CHEMISTRY, BIOLOGICAL AND PHARMACOLOGICAL PROPERTIES OF MEDICINAL PLANTS FROM THE AMERICAS

CHEMISTRY, BIOLOGICAL AND PHARMACOLOGICAL PROPERTIES OF MEDICINAL PLANTS FROM THE AMERICAS

Proceedings of the IOCD/CYTED Symposium, Panama City, Panama, 23–26 February 1997

Edited by

K. Hostettmann
Institut de Pharmacognosie et Phytochimie
Université de Lausanne
Switzerland

M.P. Gupta
College of Pharmacy, University of Panama
Republic of Panama

and

A. Marston
Institut de Pharmacognosie et Phytochimie
Université de Lausanne
Switzerland

CRC Press
Taylor & Francis Group
Boca Raton London New York

CRC Press is an imprint of the
Taylor & Francis Group, an **informa** business
A TAYLOR & FRANCIS BOOK

CRC Press
Taylor & Francis Group
6000 Broken Sound Parkway NW, Suite 300
Boca Raton, FL 33487-2742

First issued in hardback 2019

Cover: Diagram of *Lisianthius jefensis*: A. Robyns, T.S. Elias, *Ann. Missouri Bot. Gard.*, 55, 60-63 (1968).

ISBN-13: 978-90-5702-397-2 (hbk)

Library of Congress Cataloging-in-Publication Data

Catalog record is available from the Library of Congress

Visit the Taylor & Francis Web site at
http://www.taylorandfrancis.com

and the CRC Press Web site at
http://www.crcpress.com

CONTENTS

PREFACE

The first symposium of the Working Group on Plant Chemistry of the International Organization for Chemical Sciences in Development (IOCD) was held in Zimbabwe during 1996 and addressed the status of research into African medicinal plants.

Following the success of this meeting, it was decided to hold a second symposium, this time in conjunction with CYTED (Programa Iberoamericano de Ciencia y Tecnologia para el Desarrollo). The symposium was held in Panama City, Republic of Panama, from 23 to 26 February 1997 and this volume is based on the contributions from the speakers presenting plenary lectures during the event.

The aim of the symposium, similar to the first IOCD symposium, was to give an opportunity to as many scientists as possible from the Americas to update themselves on research on local plants. It is often difficult for central and southern American scientists to get to congresses in Europe and even north America. For this reason, an attempt was made to provide a local forum for the exchange of ideas on work concerning American plants. That the idea received the necessary resonance can be judged from the fact that more than 160 participants from Europe and all over the Americas attended.

The Americas are home to an incredible diversity of plant species and include, of course, the untapped treasures of the Amazon. As has been repeated a multitude of times, the tropical forests are amazing potential sources of new pharmaceutical products and have the possibility of providing both important revenue and job prospects for the countries involved. This is not to forget what can be found and is found in plant material from other climatic regions.

Local populations of the Americas make rich use of plants for the treatment of ailments and as healing drugs. The science of ethnomedicine is being applied to unravel the mysteries of these remedies and, together with work by phytochemists and pharmacologists, much is now known about the chemistry and biological activities of their constituents. However, time is very short, in the face of over-exploitation and erosion of natural resources. For this reason, it is essential to study this disappearing plant material before the chance is completely lost. This requires an all-out collaborative effort by botanists, ethnobotanists, chemists, phytochemists, medicinal chemists, biologists and pharmacologists from both academia and industry. Much of the work covered by these disciplines is treated in this book: the latest results from research on the chemistry and bioactivity of plants from the Americas are presented, and an idea is given of how useful the plant-based products may be.

The authors would like to express their gratitude to all the speakers at the Panama symposium who have contributed to this volume. Thanks are also due to the sponsors of the meeting and to the local organizers in Panama.

Finally, it is hoped that readers of this book will be stimulated to undertake new research on some of the exciting new material which remains to be investigated in both the North and South continents of America.

K. Hostettmann
M.P. Gupta
A. Marston

R. Evans
Missouri Botanical Garden
P.O. Box 299
St. Louis, MO 63166-0299
USA

L. Famolare
Conservation International
2501 M Street, Suite 200
Washington, DC 20037
USA

S. Farmer
Dept. of Microbiology
University of British Columbia
Vancouver, B.C.
Canada V6T 1Z3

E. Gibbons
Veterinary Infectious Disease Organization
Saskatoon, Saskatchewan
Canada

A.A.L. Gunatilaka
Dept. of Chemistry
Virginia Polytechnic Institute and State University
Blacksburg, VA 24061-0212
USA

M.P. Gupta
International Coordinator, CYTED
Centre for Pharmacognostic Research on Panamanian Flora
College of Pharmacy
University of Panama and Smithsonian Tropical Research Institute
Panama
Republic of Panama

K. Hostettmann
Institut de Pharmacognosie et Phytochimie
Université de Lausanne
CH-1015 Lausanne
Switzerland

M. Jang
PCRPS, and Dept. of Medicinal Chemistry and Pharmacognosy
College of Pharmacy
University of Illinois at Chicago
Chicago, IL 60612
USA

H.-J. Jeong
PCRPS, and Dept. of Medicinal Chemistry and Pharmacognosy
College of Pharmacy
University of Illinois at Chicago
Chicago, IL 60612
USA

A.D. Kinghorn
PCRPS, and Dept. of Medicinal Chemistry and Pharmacognosy
College of Pharmacy
University of Illinois at Chicago
Chicago, IL 60612
USA

D.G.I. Kingston
Dept. of Chemistry
Virginia Polytechnic Institute and State University
Blacksburg, VA 24061-0212
USA

H. Klingemann
Terry Fox Labs
B.C. Cancer Research Center
Vancouver, B.C.
Canada

A.J. Lapa
Natural Products Section
Dept. of Pharmacology
Escola Paulista de Medicina/UNIFESP
04044-020 São Paulo, SP
Brazil

S.K. Lee
PCRPS, and Dept. of Medicinal Chemistry and Pharmacognosy
College of Pharmacy
University of Illinois at Chicago
Chicago, IL 60612
USA

A. Lobstein
Laboratoire de Pharmacognosie
Faculté de Pharmacie, B.P. 24
67401 Illkirch Cédex
France

S. Malone
Conservation International Suriname
Burenstraat 7
Paramaribo
Suriname

A. Marston
Institut de Pharmacognosie et Phytochimie
Université de Lausanne
CH-1015 Lausanne
Switzerland

R. Mata
Facultad de Química
Universidad Nacional Autónoma de México
Coyoacán 04510
México DF
México

E. Mata-Greenwood
PCRPS, and Dept. of Medicinal Chemistry and Pharmacognosy
College of Pharmacy
University of Illinois at Chicago
Chicago, IL 60612
USA

H. Matsuura
Dept. of Botany
The University of British Columbia
Vancouver, B.C.
Canada V6T 1Z4

A.R. McCutcheon
Dept. of Botany
The University of British Columbia
Vancouver, B.C.
Canada V6T 1Z4

R.G. Mehta
Dept. of Surgical Oncology
College of Medicine
University of Illinois at Chicago
Chicago, IL 60612
USA

J. Memelink
Institute for Plant Molecular Sciences
Leiden University
The Netherlands

R. Mittermeier
Conservation International
2501 M Street, Suite 200
Washington, DC 20037
USA

K. Nakanishi
Dept. of Chemistry
Columbia University
New York, NY 10027
USA

J. Page
Dept. of Botany
The University of British Columbia
Vancouver, B.C.
Canada V6T 1Z4

J.M. Pezzuto
PCRPS, and Dept. of Medicinal Chemistry and Pharmacognosy
College of Pharmacy
University of Illinois at Chicago
Chicago, IL 60612
USA

E. Pisha
PCRPS, and Dept. of Medicinal Chemistry and Pharmacognosy
College of Pharmacy
University of Illinois at Chicago
Chicago, IL 60612
USA

J.F. Rivero-Cruz
Facultad de Química
Universidad Nacional Autónoma de México
Coyoacán, 04510
México DF
México

T.E. Roberts
Veterinary Infectious Disease Organization
Saskatoon, Saskatchewan
Canada

A. Rojas
Facultad de Química
Universidad Autónoma de Querétaro
Centro Universitario
Querétaro
México

M. Sauvain
ORSTOM UR 45
Faculté de Pharmacie
51 Rue Cognacq-Jay
51096 Reims Cédex
France

G. Saxena
TerraGen Diversity Inc.
Suite 300-2386 East Mall
University of British Columbia
Vancouver, B.C.
Canada V6T 1Z3

L.A. Shamon
PCRPS, and Dept. of Medicinal Chemistry and Pharmacognosy
College of Pharmacy
University of Illinois at Chicago
Chicago, IL 60612
USA

P. Sokol
Dept. of Microbiology and Infectious Diseases
University of Calgary
Alberta
Canada

L.L. Song
PCRPS, and Dept. of Medicinal Chemistry and Pharmacognosy
College of Pharmacy
University of Illinois at Chicago
Chicago, IL 60612
USA

C. Souccar
Natural Products Section
Dept. of Pharmacology
Escola Paulista de Medicina/UNIFESP
04044-020 São Paulo, SP
Brazil

R.W. Stokes
Infectious and Immunological Diseases
B.C. Children's Hospital
Vancouver, B.C.
Canada

H.J.G. ten Hoopen
Dept. of Biochemical Engineering
Delft University of Technology
Delft
The Netherlands

L.M. Thorson
Infectious and Immunological Diseases
B.C. Children's Hospital
Vancouver, B.C.
Canada

G.H.N. Towers
Dept. of Botany
The University of British Columbia
Vancouver, B.C.
Canada V6T 1Z4

R. van der Heijden
Division of Pharmacognosy
Leiden/Amsterdam Center for
Drug Research
Leiden University
P.O. Box 9502
2300 RA Leiden
The Netherlands

H. Van der Werff
Missouri Botanical Garden
P.O. Box 299
St. Louis, MO 63166-0299
USA

R. Verpoorte
Division of Pharmacognosy
Leiden/Amsterdam Center for
Drug Research
Leiden University
P.O. Box 9502
2300 RA Leiden
The Netherlands

B. Weniger
ORSTOM
Universidad del Valle
Cali
Colombia

J.H. Wisse
Bedrijf Geneesmiddelen
Voorziening Suriname
Commissaris Roblesweg 156
Paramaribo
Suriname

J.-L. Wolfender
Institut de Pharmacognosie et
Phytochimie
Université de Lausanne
CH-1015 Lausanne
Switzerland

B.-N. Zhou
Dept. of Chemistry
Virginia Polytechnic Institute and
State University
Blacksburg, VA 24061-0212
USA

1. LESSONS FROM NATURAL MEDICINES

KOJI NAKANISHI

Department of Chemistry, Columbia University, New York, N.Y. 10027, USA

The chemistry dealing with natural products is the oldest branch of organic chemistry. Natural products chemistry started with the curiosity regarding odor, taste, color, folk medicinal cures, etc. In early days what is now known as natural products chemistry was focused on isolating the more readily available constituents and determining their structures. In certain cases this was followed by synthesis and elucidation of biosynthetic routes. With the rapid advancement in isolation techniques and spectroscopy, structure determination became increasingly routine in most cases; the trend also shifted towards activity-monitored isolation and structural studies. However, in more recent years, it has become possible for organic chemists to direct their attention towards the mode of action, which almost invariably involves the interaction between a ligand and its biopolymeric receptor; a decade ago, such studies were impossible to be performed on a molecular structural level. Organic chemists, especially those involved in structural studies have the techniques, imagination, and knowledge with respect to the approach in such studies. However, nature is far more complex, and it is only with multidisciplinary collaborative research encompassing many disciplines that such targets can be successfully investigated. Why does sugar taste sweet or how do we perceive color? These are extremely challenging problems which at present cannot be answered even with a major multidisciplinary effort. We are at the starting point of a new field which is filled with exciting problems that have an immediate bearing directed towards a better understanding of Nature.

To understand the mode of action of folk herbs and related products from nature is even more complex because unfractionated or partly fractionated extracts are used. Nevertheless, the efficacy from this source has been proven over the years, in some cases covering thousands of years. The fact that natural medicines are attracting renewed attention is encouraging. Clarification of the constituents involved and understanding the mode of actions are clearly challenging but all the more worthwhile. In the following, some of our past studies in this area are outlined.

MICROBIAL PRODUCTS

Monascorubrin and Monascoflavin (Figure 1)

These orange and yellow pigments, respectively, isolated in 1926 by the late Eijiro Nishikawa (Tottori University) from *Monascus purpureus* Mentii, are widely used for coloring foodstuff in Asia and also are the main pigments of the popular Chinese wine "chao xing chiu". M. Ohashi and others started working on these pigments collected

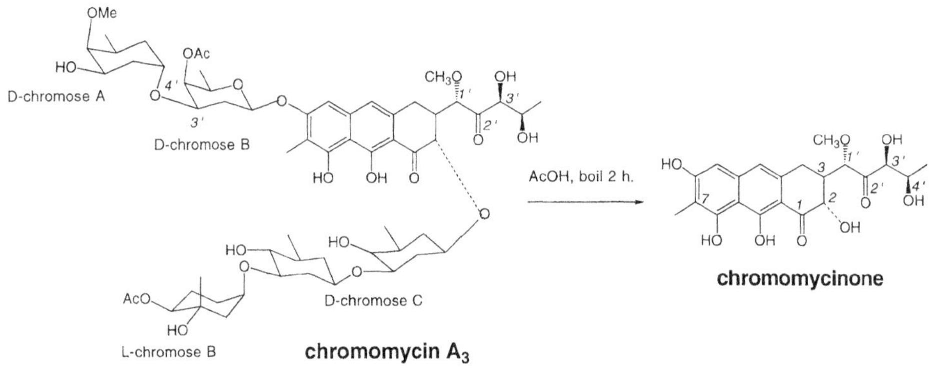

Figure 3 Chromomycin A_3 and chromomycinone

most extensive in those days, clarified all connectivities to disclose the full structure which to our surprise was found to contain one more carbon than originally thought.

The original sugar connectivities in chromomycin A_3 (Miyamoto *et al.* 1964, 1967) were revised later on the basis of extensive measurements by 300 MHz NMR (Thiem and Meyer 1979); in the original proposal the linkages at 3′ and 4′ of the D-chromose D moiety were interchanged and the stereochemistries at glycosidic centers were undetermined. The absolute stereochemistry was found (Harada *et al.* 1969) by the exciton chirality method in which the split circular dichroic curve resulting from coupling between the naphthalenoid transition and that of a benzoate chromophore at C-1′ was analysed. In the chromomycin studies, the Takeda team carried out beautiful and very extensive degradative studies while we were mainly involved in interpretations of spectral data. A chromomycin derivative has since been used as a template for application of red-shifted chromophores in the exciton chirality method (Cai *et al.* 1993).

Phytoalexins Produced by Sweet Potato (Figure 4)

Infection of plants by pathogens leads to the rapid production of antifungal compounds for self-defense. The molecular mechanism of this rapid and efficient defense is an important problem but is still obscure. When the sweet potato *Ipomea batatas* is invaded by the pathogenic black rot fungus *Ceratocystis fimbriata* and other fungi, ipomeamarone, a bitter sesquiterpenoid, rapidly accumulates in the infected tissue. The structure of this first phytoalexin was elucidated in 1956 by Kubota and Matsuura. As shown in Figure 4, inoculation of the roots with fungi or treatment with $HgCl_2$, incubation for 3 days and extraction yielded ten new farnesol-derived ipomeamarones and 7-hydroxycostal, a new class of phytoalexins (Schneider *et al.* 1984a). The 7-deoxyaldehyde was also produced but this was devoid of activity. It was therefore proposed that the antifungal activity of 7-hydroxycostal is due to the production of acrolein; indeed, a one week incubation led to a 10% production of the bicyclic ketone. In view of the multitude of phytoalexins produced upon inoculation or injuring, this may provide an attractive and simple means of producing new antifungal agents in general. Biosynthetic studies employing ^2H NMR showed the occurrence of an unusual 1,3-hydride shift

1. LESSONS FROM NATURAL MEDICINES

KOJI NAKANISHI

Department of Chemistry, Columbia University, New York, N.Y. 10027, USA

The chemistry dealing with natural products is the oldest branch of organic chemistry. Natural products chemistry started with the curiosity regarding odor, taste, color, folk medicinal cures, etc. In early days what is now known as natural products chemistry was focused on isolating the more readily available constituents and determining their structures. In certain cases this was followed by synthesis and elucidation of biosynthetic routes. With the rapid advancement in isolation techniques and spectroscopy, structure determination became increasingly routine in most cases; the trend also shifted towards activity-monitored isolation and structural studies. However, in more recent years, it has become possible for organic chemists to direct their attention towards the mode of action, which almost invariably involves the interaction between a ligand and its biopolymeric receptor; a decade ago, such studies were impossible to be performed on a molecular structural level. Organic chemists, especially those involved in structural studies have the techniques, imagination, and knowledge with respect to the approach in such studies. However, nature is far more complex, and it is only with multidisciplinary collaborative research encompassing many disciplines that such targets can be successfully investigated. Why does sugar taste sweet or how do we perceive color? These are extremely challenging problems which at present cannot be answered even with a major multidisciplinary effort. We are at the starting point of a new field which is filled with exciting problems that have an immediate bearing directed towards a better understanding of Nature.

To understand the mode of action of folk herbs and related products from nature is even more complex because unfractionated or partly fractionated extracts are used. Nevertheless, the efficacy from this source has been proven over the years, in some cases covering thousands of years. The fact that natural medicines are attracting renewed attention is encouraging. Clarification of the constituents involved and understanding the mode of actions are clearly challenging but all the more worthwhile. In the following, some of our past studies in this area are outlined.

MICROBIAL PRODUCTS

Monascorubrin and Monascoflavin (Figure 1)

These orange and yellow pigments, respectively, isolated in 1926 by the late Eijiro Nishikawa (Tottori University) from *Monascus purpureus* Mentii, are widely used for coloring foodstuff in Asia and also are the main pigments of the popular Chinese wine "chao xing chiu". M. Ohashi and others started working on these pigments collected

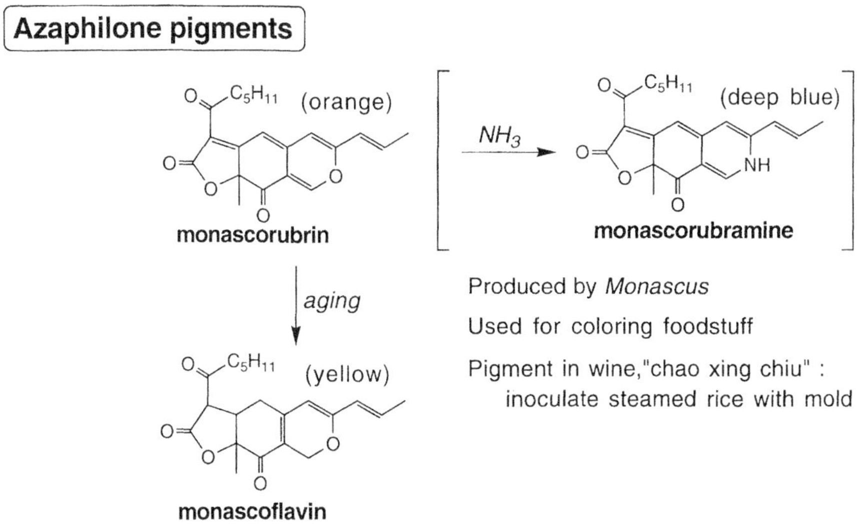

Figure 1 Monascorubrin and monascoflavin, azaphilones

earlier by Nishikawa. They belong to a group of pigments studied intensively by the Liverpool school of R. Robertson, B. Whalley and coworkers, and were generically called azaphilones because ammonia readily replaces the pyran oxygen with nitrogen yielding the violet nitrogen analog. The microorganism is spread on warm steamed rice to first produce the orange pigment; upon further fermentation, the pigment changes into the yellow monascoflavin, which gives rise to a higher quality wine. Besides structural studies, we performed our first biosynthetic studies to show the polyketide origin (Ohashi *et al.* 1960; Kumasaki *et al.* 1962; Kurono *et al.* 1963a).

Illudin S (Figure 2)

Our studies on illudin-S (lampterol) started in Tokyo in 1962 and were completed in Sendai by Mamoru Ohashi, Masaru Tada and Yasuji Yamada (Nakanishi *et al.* 1965). It is a constituent of the bioluminescent and toxic "tsuki-yo-take" (*Lampteromyces japonicus*) that grows on rotten beech trees; the mushroom has been the cause of fatal accidents because of its similarity to edible mushrooms. Independent structural studies were carried out by Marjorie Anchel and Trevor McMorris at New York Botanical Garden (from *Clitocybe illudens* or Jack-o-lantern) (McMorris and Anchel 1963) and by Takeshi Matsumoto's group at Hokkaido University (same source as ours) (Matsumoto *et al.* 1965). A large number of sesquiterpenoids with the illudan framework has since been identified. We were originally interested in the bioluminescent factor and toxin, but when an intensive screening test carried out by Takeda Company showed the mushroom to possess strong antitumor activity, extractions of the toxic and antitumor factors were carried out in parallel; the two turned out to be the same. We combined the collection of 300 kg of mushrooms with a fall group picnic. In groups of 2–3 we spread out into the fields of beech trees, collected the mushrooms in knapsacks,

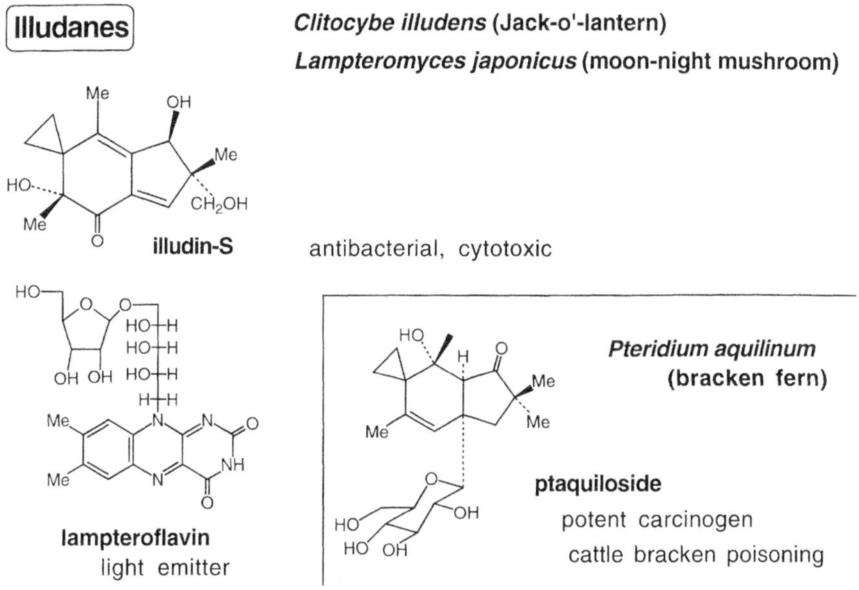

Figure 2 Illudanes

returned to the road and left them at the roadside to be picked up by a mini-truck; this was repeated several times until evening. The glow in the dark from the mushrooms dipped in methanol was an eery sight indeed. Although we attempted to isolate the bioluminescent factor, I had to give up after cultivating a fluorescent fungi (Endo *et al.* 1970), because of my transfer to Columbia University; the factor was successful characterized in 1988 by Isobe and Goto (Isobe *et al.* 1988). The illudan skeleton attracted wide interest when a heat-unstable derivative of it was found to be responsible for the potent carcinogenecity of unboiled "warabi" (Hirono *et al.* 1984).

Chromomycins (Figure 3)

One of the earliest example of applications of exhaustive NMR decoupling coupled with extensive chemical studies established the structure of this group of antibiotics comprising a tricyclic skeleton and four new 2,6-deoxysugars. They were important clinical anticancer drugs; extensive studies are currently being performed at various institutes to understand their mode of interaction with DNA. The principal antibiotic chromomycin A_3 was sold under the name of toyomycin; a number of similar compounds were isolated in the US, Russia, Italy and Japan but structures were unknown. A_3 could be hydrolyzed with aqueous acetic acid to give four new deoxy sugars, the chromoses and the aglycone chromomycinone (CHR). Detailed 100 MHz NMR measurements, including double and triple resonance and solvent shift techniques to separate overlapping peaks, were carried out on the new sugars, the aglycone and derivatives. They were performed by Norman Bhacca (currently at Louisiana State University), a visiting scientist from Varian. The NMR studies carried out by Norman, some of the

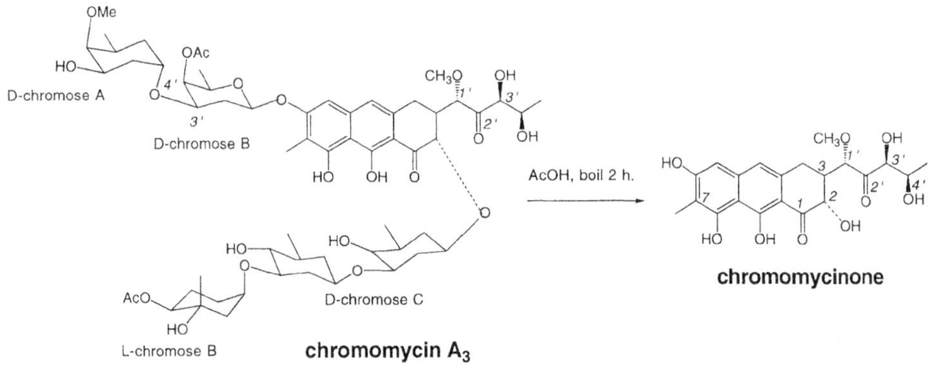

Figure 3 Chromomycin A_3 and chromomycinone

most extensive in those days, clarified all connectivities to disclose the full structure which to our surprise was found to contain one more carbon than originally thought.

The original sugar connectivities in chromomycin A_3 (Miyamoto *et al.* 1964, 1967) were revised later on the basis of extensive measurements by 300 MHz NMR (Thiem and Meyer 1979); in the original proposal the linkages at 3′ and 4′ of the D-chromose D moiety were interchanged and the stereochemistries at glycosidic centers were undetermined. The absolute stereochemistry was found (Harada *et al.* 1969) by the exciton chirality method in which the split circular dichroic curve resulting from coupling between the naphthalenoid transition and that of a benzoate chromophore at C-1′ was analysed. In the chromomycin studies, the Takeda team carried out beautiful and very extensive degradative studies while we were mainly involved in interpretations of spectral data. A chromomycin derivative has since been used as a template for application of red-shifted chromophores in the exciton chirality method (Cai *et al.* 1993).

Phytoalexins Produced by Sweet Potato (Figure 4)

Infection of plants by pathogens leads to the rapid production of antifungal compounds for self-defense. The molecular mechanism of this rapid and efficient defense is an important problem but is still obscure. When the sweet potato *Ipomea batatas* is invaded by the pathogenic black rot fungus *Ceratocystis fimbriata* and other fungi, ipomeamarone, a bitter sesquiterpenoid, rapidly accumulates in the infected tissue. The structure of this first phytoalexin was elucidated in 1956 by Kubota and Matsuura. As shown in Figure 4, inoculation of the roots with fungi or treatment with $HgCl_2$, incubation for 3 days and extraction yielded ten new farnesol-derived ipomeamarones and 7-hydroxycostal, a new class of phytoalexins (Schneider *et al.* 1984a). The 7-deoxyaldehyde was also produced but this was devoid of activity. It was therefore proposed that the antifungal activity of 7-hydroxycostal is due to the production of acrolein; indeed, a one week incubation led to a 10% production of the bicyclic ketone. In view of the multitude of phytoalexins produced upon inoculation or injuring, this may provide an attractive and simple means of producing new antifungal agents in general. Biosynthetic studies employing ^2H NMR showed the occurrence of an unusual 1,3-hydride shift

Ginkgo biloba, the ginkgo tree

One of the first trees, *Ginkgoales*, appeared 250 million years ago.

All species vanished, except one, *Ginkgo biloba*, which has remained unchanged for 150 million years.

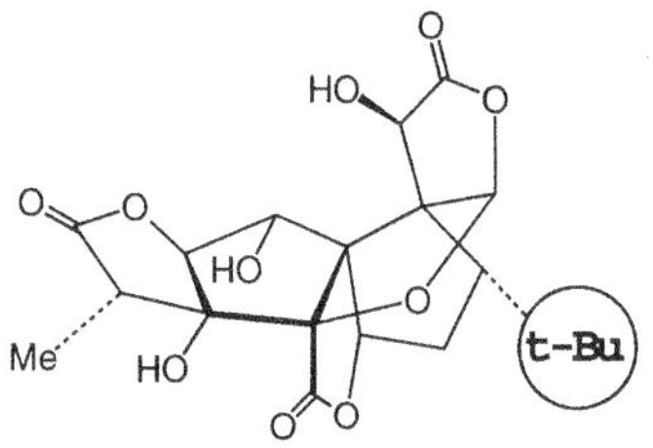

ginkgolide B

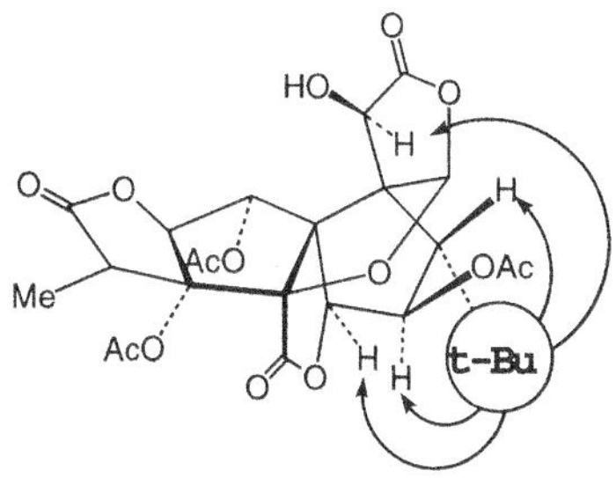

nuclear Overhauser effect

Mentioned in Chinese materia medica, 2800 B.C.

Inhibitor of platelet activating factor.

$ 2-3 billion of *Ginkgo* tree extract sold over-the-counter annually.

Figure 8 Ginkgolides and nuclear Overhauser effect

have shown that these ancient diterpenoids do not follow the conventional mevalonate biosynthetic pathway; instead they follow the nonmevalonate sugar route of biosynthesis (Schwarz 1994).

One of the ginkgolides, ginkgolide B (shown in Figure 8), has since been found to be a potent inhibitor of platelet activating factor. The extract of *G. biloba* leaves, a mixture of flavonoids and ginkgolides, is said to improve memory and have some positive effect in Alzheimer's disease and is selling as a dietary supplement at an explosive rate throughout the world.

Cabenegrins, Anti-snake Venom Factors from *Cabeca de negro* (Figure 9)

This problem evolved in a most unusual style. Dr. Laszlo Darko whom I had never met walked into my office and produced a bottle containing a colorless solution claiming it was an extract of an Amazonian plant "Cabeca de negro", and was used by plantation workers in the jungle as an antidote against snake and spider venoms. He had obtained it from a hunter who did not want to disclose the identity of the plant. Two-years of extraction, monitored by injections into mice of the 2.5-fold lethal dose of the venom from a common South American snake *Fer de lance* (*Bothrops atrox*) indeed yielded two phytoalexins, cabenegrins A-I and II (Nakagawa *et al.* 1982). The minimal dosage required for survival was 2.0–2.8 mg/kg of A-I or A-II.

Injection of 2.5 mg/kg of the snake venom into a 9 kg male beagle dog led to hypotension, followed by respiratory and cardiac arrest. Injection of 1 mg/kg of A-I

Anti-snake venom from *Cabeca de negro*

aqueous ethanol extract of root, 135 ml

cabenegrin A-I, 44 mg | **cabenegrin A-II,** 1 mg

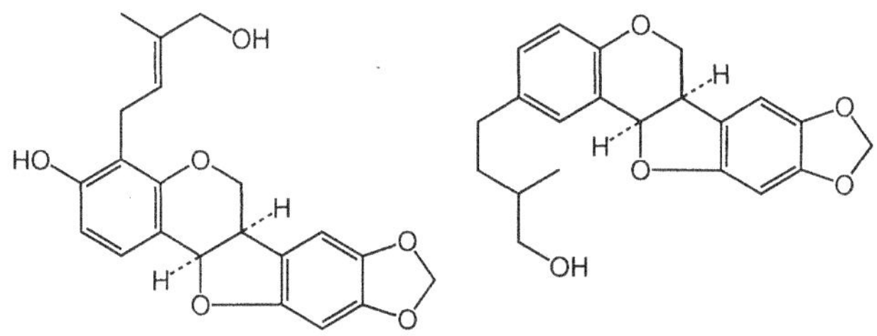

antidote against Fer de lance (*Bothrops atrox*) snake venom.

injection of 2.5 mg / kg venom (lethal dose) to 9 kg beagle dog leads to hypotension, respiratory and cardiac arrest.

injection of 1.0 mg / kg A-I, 15 min before or after venom injection restores or reverses effect in 90 min.

Figure 9 Anti-snake venoms from a tropical plant

15 min prior to i.v. injection of the venom restored all three symptoms after 1 hour; conversely, injection of venom followed by injection of cabenegrin also reversed the toxic cardiovascular effects. However, when A-I and -II were tested against some other snake venoms (at a snake venom laboratory in Okinawa) they were ineffective. The cabenegrins must be specific for the *Fer de lance* venom. The colorless solution contained a more potent antidote, but the studies were discontinued because Dr. Vick who was performing the assays was retiring. Furthermore, we found that at least ten plants exist in South America called by the name of "Cabeca de negro". Studies of the mode of action involving the specific snake venom as opposed to others would be interesting and important. However, from a more pragmatic view point, further investigations to collect more material and characterize the more potent antidote(s) are necessary.

Antheridiogen, a fern hormone (Figure 10)

This diterpene, produced at the prothali stage of ferns, is the hormone that induces formation of antheridium, the male sex organ. The project to characterize it was brought to us by Ulrich Näf, Manhattan College, a plant physiologist who had been working on this factor since the mid 1950's. A large scale culture of the fern *Anemia phylliudis* (*Schizaiaceae*), in 1500 flasks, each containing 38 ml of culture medium and cultivated for 53 days, and followed by submicroscopic observation of antheridia formation, gave 18 mg of pure antheriodogen-An (A_{An}). Structural studies performed

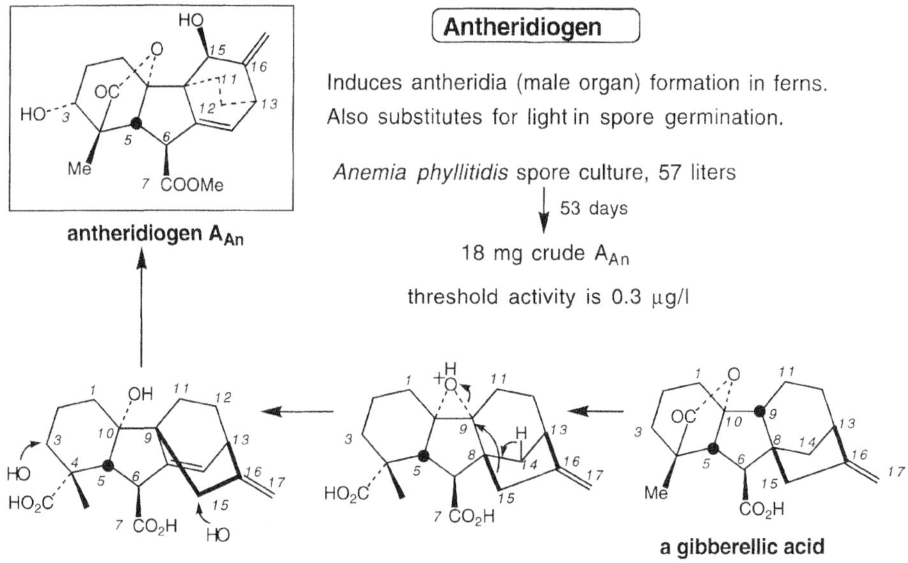

Figure 10 Antheridia forming factor

by M. Endo on 3 mg of the factor led to the structure having a rearranged gibberellin skeleton (Nakanishi *et al.* 1971). Our attempts to isolate a second anthridiogen were not successful but subsequently several antheridiogens have been characterized. A_{An} has been synthesized independently by Corey (Corey *et al.* 1988) and by Mander (Farber and Mander 1988), who have shown that the 3-OH is a and not b as in the original structure.

Insect antifeedants (Figure 11)

An obvious defense mechanism for plants against insects, birds, and animals is to produce feeding deterrents or antifeedants (Kubo and Nakanishi 1977). It is not surprising that plants growing in tropical rain forests are exposed to many more natural hazards and hence are better developed in producing antifeedants. During my connection with the International Centre of Insect Physiology and Ecology, Nairobi, Kenya (1969–1977), we carried out searches for insect antifeedants, which developed into an intensive survey by Isao Kubo. Of the ca. 30 antifeedants isolated, warburganal and azadirachtin are the two most potent.

Warburganal was isolated from the leaves and bark of *Warburgia stuhlmannii* and *W. ugandensis*, plants that are widely used as food spice and folk medicine in East Africa. Electrophysiological studies show that it irreversibly blocks the taste buds of *Spodoptera exempta* (African army-worm) (Nakanishi and Kubo 1977) so that when the caterpillar is placed onto a corn leaf treated with warburganal for 1 hour and then transferred to an untreated leaf, it starves to death. Warburganal exhibits potent molluscidal and antibiotic activity (e.g. *Candida utilis* 3.1 μg/ml). It is contained in "TADE", a spice traditionally

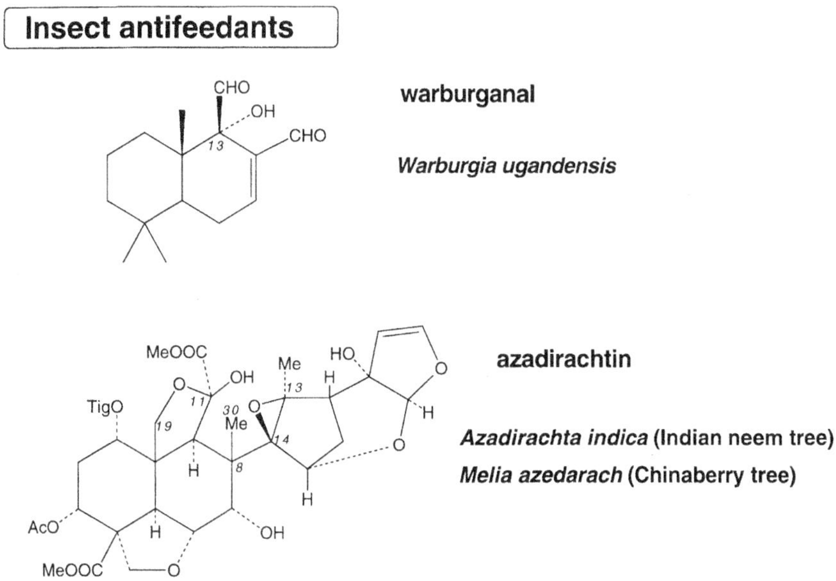

Figure 11 Insect antifeedants from plants

accompanying sashimi, but upon injection exhibits acute toxicity (LD_{50} 20.4 mg/kg against mice) and cytotoxicity (KB 0.01 μg/ml). After Steve Tanis in our lab synthesized warburganal, many syntheses have followed. Electrophysiology tests for insect antifeedants are easy to set up and readily give impulse signals that reflect the responses visually and semiquantitatively.

Azadirachtin, isolated from the Indian neem tree *Azadirachta indica* A. Juss is more than an antifeedant and exerts profound physiological effects on insects. It is the best known antifeedant, and because of its ready availability from the plant, it is attracting great interest; companies dealing with azadirachtin have been formed in India and the US. A structure proposed by us in 1975 (Zanno *et al.*) was revised by the groups of W. Kraus (1986), S. Ley, and ourselves (1987); the papers were published together in 1987, Bilton *et al.*, Kraus *et al.*, Nakanishi *et al.*, preceeded by a comment (Taylor 1987).

CAPE, an Antiinflammatory and Preferential Cytotoxic Agent from Propolis (Figure 12)

Propolis, the brown gum made by honeybees on beehives has long been a popular folk medicine in Eastern Europe, the Near East and other countries, and is alleged to possess antibiotic, antiinflammatory and tumor-growth-arresting properties. Fractionation monitored by cytotoxicity assays yielded the simple but unknown caffeic acid phenethyl ester (CAPE) (Grünberger *et al.* 1988). CAPE exhibits preferential cytotoxicity on tumor cells and is a potent antiinflammatory agent. CAPE should be an intriguing subject for structure/activity and mechanistic studies. Is the role of the phenethyl moiety simply for increasing the permeability of caffeic acid, which is a well-documented

CAPE Caffeic Acid Phenethyl Ester

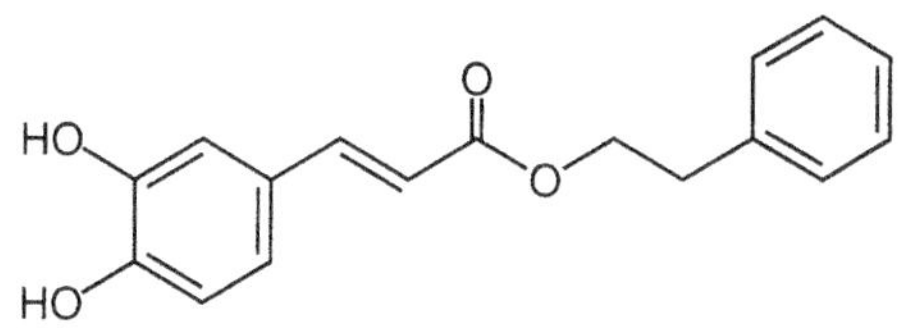

from **propolis**, dark-brown wax of beehives

differential cytotoxicity observed in
melanoma and breast carcinoma of:
normal rat/human *versus* transformed rat/human cell lines

potent anti-inflammatory agent

Figure 12 A propolis constituent, caffeic acid phenethyl ester

antibiotic? What is the basis of its antiinflammatory activity? It should be noted that since propolis is collected by honeybees from plants, the ingredients are not constant and would differ depending on the season, the region of collection, etc. Namely, honey bees are playing the role of collectors.

INSECT/CRUSTACEAN MOLTING HORMONES, MOLT INHIBITORS etc.

Phytoecdysteroids (Figure 13)

The structure determination and isolation of ecdysone, the precursor, and 20-hydroxy-ecdysone, the insect molting hormone was a brilliant and exciting achievement performed by the Butenandt school, between 1965 and 1966; Horn and coworkers also isolated 2 mg of 20-hydroxyecdysone from 1 ton of crayfish waste and showed it to be the crustacean molting hormone. At that time our group and that of Takemoto's had isolated from Chinese folk medicinal herbs, large quantities of similar compounds which exhibited potent molting activity, the ponasterones and inokosterone (Nakanishi *et al.* 1966; Takemoto *et al.* 1967). Over 70 ecdysteroids have since been isolated from plants. The discovery of phytoecdysteroids made a big impact on insect physiology because of their abundance and variations in structures and activities. It is thought that plants contain these ecdysteroids as a defense against predators. The recent finding that certain ecdysteroids have been found to induce gene expression in mammalian cells and transgenic mice is attracting great interest (No *et al.* 1996).

Ecdysone Biosynthesis Inhibitor (EBI) of Crustaceans (Figure 14)

Since the early 1900's it has been known that removal of eyestalks from crustaceans leads to precocious ecdysis, thus suggesting the presence of a "molt-inhibiting hormone (MIH)". Four thousand blue crabs (*Callinectes sapidus*), were collected in Long

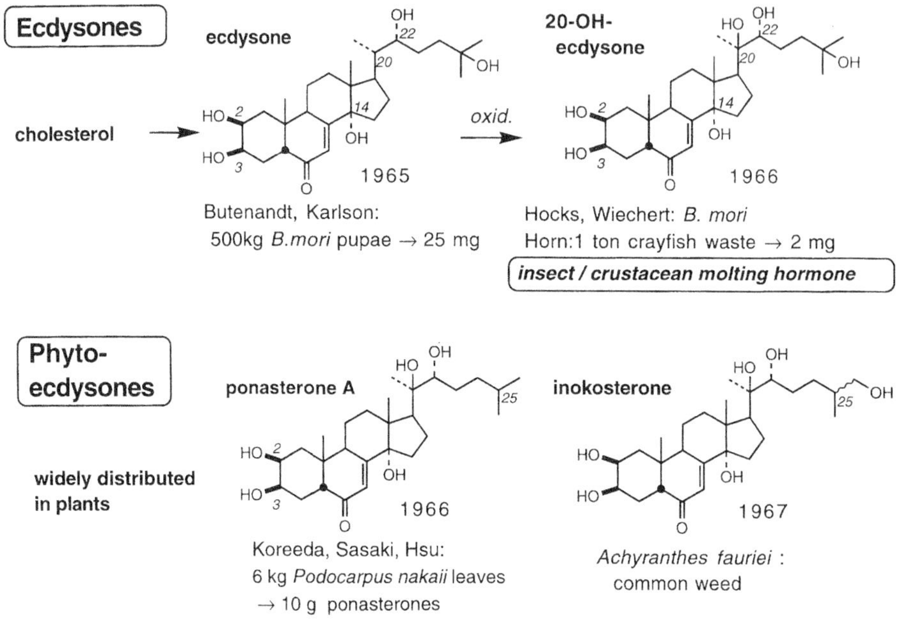

Figure 13 Ecdysteroids from insects, crustaceans and plants

Crustacean molt regulators

Removal of eyestalks (ES) → precocious ecdysis or shedding of shell (1905).

Y-organs at stem of ES : biosynthesize ecdysone (Chang, O'Connor, 1976).

X-organs in ES : release peptidic molt-inhibiting hormone (MIH, 1986-90).

Isolation of ecdysone biosynthesis inhibitor (EBI)

ES of blue crabs, 16 g from 660 X-organs

↓ *fractionation monitored by ecdysone titer assay using Y-organ homogenate*

0.7 mg of **EBI** or **3-hydroxykynurenine** (3-OHK) (1988)

0.3 mg of **xanthurenic acid** (XA) isolated (1989).

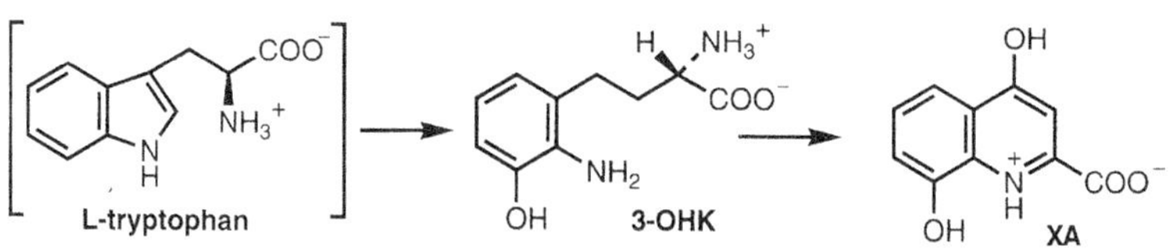

Figure 14 Crustacean molt inhibitor

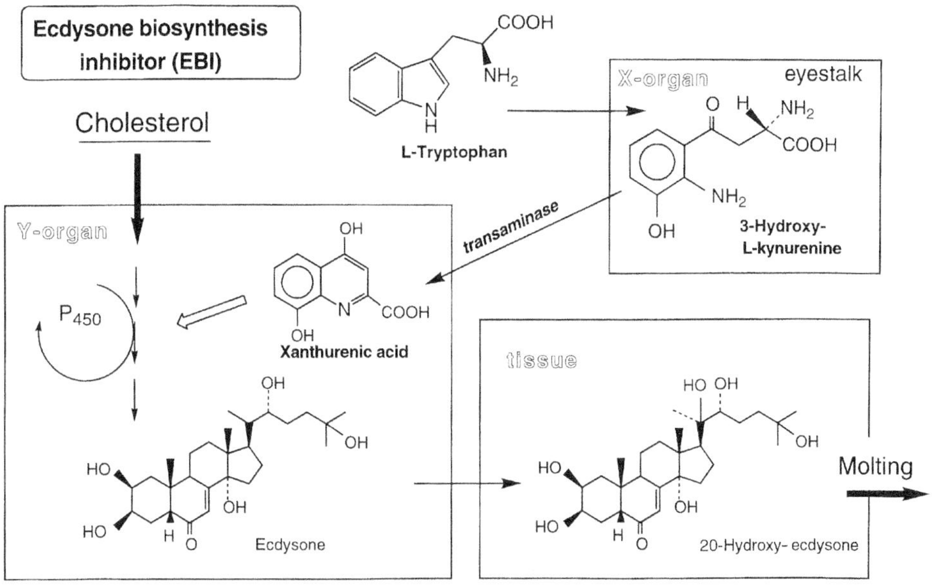

Figure 15 Ecdysone biosynthesis inhibitor of crustaceans

Island and shipped to the Suntory Institute for Bioorganic Research for isolation of this hormone. However, Naya and coworkers experienced many frustrations, because the bioassay, which monitors the production/inhibition of ecdysone by the Y organs, would not work during the summer or even over periods longer than a year, whether the crabs were collected in the United States or in Japan. Naya finally succeeded in isolating 0.7 mg of the molt inhibitor and identified this endogenous compound contained in the X organ of the eyestalk as 3-hydroxy-L-kynurenine (3-OH-K), a key metabolite of L-tryptophan (Naya *et al.* 1988).

Ecdysone Biosynthesis Inhibitor (EBI) in Crustaceans (Figure 15)

Naya and co-workers further found that during the flow of 3-OH-K to the Y-organ complex (YOC) located at the stem of the eyestalk, the site of ecdysone biosynthesis, it was converted into xanthurenic acid, the real ecdysone biosynthesis inhibitor. Xanthurenic acid interferes with the hydroxylation of cholesterol to ecdysone by interacting with the P-450 system, but does not block the hydroxylation step leading from ecdysone to 20-hydroxyecdysone.

ACKNOWLEDGEMENTS

Part of the studies have been supported by NIH AI 10187.

REFERENCES

Bilton, J.N., Broughton, H.B., Jones, P.S., Ley, S.V., Lidert, Z., Morgan, E.D., Rzepa, H.S., Sheppard, R.N., Slawin, A.M.Z., Williams, D.J. (1987) An X-ray crystallographic, mass spectroscopic, and NMR study of the limonoid insect antifeedant azadirachtin and related derivatives. *Tetrahedron*, **43**, 2805–2815.

Cai, G., Bozhkova, N., Odingo, J., Berova, N., Nakanishi, K. (1993) CD exciton chirality method. New red-shifted chromophores for hydroxyl groups. *J. Am. Chem. Soc.*, **115**, 7192–7198.

Corey, E.J., Kand, M., Desai, M.C., Ghosh, A.K., Houpis, I.N. (1988) Total synthesis of ginkgolide B. *J. Am. Chem. Soc.*, **110**, 649–651.

Endo, M., Kajiwara, M., Nakanishi, K. (1970) Fluorescent constituents and cultivation of *Lampteromyces japonicus. J. Chem. Soc., Chem. Commun.*, 309–310.

Eya, B.K., Kenny, P.T.M., Tamura, S.Y., Ohnishi, M., Naya, Y., Nakanishi, K., Sugiura, M. (1989) Chemical association in symbiosis: sterol donors in planthoppers. *J. of Chemical Ecology*, **15**, 373–380.

Eyre, D.H., Harrison, J.W., Scrowston, R.M., Lythgoe, B. (1963) The constitution of taxicin-I and -II. *Proc. Chem. Soc.*, 271–272.

Farber, M. and Mander, L.N. (1988) Synthesis and confirmation of the structure of antheridium-inducing factor from the fern *Antheridium mexicana. J. Am. Chem. Soc.*, **110**, 4084–4085.

Fredenhagen, A., Tamura, S.Y., Kenny, P.T.M., Komura, H., Naya, Y., Nakanishi, K. (1987) A new peptide antibiotic produced by an intracellular bacterial symbiont isolated from a brown planthopper. *J. Am. Chem. Soc.*, **109**, 4409–4411.

Grünberger, D., Banerjee, R., Eisinger, K., Oltz, E.M., Efros, L., Caldwell, M., Estevez, V., Nakanishi, K. (1988) Preferential cytotoxicity on tumor cells by caffeic acid phenethyl ester isolated from propolis. *Experientia*, **44**, 230–232.

Harada, R., Kakisawa, H., Kobayashi, S., Musya, M., Nakanishi, K., Takahashi, Y. (1962) Structure of pristimerin, a quinonoid triterpene. *Tetrahedron Lett.*, 603–607.

Harada, N., Nakanishi, K., Tatsuoka, S. (1969) A method for determining the chirality of two aromatic chromophores and the absolute configurations of chromomycin A_3 and related antibiotics. *J. Am. Chem. Soc.*, **91**, 5896–5898.

Hirono, I., Yamada, K., Niwa, H., Shizuki, Y., Ojika, M., Hosaka, S., Yamaji, T., Wakamatsu, T., Kigoshi, H., Niiyama, K., Uosaki, Y. (1984) Separation of carcinogenic fraction of bracken fern. *Cancer Lett.*, **21**, 239–246.

Isobe, M., Uyakul, D., Goto, T. (1988) Lampteromyces bioluminescence. 2. Lamptoflavin, a light emitter in the luminous mushroom *Lampteromyces japonicum. Tetrahedron Letters.* 1169–1172.

Kraus, W., Bokel, M., Bruhn, A., Cramer, R., Klaiber, I., Klenk, A., Nagl, G., Pöhnl, H., Sadlo, H., Vogler, B. (1987) Structure determination by NMR of azadirachtin and related compounds from *Azadirachta indica* A. JUSS (Meliaceae). *Tetrahedron*, **43**, 2805–2815.

Kubo, I. and Nakanishi, K. (1977) Insect antifeedants and repellents from African plants. In P. Hedin, (ed.), *Host Plant Resistance to Pests*, *Am. Chem. Soc. Symposium Series*, **62**, 165–178.

Kumasaki, S., Nakanishi, K., Nishikawa, E., Ohashi, M. (1962) Structure of monascorubrin. *Tetrahedron*, **18**, 1171–1184.

Kurono, M., Nakanishi, K., Shindo, K., Tada, M. (1963a) Biosynthesis of monascorubrin and monascoflavin. *Chem. Pharm. Bull.*, **11**, 359–362.

Kurono, M., Nakadaira, Y., Onuma, S., Sasaki, K., Nakanishi, K. (1963b) Taxinine. *Tetrahedron Lett.*, 2153–2160.

Maruyama, M., Terahara, A., Itagaki, Y., Nakanishi, K. (1967a) The ginkgolides I: Isolation and characterization of the various groups. *Tetrahedron Lett.*, 299–302.

Maruyama, M., Terahara, A., Itagaki, Y., Nakanishi, K. (1967b) The ginkgolides II: Derivation of partial structures. *Tetrahedron Lett.*, 303–308.

Maruyama, M., Terahara, A., Nakadaira, Y., Woods, M.C., Nakanishi, K. (1967c) The ginkgolides III: The structure of the ginkgolides. *Tetrahedron Lett.*, 309–314.

Maruyama, M., Terahara, A., Nakadaira, Y., Woods, M.C., Takagi, Y., Nakanishi, K. (1967d) The ginkgolides IV: Stereochemistry of the ginkgolides. *Tetrahedron Lett.*, 315–320.

Matsumoto, T., Shirahama, H., Ichihara, Y., Fukuoka, Y., Takahashi, Y., Mori, Y., Watanabe, M. (1965) Structure of lampterol (illudin S). *Tetrahedron*, **21**, 2671–2676.

McMorris, T.C. and Anchel, M. (1963) The structures of the Basidiomycete metabolites illudin S and illudin M. *J. Am. Chem. Soc.*, **85**, 831–832.

McMorris, T.C. and Anchel, M. (1965) Fungal metabolites. The structures of the novel sesquiterpenoids illudin-S and -M. *J. Am. Chem. Soc.*, **87**, 1594–1600.

Miyamoto, M., Morita, K., Kawamatsu, Y., Noguchi, S., Marumoto, R., Tanaka, K., Nakadaira, Y., Bhacca, N.S. (1964) Chromomycinone, the aglycone of chromomycin A_3. *Tetrahedron Lett.*, 2355–2365.

Miyamoto, M., Kawamatsu, Y., Kawashima, K., Shinohara, M., Tanaka, K., Tatsuoka, S., Nakanishi, K. (1967) Chromomycin A_2, A_3 and A_4. *Tetrahedron*, **23**, 412–437.

Nakagawa, M., Nakanishi, K., Darko, L.L., Vick, J.A. (1982) Structures of cabenegrins A-I and A-II, potent anti-snake venoms. *Tetrahedron Lett.*, 3855–3858.

Nakanishi, K. (1967) The ginkgolides. *Pure and Appl. Chem.*, **14**, 89–113.

Nakanishi, K. (1992) Past and present studies with ponasterones, the first insect molting hormones from plants. *Steroids*, **57**, 649–657.

Nakanishi, K. and Kubo, I. (1977) Studies on warburganal, muzigadial and related compounds. *Israel J. Chem.*, **16**, 28–31.

Nakanishi, K., Kakisawa, H., Hirata, Y. (1956) Structures of pristimerin and celastrol. *Bull. Chem. Soc., Japan*, **29**, 7–15.

Nakanishi, K., Ohashi, M., Tada M., Yamada, Y. (1965) Illudin S (lampterol). *Tetrahedron*, **21**, 1231–1246.

Nakanishi, K., Koreeda, S., Sasaki, S., Chang, M.L., Hsu, H.Y. (1966) Insect hormones I: The structure of ponasterone A, an insect molting hormone from the leaves of *Podocarpus nakaii* Hay. *J. Chem. Soc., Chem. Commun.*, **1966**, 915–916.

Nakanishi, K., Endo, M., Näf, U., Johnson, F.F. (1971) Structure of the antheridium inducing factor of the fern. *J. Am. Chem. Soc.*, **93**, 5579–5581.

Naya, Y., Kishida, K., Sugiyama, M., Murata, M., Miki, W., Ohnishi, M., Nakanishi, K. (1988) Endogenous inhibitor of ecdysone synthesis in crabs. *Experientia*, **44**, 50–52.

Naya, Y., Miki, W., Ohnishi, M., Ikeda, M., Nakanishi, K. (1989) Endogenous xanthurenic acid as a regulator of the crustacean molt cycle. *Pure and Appl. Chem.*, **61**, 465–468.

No, D., Yao, T.-P., Evans, R.M. (1996) Ecdysone-inducible gene expression in mammalian cells and transgenic mice. *Proc. Natl. Acad. Sci. USA*, **93**, 3346–3351.

Ohashi, M., Yamamura, S., Terahara, A., Nakanishi, K. (1960) Monascoflavin. *Bull. Chem. Soc. Jap.*, **33**, 1630.

Schneider, J.A., Yoshihara, K., Nakanishi, K. (1983) A new class of sweet potato phytoalexins *J. Chem. Soc., Chem. Commun.*, 352–353.

Schneider, J.A., Lee, J., Naya, Y., Nakanishi, K., Oba, K., Uritani, I. (1984a) The fate of the phytoalexin ipomeamarone: furanoterpenes and butenolides from *Ceratocystis fimbriata*-infected sweet potatoes. *Phytochemistry*, **23**, 759–764.

Schneider, J., Lee, J., Yoshihara, K., Mizukawa, K., Nakanishi, K. (1984b) Biosynthetic studies of ipomeamarone. *J. Chem. Soc., Chem. Commun.*, 372–374.

Schwarz, M.K. (1994) Ph.D. Thesis, Eidgenössische Technische Hochschule, Zürich, Switzerland.

Shiro, M., Sato, T., Koyama, H., Maki, Y., Nakanishi, K., Uyeo, S. (1966) The stereochemistry of taxinine: X-ray analysis of 2,5,9,10-tetra-*O*-acetyl-14-bromotaxinol. *J. Chem. Soc., Chem. Commun.*, 97–98.

Takemoto, T., Ogawa, S., Nishimoto, N. (1967) Inokosterone, an insect molting hormone from *Achyranthes fauriei. Yakugaku Zasshi*, **87**, 325–326.

Taylor, D.A.H. (1987) Azadirachtin. A study in the metholody of structure determination.

Thiem, J. and Meyer, B. (1979) Studies on the structure of chromomycin A_3 by ^1H and ^{13}C nuclear magnetic resonance spectroscopy. *J. Chem. Soc., Perkin Trans. II*, 1331–1336.

Turner, C.J., Tempesta, M.S., Taylor, R.B., Zagorski, M.G., Termini, J.S., Schroeder, D.R., Nakanishi, K. (1987) An NMR spectroscopic study of azadirachtin and its trimethyl ether. *Tetrahedron*, **43**, 2789–2803.

Woods, M.C., Miura, I., Nakadaira, Y., Terahara, A., Maruyama, M., Nakanishi, K. (1967) The ginkgolides V: Some aspects of their NMR specta. *Tetrahedron Lett.*, 321–326.

Woods, M.C., Chiang, H.C., Nakadaira, Y., Nakanishi, K. (1968) The nuclear Overhauser effect, a unique method of defining the relative stereochemistry and conformation of taxane derivatives. *J. Am. Chem. Soc.*, **90**, 522–523.

Zanno, P., Miura, I., Nakanishi, K., Elder, D. (1975) Structure of the insect phagorepellent azadirachtin. Application of PRFT/CWD carbon-13 nuclear magnetic resonance. *J. Am. Chem. Soc.*, **97**, 1975–1976.

2. APPLICATION OF LC/MS AND LC/NMR IN THE SEARCH FOR NEW BIOACTIVE COMPOUNDS FROM PLANTS OF THE AMERICAS

KURT HOSTETTMANN and JEAN-LUC WOLFENDER

Institut de Pharmacognosie et Phytochimie, Université de Lausanne, BEP, CH-1015 Lausanne, Switzerland

INTRODUCTION

Efficient detection and rapid characterisation of natural products play an important role as an analytical support in the work of phytochemists. The identification of a metabolite at the earliest stage of separation is a strategic element for guiding an efficient and selective isolation procedure. In this respect, the role of hyphenated techniques such as high performance liquid chromatography coupled to mass spectrometry (LC/MS) or to nuclear magnetic resonance (LC/NMR) has become of great interest.

The plant kingdom represents an extraordinary reservoir of novel molecules. Of the estimated 400,000–500,000 plant species around the globe, only a small percentage has been investigated phytochemically and the fraction submitted to biological or pharmacological screening is even lower. Since plants may contain hundreds, or even thousands of metabolites, there is currently a resurge of interest in the vegetable kingdom as a possible source of new lead compounds for introduction into therapeutical screening programmes. The rapid disappearance of tropical forests and other important areas of vegetation has meant that it is essential to have access to methods which lead to the rapid isolation and identification of bioactive natural products.

The approach adopted to obtain an exploitable pure plant constituent involves interdisciplinary work in botany, pharmacognosy, pharmacology, chemistry and toxicology. The plant material is extracted by solvents of increasing polarity. The obtained extracts are screened with different bioassays and submitted to fractionation with chromatographic techniques. This process is repeated until the isolation of a pure active constituent which is finally identified by spectroscopical methods (bioactivity-guided isolation).

By following this approach alone, there is a risk of unnecessarily isolating known plant constituents. Furthermore, interesting lead compounds which do not exhibit the tested activity will simply be missed. In order to avoid the time-consuming isolation of known constituents, hyphenated techniques such as LC/MS, LC/UV and recently even LC/NMR can be used at the earliest stage of separation to screen chemically the crude extracts (Figure 1)(Wolfender and Hostettmann 1995).

In this chapter, the role of LC chemical screening of crude extracts used as a complement to the biological screening will be particularly emphasised and mainly examples of analysis of plants from South America will be discussed.

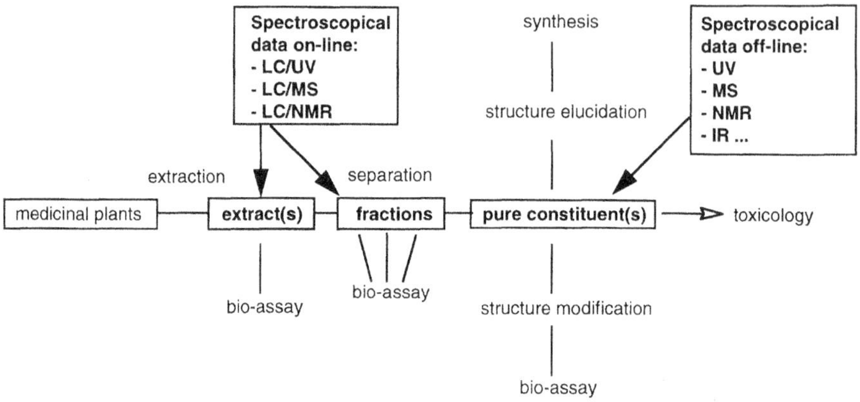

Figure 1 Procedure for obtaining the active principles from plants and use of LC hyphenated techniques as strategic analytical screening tools during the isolation process of a plant extract

HYPHENATION IN HIGH PERFORMANCE LIQUID CHROMATOGRAPHY (HPLC)

High performance liquid chromatography (HPLC) is used routinely in phytochemistry to "pilot" the preparative isolation of natural products (optimisation of the experimental conditions, checking of the different fractions throughout the separation) and to control the final purity of the isolated compounds (Kingston 1979). For chemotaxonomic purposes, the botanical relationships between different species can be shown by chromatographic comparison of their chemical composition. Comparison of chromatograms, used as fingerprints, between authentic samples and unknown compounds permits identification of drugs and/or search for adulteration. HPLC is thus the most well fitted technique for an efficient separation of crude plant extracts and can be coupled with different spectroscopic detection methods.

Liquid chromatography/UV photodiode array (LC/DAD-UV)

HPLC coupled with UV photodiode array detection (LC/UV) has been used since more than a decade by phytochemists for screening extracts (Hostettmann *et al.* 1984) and is now widely spread in many laboratories. The UV spectra of natural products give useful information on the type of constituents and also, as it is the case for polyphenols, information on the oxidation pattern. New instruments allow the recording of UV spectra of reference compounds in databases and computer matching can be realised automatically when screening for known constituents.

LC/DAD-UV however is limited to the detection of UV active constituents. The information recorded on-line are generally not sufficient for a complete identification and other on-line complementary spectroscopic information is needed.

Liquid chromatography/mass spectrometry (LC/MS)

If LC/UV has often been used for the analysis of crude plant extracts, LC/MS has been introduced more recently and is still not widely spread in the phytochemical community (Wolfender *et al.* 1994). At present, MS is one of the most sensitive methods of molecular analysis. Moreover, it has the potential to yield information on the molecular weight as well as on the structure of the analytes. Due to its high power of mass separation, very good selectivities can be obtained.

The coupling between LC and MS has not been straightforward since the normal operating conditions of a mass spectrometer (high vacuum, high temperatures, gas-phase operation, and low flow rates) are diametrically opposed to those used in high performance liquid chromatography (HPLC), namely liquid-phase operation, high pressures, high flow rates, and relatively low temperatures (Garteiz and Vestal 1985). Because of the basic incompatibilities between HPLC and mass spectrometry (MS), on-line coupling of these instrumental techniques has been difficult to achieve.

To cope with these different problems, a considerable number of LC/MS interfaces has been built up. These LC/MS interfaces must accomplish nebulisation and vaporisation of the liquid, ionisation of the sample, removal of the excess solvent vapour, and extraction of the ions into the mass analyser. Thus, different techniques may involve variations in the methods by which these steps are accomplished. The most commonly used LC/MS interfaces are particle beam (PB) (Wiloughby and Browner 1984), thermospray (TSP) (Blakley and Vestal 1983), atmospheric pressure chemical ionisation (APCI) (Bruins *et al.* 1987) and electrospray (ES) (Whitehouse *et al.* 1985). Each of these interfaces has its own characteristics and range of applications. At present, with the appropriate LC/MS configuration, the analysis of small non-polar molecules to very large polar molecules such as proteins is possible.

A number of LC/MS interfaces are suitable for the analysis of plant secondary metabolites. In our approach to LC/MS, mainly used for the HPLC screening of crude plant extracts, three interfaces, thermospray (TSP) (Blakley and Vestal 1983), continuous flow FAB (CF-FAB) (Caprioli *et al.* 1986) and electrospray (ES) (Whitehouse *et al.* 1985) have been investigated (Wolfender *et al.* 1995). They cover the ionisation of relatively small non-polar products (aglycones, 200 u) to high polar molecules (glycosides, 2000 u) (Figure 2). In our laboratory, thermospray is the most widely used interface.

Liquid chromatography/nuclear magnetic resonance (LC/NMR)

Nuclear magnetic resonance (NMR) spectroscopy is by far the most powerful spectroscopic technique for obtaining detailed structural information about organic compounds in solution. Its particular strength lies in its ability to differentiate between most structural, conformational and optical isomers. Given a molecular mass, NMR spectroscopy can usually provide all the additional information to unambiguously identify a completely unknown compound (Albert 1995). Thus LC/NMR represents potentially an important complementary hyphenated technique to LC/UV and LC/MS.

Proton detected high-performance liquid chromatography/nuclear magnetic resonance (LC/^{1}H-NMR), despite being known for over fifteen years (Watanabe *et al.* 1979), has not yet been a widely accepted technique, mainly because of its lack of sensitivity.

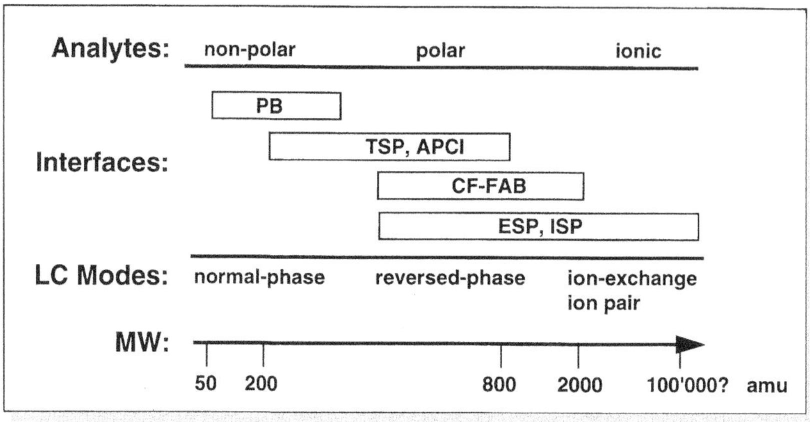

Figure 2 Summary of the ionisation range of various LC/MS interfaces

However, the recent progress in pulse field gradients and solvent suppression, the improvement in probe technology and the construction of high field magnets have given a new impulse to this technique.

While LC coupling itself was rather straightforward compared to LC/MS (Niessen and van der Greef 1992), one of the main problems of LC/NMR was the difficulty of observing analyte resonances in the presence of the much larger resonances of the mobile phase. This problem was even worsened in the case of typical LC reversed phase operating conditions, where more than one protonated solvent was used and where the resonances changed frequencies during the analysis in the gradient mode. Furthermore, the continuous flow of sample in the detector coil complicated solvent suppression. These problems have now been overcome thanks to the development of fast reliable and powerful solvent suppression techniques such as WET (Smallcombe *et al.* 1995), which produced in both on-flow and stopped-flow modes high quality spectra. These techniques consist of a combination of pulsed field gradients, shaped rf pulses, shifted laminar pulses and selective ^{13}C decoupling and are much faster than classical presaturation techniques previously used in this field (Smallcombe *et al.* 1995). Thus, in reversed-phase HPLC conditions, non-deuterated solvents such as MeOH or MeCN can be used, while water is replaced by D_2O.

In Figure 3, a comparison of the ^{1}H-NMR spectra of a naphthoquinone recorded in a MeCN-D_2O mixture with and without solvent suppression is presented. In the spectrum without solvent suppression (Figure 3B), the peaks of MeCN and HOD are of several order of magnitude higher than those of the analyte, but are selectively eliminated after application of the WET sequence.

In contrast with conventional NMR probes, in LC/NMR the sample to be analysed is not placed in a rotating 5 mm i.d. tube. The mobile phase is flowing continuously through a non-rotating glass tube connected at both ends with HPLC tubing. The detection is made, in most LC/NMR systems, in flow cells of 60–180 μl with 2–4 mm i.d. placed in the active NMR coil region of a conventional probe body (Albert 1995).

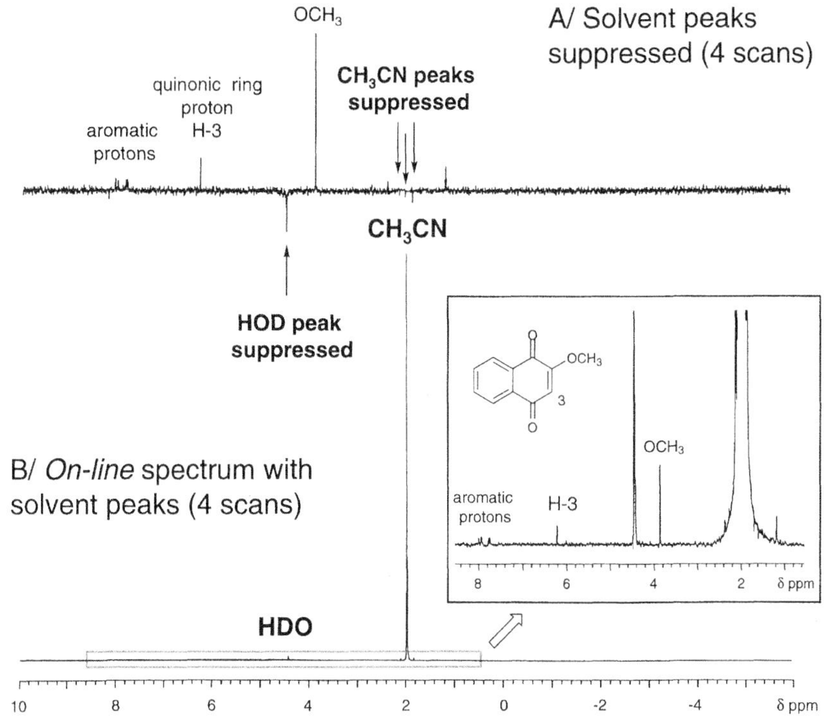

Figure 3 Example of the typical level of on-line solvent suppression obtained using the WET sequence. Spectra A and B have been measured for a naphthoquinone in the LC/^{1}H-NMR analysis of *Swertia calycina* (Wolfender *et al.* 1997). Note the total suppression of the residual HOD signal, together with the huge CH_3 CN peaks and its two ^{13}C satellites

The choice of the volume of the flow cell is dependent on the sensitivity and/or the LC resolution required for each specific application.

Setup used for the LC/UV/MS and LC/NMR analyses

In order to perform LC/UV/MS and LC/NMR analyses of crude plant extracts, a setup allowing the use of standard HPLC conditions (1 ml/min, gradient with aqueous solvent systems) with all the detectors has been used. An HPLC pump equipped with a gradient controller provides the eluent for the HPLC separation of the plant extracts (Figure 4). A reversed-phase column (i.d. about 4 mm) is most often used. At the column outlet, the eluent passes through a photodiode array detector equipped with a high pressure cell.

For LC/MS and with the aim of using the same columns without changing the chromatographic conditions, the buffer needed for the various LC/MS interfaces is added post-column after the UV detector. For LC/TSP-MS, all the eluent is entering into the MS, while for CF-FAB and ES, the eluent is splitted (Figure 4).

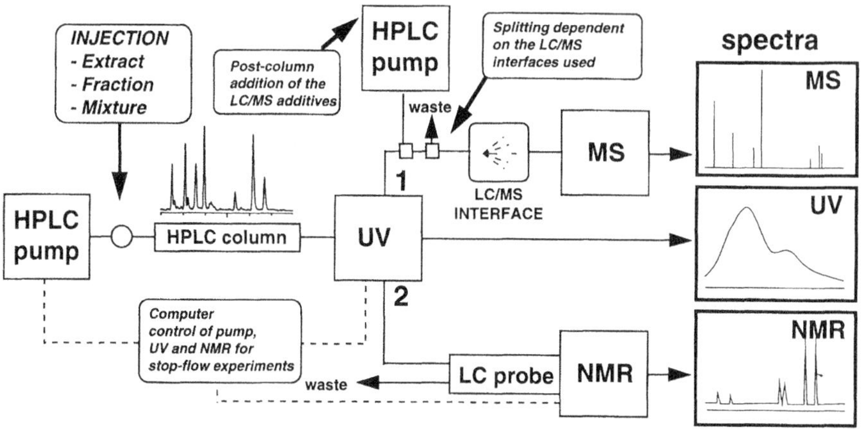

Figure 4 Schematic representation of the experimental setup used for LC/UV/MS (1) and LC/UV/NMR (2) analyses

For LC/NMR, no post-column addition is necessary and the water is replaced by D_2O. As for LC/MS, the separation is first monitored by a UV detector. In the online mode, all the eluent passes through the UV detector and then through the LC/NMR probe without splitting. In the stopped-flow mode, the HPLC pump is stopped as soon as a peak of interest reaches the centre of the NMR detection cell. A computer controlling the LC pump, the UV detector and the NMR is used to synchronise this operation. The UV monitor is used to detect the peak and the HPLC is stopped after a delay corresponding to the time needed by this peak to reach the NMR cell.

CHEMICAL SCREENING OF GENTIANACEAE EXTRACTS BY COMBINED LC/UV, LC/MS AND LC/NMR

Numerous phytochemical studies have been performed on species of the family Gentianaceae in our laboratories (Wolfender and Hostettmann 1993; Rodriguez *et al.* 1995). In Europe, gentians and especially the yellow gentian (*Gentiana lutea*) are well known for their bitterness and are used to cure digestive problems. This bitter character is mainly due to their content in monoterpene glycosides of the secoiridoid class. In the traditional Indian medicine, these plants are used to cure depression. Research started in 1972 by Indian scientists (Bhattacharya *et al.* 1972) showed that the gentian extracts, especially those of *Canscora decussata*, stimulated the central nervous system. Our group, together with Suzuki (Suzuki *et al.* 1981), demonstrated that xanthones, rare polyphenols occurring in these plants, presented a remarkable *in vitro* inhibition of monoamine oxidase (IMAO), an enzyme which plays an important role in the regulation of neurotransmitters in the central nervous system. As it is known that patients suffering from depression lack some neurotransmitters, xanthones are potentially interesting as new antidepressive drugs.

In order to find new xanthones, numerous gentians have been screened chemically by both LC/UV and LC/MS. Minute amounts of plant material were required for these analyses and one plant specimen was often sufficient. The different organs of the gentians were extracted successively with dichloromethane and methanol. Without prepurification, the crude extracts (ca. 200 µg) were directly separated on a 4 mm i.d. reversed-phase column with an acidic acetonitrile-water gradient (1 ml/min). Under these conditions it was possible to obtain a satisfactory separation of the main constituents. The UV spectra recorded on-line allowed the distinction of xanthones from the other constituents. These polyphenols show characteristic UV spectra usually with four bands of decreasing intensity between 200–400 nm. The other metabolites were generally secoiridoids (one UV maximum at ca. 240 nm) or flavones (two UV maxima at 200–400 nm). For the LC/MS analysis of these extracts, the TSP interface which is well adapted for the ionisation of moderately polar molecules, was chosen.

For example, the LC/UV/MS analysis of a methanolic root extract of a Gentianaceae from Argentina, *Gentianella cabrerae* presented different peaks characteristic of secoiridoids, flavone C-glycosides and xanthones. In order to show the type of information recorded on-line, the LC/TSP-MS and LC/UV spectra of two polyphenols have been selected (Figure 5). The UV spectrum of **2** was characteristic for a xanthone substituted in 1, 3, 5, 8. The LC/TSP-MS spectrum of this peak exhibited a protonated molecular $[M+H]^+$ ion at 275 u. The molecular weight of this xanthone was thus 274 u. As the mass of a bare xanthone nucleus is 196 u, this indicated that the polyphenol was tetrasubstituted by three hydroxyls (3 × 16 u) and one methoxyl group (1 × 30 u). A comparison of the on-line UV spectra with those of tetrasubstituted xanthones from an in-house UV data base confirmed that this compound was indeed bellidifolin, a widespread xanthone in the genus *Gentianella*. Compound **1** presented a UV spectrum characteristic for a flavonoid compound. Its LC/TSP-MS spectrum was more complicated than that recorded for **2** and showed different fragments. **1** exhibited a protonated molecular ion $[M+H]^+$ at m/z 463 and fragments $[M+H-90]^+$ and $[M+H-120]^+$ at m/z 373 and 343 which are characteristic for the cleavage of flavone C-glycosides. According to these data, **1** was most probably a flavone C-glycoside substituted by three hydroxyl and one methoxyl groups. Two isomeric flavones occurring in Gentianaceae, isoscoparin and swertiajaponin, fitted with such data. An LC/TSP-MS/MS was finally performed in order to confirm that **1** was indeed swertiajaponin. Examples of LC/MS/MS of flavones will be explained in more detail in the case of *G. ottonis* (see below).

LC/UV/MS ANALYSES OF DIFFERENT *HALENIA* SPECIES

Following the same analytical approach as for *G. cabrerae*, other South American Gentianaceae have been efficiently screened by LC/UV/MS. Among them various species of *Halenia* were analysed: *H. woodsoniana* from Panama, *H. campanulata* and *H. asclepiadea* both from Columbia. The compositions of these Gentianaceae plants were compared with that of another species: *Halenia corniculata* from Mongolia. Plants of the genus *Halenia* (ca. 80 species) were particularly interesting because very little is known on their constituents (Recio-Iglesias *et al.* 1992).

As shown on Figure 6, the LC/UV analyses of both methanolic and dichloromethanic extracts of *Halenia corniculata* revealed the presence of numerous xanthones

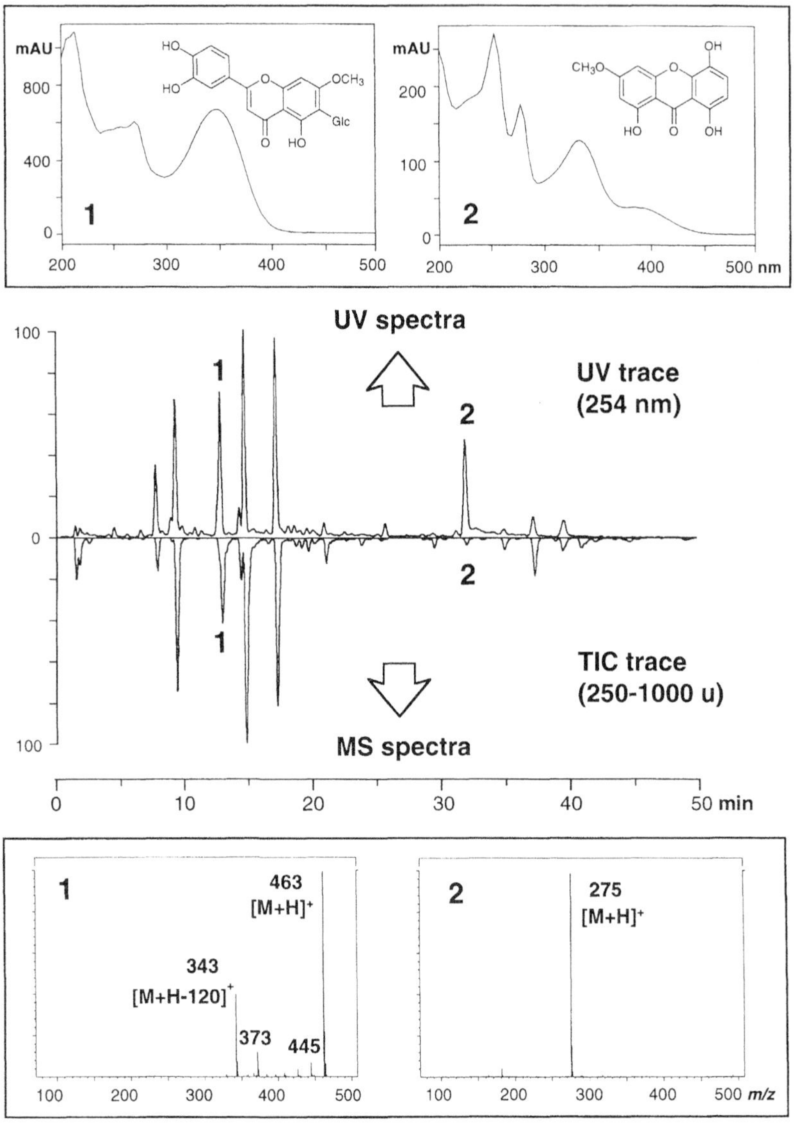

Figure 5 LC/UV/MS analysis of the methanolic extract of *Gentianella cabrerae* (Gentianaceae). The UV and TSP-MS spectra of a flavone C-glycoside, swertiajaponin (**1**) and a xanthone, bellidifolin (**2**) are displayed. The UV trace was recorded at 254 nm and UV spectra from 200–500 nm. The LC/TSP-MS trace was recorded from 150–900 u. HPLC: Column, RP-18 NovaPak (4 μm, 150 × 3.9 mm i.d.); gradient, CH_3CN-H_2O (0.1% TFA) 5:95 → 65:35 in 50 min (1 ml/min), NH_4OAc 0.5 M (0.2 ml/min post-column). TSP: positive ion mode; filament off; vaporiser 100°C; source 280°C

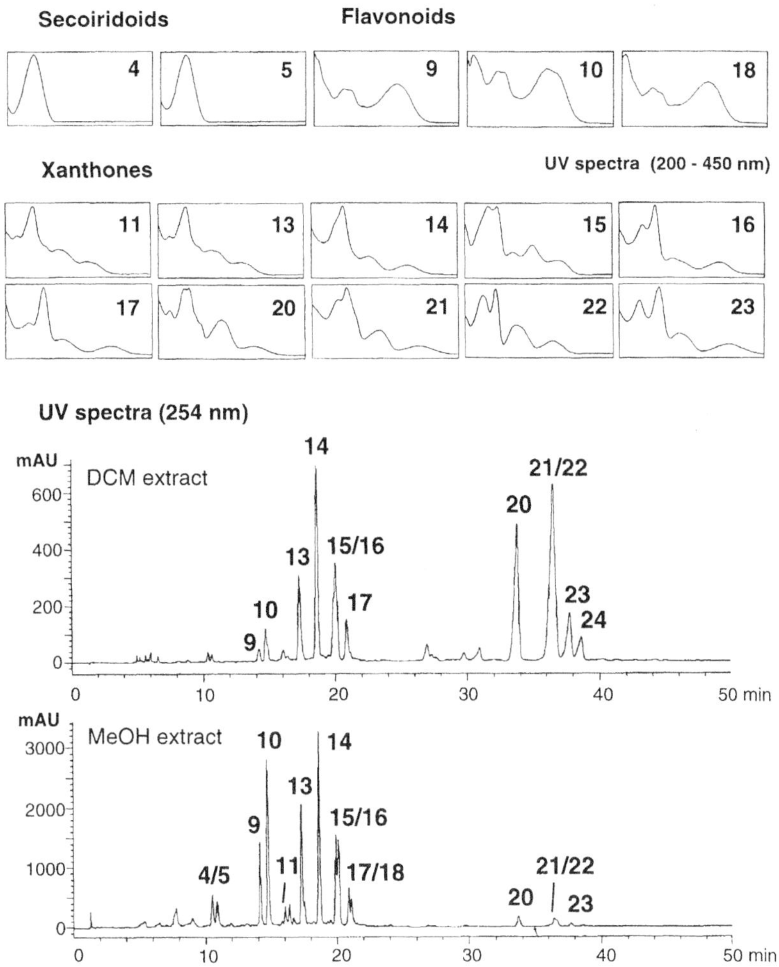

Figure 6 LC/UV analyses of the dichloromethane and methanolic extracts of *Halenia corniculata*. The UV spectra (200–450 nm) of the different types of constituents obtained on-line are displayed separately. Same conditions as in Figure 5

exhibiting typical UV spectra with four or five absorption bands of decreasing intensity (**11**, **13**–**17**, **20**–**23**). Some flavonoids **9**, **10** and **18** were also present together with bitter principles **4** and **5**. The LC/TSP-MS analyses of these extracts allowed the molecular weight determination of all these constituents and was particularly efficient for the localisation of the different xanthone aglycones and glycosides present.

Two types of xanthone O-glycosides, together with xanthone aglycones, were detected in this extract. As LC/MS can also be used as a very selective detector, a display of the selective ion traces of the molecular ions of the aglycones and glycosides of these xanthones was used for their specific detection. As shown in Figure 7 for example,

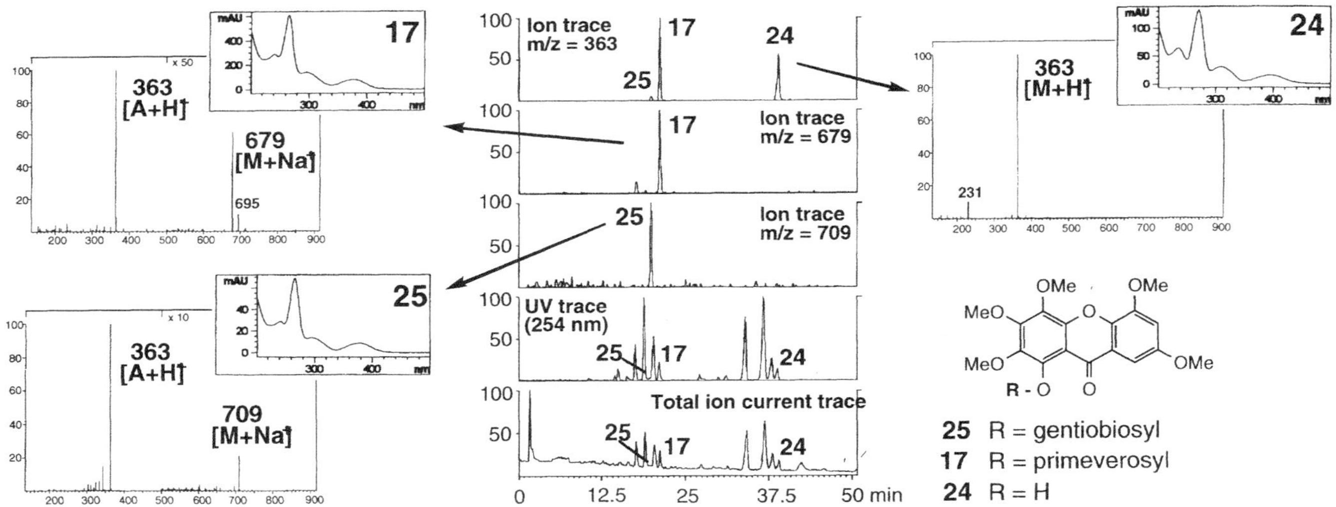

Figure 7 LC/UV/MS analysis of the crude dichloromethane extract of *Halenia corniculata*. The UV and MS spectra of three related xanthones (aglycone **24** and its two glycoside derivatives **17** and **25**) are displayed. The ion m/z 363 is the main fragment and corresponds to the aglycone moiety $[A+H]^+$. The ion traces displayed at m/z 679 and 709 correspond to the molecular adduct ions $[M+Na]^+$ of primeveroside **17** and gentiobioside **25** respectively. Same conditions as in Figure 5

three xanthones of *H. corniculata*, **17**, **24** and **25** appeared in the single ion trace *m/z* 363, which indicated the presence of a common hexasubstituted aglycone (one hydroxyl and five methoxyl groups) in each case. This was confirmed by similar UV spectra for all the three xanthones. The glycosides **25** and **17** were detected separately at *m/z* 709 and 679, respectively. These ions corresponded to weak molecular sodium adducts $[M+Na]^+$ of the xanthone glycosides. Important fragments corresponding to the aglycone moiety $[A+H]^+$ were also observed in the LC/TSP-MS spectra at *m/z* 363 (Figure 7). The difference of 316 u between the ions $[M+Na]^+$ and $[A+H]^+$ in the TSP spectrum of **17** was characteristic of the loss of a disaccharide moiety constituted of hexosyl and pentosyl units (-pentosyl (132) -hexosyl (162) -Na (23) +H (1)). Likewise, a difference of 346 u (-hexosyl (162) -hexosyl (162) -Na (23) +H (1)) for **25** indicated the presence of a dihexosyl derivative. As primeverosyl and gentiobiosyl are the only disaccharide residues corresponding to these masses which have been found to date in the Gentianaceae family, **25** and **17** were respectively the gentiobiosyl and the primeverosyl derivatives of the corresponding free aglycone **24**. The UV spectra of the three compounds were very similar but did not correspond to any of the available data. Thus, a rare oxidation pattern was indicated. Following this approach, other series of aglycones and corresponding glycosides (**11**, **13–17**, **20–25**) were detected on-line. The isolation of the most interesting xanthones was subsequently undertaken (Rodriguez *et al.* 1995). With the aid of different NMR experiments, the structures of six isolated xanthone glycosides were fully assigned. This information together with the on-line data obtained was used to confirm the on-line identification of eleven other xanthones of the extract. As expected from the on-line data, nine xanthones were new natural products (Rodriguez *et al.* 1995).

The extracts of the other *Halenia* species *H. woodsoniana*, *H. campanulata* and *H. asclepiadea* were analysed by LC/UV/MS following the same procedure. The LC/UV chromatograms of the methanolic extract of the different *Halenia* species are shown in Figure 8. The UV and MS information obtained for each peak allowed a very precise comparison of the extract composition of these plants. These data were very important to ascertain the peak assignment because the retention time of certain metabolites may vary a few between each analysis. Indeed all these Gentianaceae extracts were not analysed at the same time but were screened at the date of their arrival in our laboratory. The comparison of the LC/UV chromatograms of these different *Halenia* species showed an important homogeneity of chemical composition from a qualitative viewpoint. From this first chemical screening, it can be concluded that species of the genus *Halenia* contain tetra- penta- and hexa- oxygenated xanthones (**11–25**) (See structure in Figure 9). All these xanthones were O-substituted at least in positions 1, 2, 3, 5 and 1, 2, 3, 7 and most of them were highly methoxylated. The same type of xanthones was reported from *H. elliptica* (Dhasmana and Garg 1990). These xanthones can thus be considered as characteristic chemotaxonomic markers for this genus. Secoiridoids such as vogeloside (**4**) or epi-vogeloside (**5**) were present in all the species except *H. asclepiadea*. Mangiferin (**3**), xanthone **12** and its corresponding glycoside **19** were clearly discernible in *H. asclepiadea* and *H. campanulata* but were absent in *H. woodsoniana* and *H. corniculata*. Flavonoids were present in all species but their pattern was less characteristic. They were mostly present in *H. corniculata*. Two important peaks **7** and **8** appeared

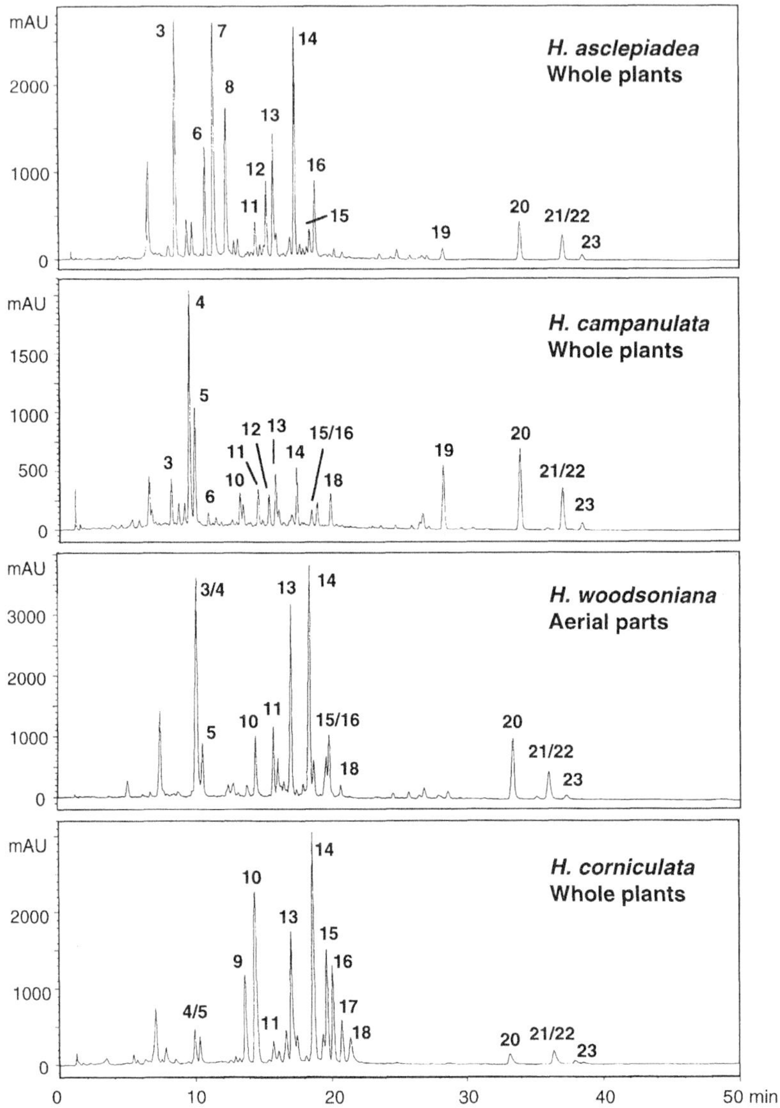

Figure 8 Comparison of the LC/UV chromatogram of the methanolic extracts of various *Halenia* species. Same conditions as in Figure 5

only in *H. asclepiadea*. These two compounds have not yet been isolated because of the restricted amount of vegetal material. According to their UV spectra (max 244 nm) and their molecular weight (MW 402), these two compounds are most probably two iridoid glycosides. Further investigations have still to be performed in order to characterise fully these two constituents. This chemotaxonomic information shows that within the Gentianaceae family, the genus *Halenia* is closely related to the genus *Frasera* which

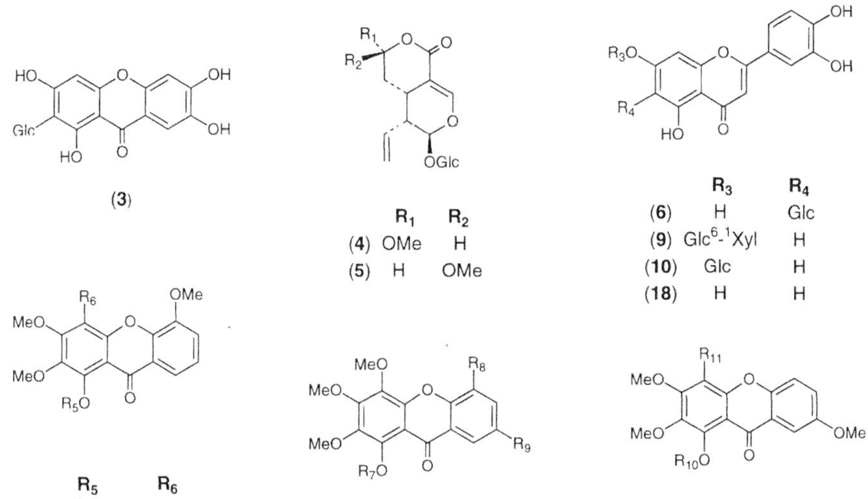

	R_1	R_2
(4)	OMe	H
(5)	H	OMe

	R_3	R_4
(6)	H	Glc
(9)	Glc^{6}-^{1}Xyl	H
(10)	Glc	H
(18)	H	H

	R_5	R_6
(11)	Glc^{6}-^{1}Xyl	H
(13)	Glc^{6}-^{1}Glc	H
(14)	Glc^{6}-^{1}Xyl	H
(20)	H	OMe
(21)	H	OMe

	R_7	R_8	R_9
(12)	Glc^{6}-^{1}Xyl	H	OH
(17)	Glc^{6}-^{1}Xyl	OMe	OMe
(19)	H	H	OH

	R_{10}	R_{11}
(15)	Glc^{6}-^{1}Xyl	H
(16)	Glc^{6}-^{1}Xyl	H
(22)	H	OMe
(23)	H	OMe

Figure 9 Structure of all the constituents identified in the different Mongolian and South American *Halenia* species

contains the same type of xanthones in addition with tri-oxygenated ones (Dreyer and Bourell 1981).

LC/TSP-MS and LC/ES-MS of *Halenia corniculata*

The use of the thermopray LC/MS interface for the analysis of polyphenols in the extracts of the different *Halenia* species has allowed the detection of most of the UV active contituents of these plants. However, other LC/MS analyses with different complementary ionisation techniques are often important for an unambiguous online determination of molecular weights. Furthermore, some constituents are not ionised with a given LC/MS interface while they give intense response with other ones. The analysis of *Halenia corniculata* for example by LC/TSP-MS and LC/ES-MS demonstrated also the importance of an LC/MS analysis with two independent methods especially for screening unknowns.

In LC/ES-MS, the best ionisation was obtained in the negative ion mode (N.I.) using TFA as buffer. Ionisation of all the glycosides found in the extract was observed in the LC/ES-MS (N.I.). In this mode, however, only some of the aglycones were correctly ionised (Wolfender *et al.* 1995). The LC/ES-MS base peak trace showed a better chromatographic resolution than the corresponding LC/TSP-MS trace. The ES spectra of the flavonoids, the xanthones as well as the secoiridoid glycosides exhibited clearly discernible TFA anion adducts $[M+CF_3COO^-]$, together with deprotonated $[M-H]^-$ ions.

No aglycone ions or other fragments were observed. For polyphenols, the TSP (P.I.) spectra recorded on-line exhibited $[M+H]^+$ ions, together with base peak $[A+H]^+$ aglycone ions. In the case of diglycosides, the loss of the terminal sugar unit was also observed. Thus according to the ionisation mode choosen, the spectra of a given compound will differ considerably. A good knowledge of the ionisation process is thus necessary for a correct peak assignment (Wolfender *et al.* 1995).

In this example, it is also shown that compound **26**, a minor secoiridoid glycoside ester of the extract, was not ionised in TSP but exhibited in LC/ES-MS a deprotonated molecular ion $[M-H]^-$ at m/z 697 and a strong TFA adduct $[M+CF_3COO^-]$ at m/z 811, indicating a molecular weight of 698 u (Figure 10). The UV spectrum of **26** (maxima

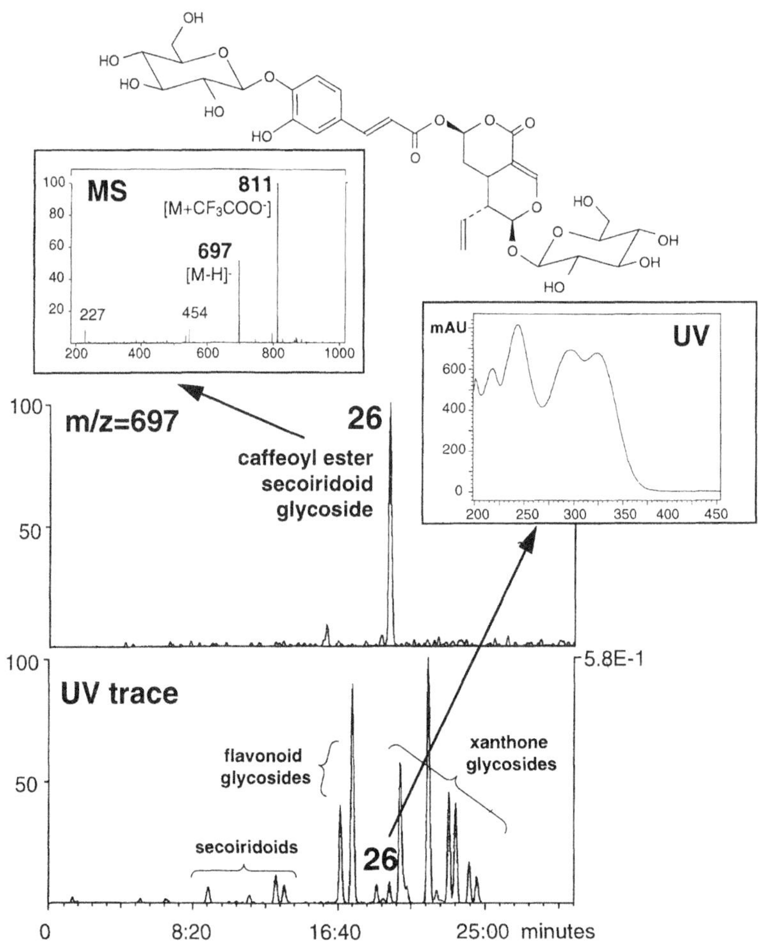

Figure 10 LC/ES-MS (N.I.) of the MeOH extract of *H. corniculata* (Gentianaceae) and spectrum of the secoiridoid glycoside corniculoside **26** not ionised by LC/TSP-MS. Same HPLC conditions as in Figure 5

at 219, 244, 297, 321 nm) differed from those usually encountered in Gentianaceae species extracts. The two UV maxima at 219, and 244 nm, together with the retention time (ca. 16 min) suggested that **26** was an iridoid-type compound. The two other maxima at 297 and 321 nm indicated the presence of an aromatic part. No previous references were found in the literature for such a compound corresponding to this molecular weight. Thus a targeted isolation of **26** was undertaken (Rodriguez *et al.* 1996). Different hydrolyses of **26**, followed by the HPLC of the degradation products, showed that it was constituted of a caffeoyl, a secoiridoid and two hexosyl moieties. Its structure was fully established by different NMR experiments as 7-β-[4′-O-(β-D-glucopyranosyl)-*trans*-caffeoyloxy] sweroside, and this new natural product was called corniculoside (Rodriguez *et al.* 1996). This example shows well that, with the use of LC/TSP-MS only, such a compound would have been missed during the chemical screening of the extract.

LC/UV/MS SCREENING OF *LISIANTHIUS* SPECIES

Continuing our studies on the chemical composition of South American Gentianaceace, we were interested by species of the genus *Lisianthius*. Plants of this genus are geographically limited to tropical America. The extracts of *L. jefensis, L. seemanii* and *L. skinneri* were analysed by LC/UV/MS. The chromatographic profiles of these different species were very similar. Contrary to the extracts of *Halenia*, no xanthones were found and different peaks exhibiting UV spectra with one maximum at 240 nm attributable to secoiridoids were present. A very major 'secoiridoid-like' constituent (**28**) was present in all extracts. Its chromatographic behaviour denoted of a rather non polar secoiridoid. The LC/TSP-MS spectrum of **28** recorded with ammonium acetate as buffer, exhibited an intense pseudomolecular ion at 734 u and an important fragment at *m/z* 359 (Figure 11). This fragment was most probably characteristic of a sweroside moiety. Sweroside (**27**) (MW 358) is indeed a widespread secoiridoid in the Gentianaceae family. This compound was also found as a very minor constituent of the extract by displaying the ion trace of *m/z* 359. The pseudomolecular ion of the main constituent *m/z* 734 was relatively important and very unusual for a secoiridoid glycoside. A single LC/TSP-MS run was insufficient for the attribution of the pseudomolecular ion 734 to $[M+NH_4^+]$ or $[M+NH_4^+-H_2O]$ since no $[M+H]^+$ ion was present. Thus, a second LC/TSP-MS analysis of the extract was carried out with another buffer to confirm the molecular weight of this secoiridoid. Diaminoethane (0.2 ml/min, 0.5 M) which has a higher proton affinity than ammonium acetate was added post-column. This buffer is indeed known to produce only $[M+61]^+$ ($[M+(H_2N(CH_2)_2NH_3)^+]$) pseudomolecular ion adducts (Chace and Callery 1992). The second analysis of the extract with this buffer exhibited an important pseudomolecular ion $[M+61]^+$ at *m/z* 777 for the main constituent, confirming the molecular weight to be 716 u (Wolfender *et al.* 1994). These on-line information permitted thus to assign this main constituent named lisianthioside (**28**) to be a dimer of sweroside (**27**). The full structural determination of this novel type of secoiridoid was confirmed after isolation of the pure substance (Hamburger *et al.* 1990).

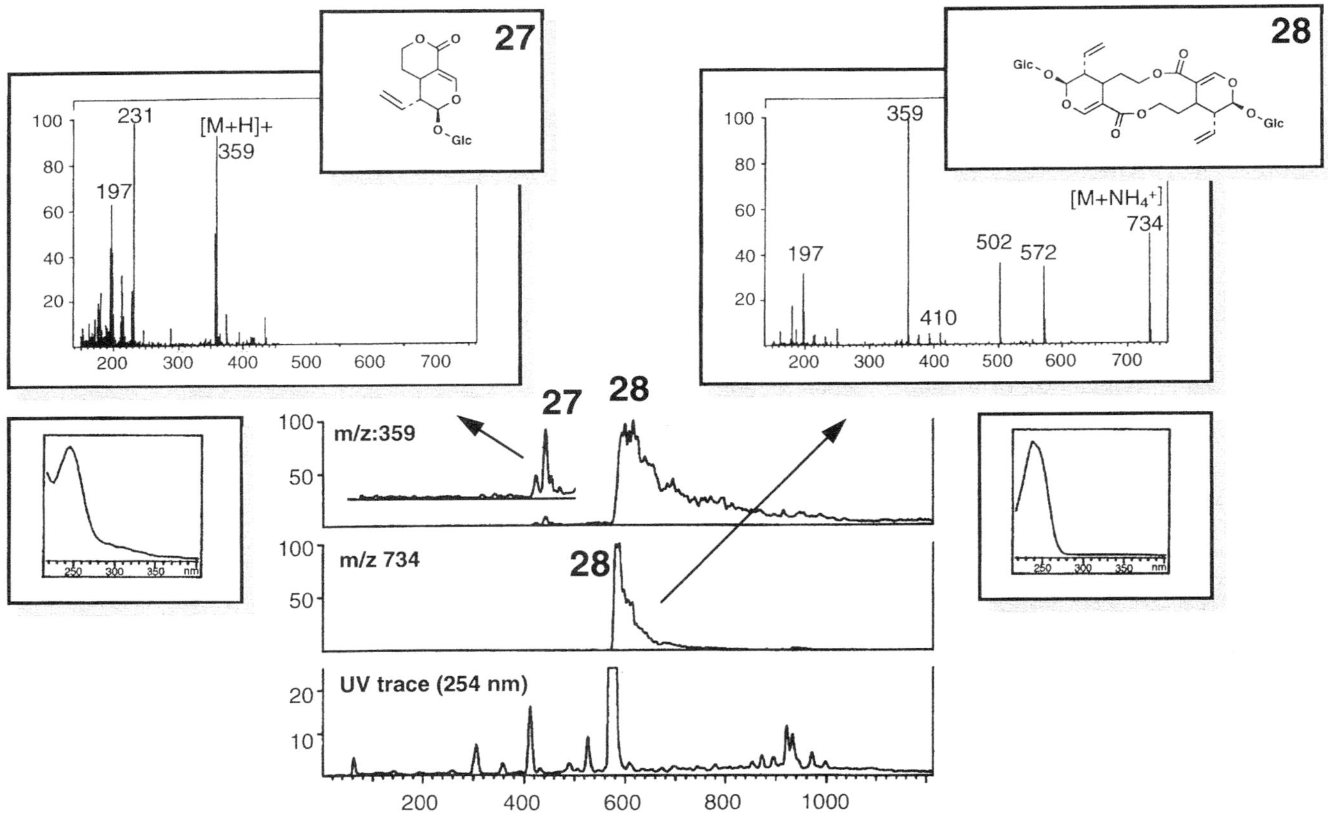

Figure 11 LC/TSP-MS analysis of the root methanolic extract of *Lisianthius seemanii* (Gentianaceae) with the spectra of lisianthioside (28) and sweroside (27). HPLC: RP-18 Nova-Pak (4 µm, 150 × 3.9 mm i.d.); gradient, CH_3CN-H_2O 5:95 → 70:30 in 50 min (1 ml/min). TSP: Vaporiser, 100°C; Source, 250°C. Ammonium acetate buffer 0.5 M (0.2 ml/min, post-column)

LC/UV/MS, LC/MS/MS AND LC/NMR ANALYSIS OF THE DICHLOROMETHANE EXTRACT OF *GENTIANA OTTONIS*

The LC/UV/MS analysis of Gentianaceae extracts gives useful structural information on-line for a partial identification of their metabolites. As shown with the extracts of *Halenia*, a full identification of some known metabolites is possible by comparison of their UV spectra and molecular weights with known standards, however for a full identification on-line complementary information are needed. In order to obtain more detailed structural information, the extracts of different Gentianaceae plants have been analysed also by LC/NMR and LC/MS/MS. In order to illustrate this approach, the analysis of the root methanolic extract of a Gentianaceae from Chile, *Gentiana ottonis*, is discussed.

The LC/UV analysis of this extract (Figure 12) presented peaks with UV spectra characteristic for secoiridoids, flavones and xanthones. The LC/TSP-MS analysis allowed the attribution of molecular weights of all the compounds. Secoiridoid **29** exhibited $[M+H^+]$ and $[M+NH_4^+]$ ions at *m/z* 375 and 392, together with a fragment at 213 characteristic for the loss of an hexosyl moiety. This MS information suggested that **29** could be swertiamarin, a widespread secoiridoid of the Gentianaceae family. Compounds **3**, **10** and **30** presented all fragments characteristic for C-glycosides (losses of 90 and 120 u). According to their UV spectra, **30** and **10** (MW 448 and 446) were respectively tri- and tetraoxygenated C-glycoside flavones. Compound **3** (MW 422) was readily identified as the widespread 1,3,6,7-tetrahydroxy-2-C-glucosylxanthone, mangiferin (Hostettmann and Hostettmann 1989). Xanthones **2**, **31–33** presented all similar UV spectra, suggesting an identical oxidation pattern. Compounds **32** and **33** presented both intense ions at *m/z* 261 characteristic for tetrahydroxylated xanthones. Substance **32** exhibited also a protonated molecular ion $[M+H]^+$ at *m/z* 423. This additional hexosyl moiety (loss of 162 u) suggested that **32** was the corresponding glucoside of aglycone **33**. The LC/TSP-MS spectra of **31** and **2** indicated the presence of another couple of related xanthones with a common aglycone ion at *m/z* 275, characteristic for substitution by one methoxyl and three hydroxyl groups.

In order to confirm these attributions and to obtain more structure information on-line, the same extract of *G. ottonis*, was submitted to an on-line LC/^{1}H-NMR analysis. The same LC conditions as for the LC/UV/MS analysis were used, except that the water of the LC gradient system was replaced by D_2O. However the quantity of extract injected onto the column was increased because of the relative insensitivity of LC/NMR compared to LC/UV and LC/MS detection. As the extract of *G. ottonis* was rather complex, only 0.4 mg could be injected onto the column without a loss of resolution. For this reason, the LC/NMR analysis was performed in the stopped-flow mode using the same gradient conditions as for the LC/UV/MS analysis but H_2O was replaced by D_2O. In this mode, the HPLC flow was stopped as soon as a UV active constituent of the extract reached the LC/NMMR cell. For simpler Gentianaceae extracts such as those of *Swertia calycina* which allow more material to be loaded on the column, on-flow LC/NMR runs are also possible (Wolfender *et al.* 1997). In this case, the measurement of the LC/^{1}H-NMR spectra of the main constituents can be performed without stopping the flow rate. The acquistion of the LC/^{1}H-NMR spectra

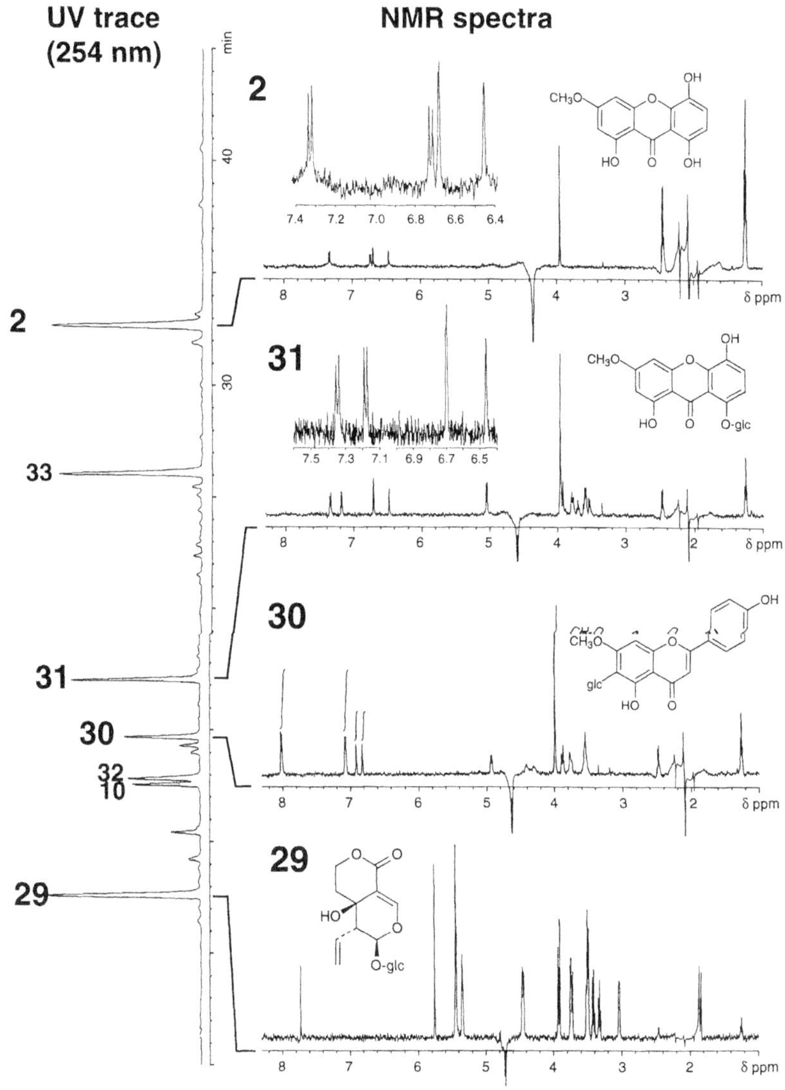

Figure 12 LC/UV chromatogram of *G. ottonis* (vertical display) with the stopped-flow LC/^{1}H-NMR spectra of 2 and 29–31. On these LC/^{1}H-NMR spectra, the coupling constants and the integration of the aromatic signals were easily measured

with solvent supression was then carried out for each constituent until that a satisfactory signal-to-noise ratio was obtained for the peak of interest (ca. 512 transients, acquisition time ca. 10 min). In order to illustrate the type of results obtained the LC/^{1}H-NMR of four types of metabolites (**2** and **29–31**) are displayed (Figure 12). The secoiridoid glycoside **29** presented signals characteristic for a glucosyl moiety between 3 and 4 ppm, together with an anomeric proton at 4.45 ppm. A singlet at 7.75 was chracteristic for the H-3 methine proton of most secoiridoids. Signals between 5 and 6 ppm were attributable to the proton of an exocyclic double bound and to H-1. A careful comparison of the LC/^{1}H-NMR spectra of these secoiridoids demonstrated clearly the absence of the H-5 signal in the case of **29**, proving an hydroxylation at C-5. A comparison of these LC/^{1}H-NMR data with those of literature allow an unambiguous identification of **29** as swertiamarin, confirming thus the attribution made according to the LC/UV/MS results.

Xanthone O-glycoside **31** (MW 436) presented signals characteristic for a O-glucosyl moiety (δ 3.3–5.1) and one methoxyl group (δ 3.96). A pair of *meta* coupled aromatic protons (1H, δ 6.70, *d*, $J=2.0$, H-4 and 1H, δ 6.47, *d*, $J=2.0$, H-2) was indicative for a 1,3-disubstituted A-ring. The B-ring protons exhibited a pair of *ortho* coupled protons (1H, δ 7.34, *d*, $J=8.8$, H-6 and 1H, 7.17, *d*, $J=8.8$, H-7). This information, together with UV data, confirmed a 1,3,5,8 substitution pattern for **31**. The LC/^{1}H-NMR of **2**, aglycone of **31**, also exhibited a pair of *meta* and *ortho* coupled protons, as well as a methoxyl group. However, with respect to **31**, one of the *ortho* protons was shifted upfield indicating a glycosylation on the B-ring (at C-5 or C-8). A careful comparison of the LC/UV data with authentic samples allowed finally the identification of **2** and **31** respectively as 1,5,8-trihydroxy-3-methoxyxanthone (bellidifolin) and 1,5-dihydroxy-3-methoxy-8-O-glucosylxanthone (8-O-glucosylbellidifolin) (Hostettmann and Jacot-Guillarmod 1978). Similar deductions allowed the identification of **33** and **32** as 1,3,5,8-tetrahydroxyxanthone (desmethylbellidifolin) and 1,3,5-trihydroxy-8-O-glucosylxanthone (8-O-glucosyldesmethybellidifolin) (Hostettmann and Jacot-Guillarmod, 1978).

The flavone C-glycoside **30** (MW 446) presented signals of 6 aromatic protons (δ 6.8–8.1) and one methoxyl group (δ 4.0), together with those of the C-glycoside moiety (3, 5-5.0). A pair of symmetric *ortho* coupled protons (2H, δ 7.06, *d*, $J=8.3$, H-3′,5′ and 2H, δ 8.00, *d*, $J=8.3$, H-2′,6′) was characteristic for a B-ring substituted in C-4′. The singlet at 6.8 was attributable to H-3. The other singlet at δ 6.9 was due to proton either at position C-6 or C-8 on the A-ring. At this stage, the LC/UV/MS together with the LC/^{1}H-NMR data did not allow a full identification of **30**. In order to ascertain the position of C-glycosylation, a LC/MS/MS experiment was performed. As shown in Figure 13, the LC/TSP-MS/MS spectrum of **30** obtained by choosing the $[M+H-120]^+$ as parent ion exhibited fragments at m/z 191 and 163 characteristic for 6-C-glycosylated flavones. Indeed it is known that TSP-MS/MS spectra of the $[M+H-120]^+$ fragments of isomeric C-glycosylflavones show different specific daughter ions (Rath *et al.* 1995). Indeed, the A-C-ring fragments issued from a retro Diels-Alder cleavage are only observable for the C-6 position isomers, as it was the case for **30**. Furthermore the fragment observed at m/z 121 was indicative for a monohydroxylated B-ring, confirming the position of the methoxyl group on the A-ring.

This UV, MS, ^{1}H-NMR and MS/MS information allowed the identification of **30** as the known 5,4′-dihydroxy-7-methoxy-6-C-glucosylflavone (swertisin). Similarly, the flavone C-glycoside **10** (MW 448) was identified as the common 5,7,3′,4′-tetrahydroxy-6-C-glucosylflavone (isoorientin).

This example showed that LC/NMR and LC/MS/MS represent an excellent complement to LC/UV/MS. The combined use of all these hyphenated techniques allow an efficient structure identification of the main constituents of an extract on-line.

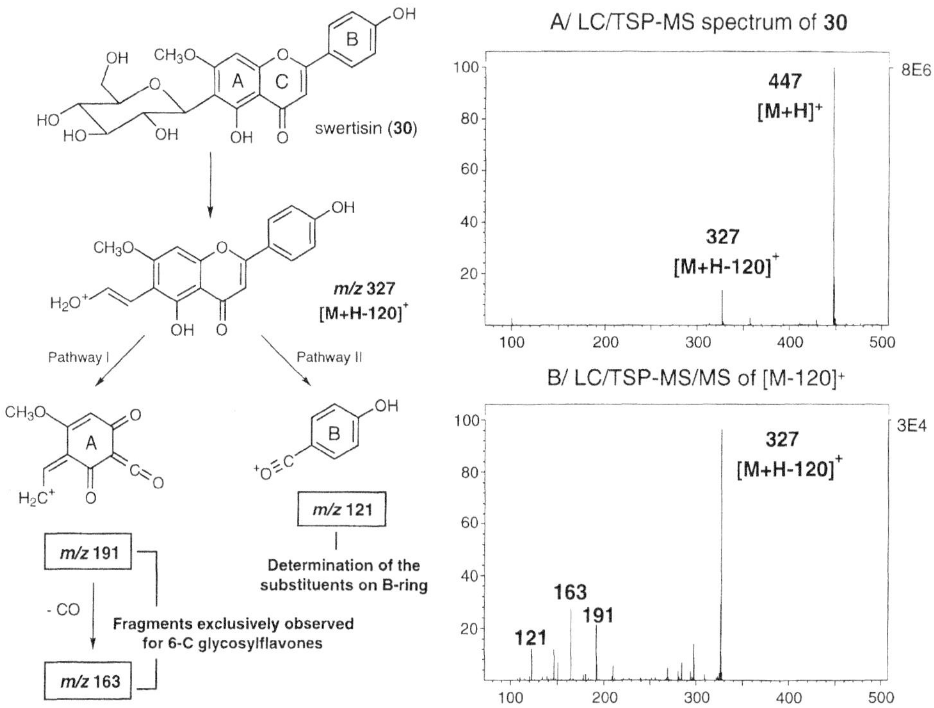

Figure 13 LC/TSP-MS (A) and LC/TSP-MS/MS (B) spectra of swertisin (**30**) in the crude extract of *G. ottonis*. The LC/MS/MS analysis was performed using the fragment $[M+H-120]^+$ (spectrum A) as parent ion. Characteristic daughter ions at *m/z* 121, 163 and 191 were observed, indicating the substitution on the A- and B-ring of the C-glycosylflavone **30**

CONCLUSION

Today the work of phytochemists is based mainly on the bioassay-guided fractionation of crude plant extracts. This type of approach has led to the isolation of numerous compounds with interesting activities. However, plants contain thousands of constituents and obviously for any successful investigation of biological material with such a wide range of properties, the future lies in the ability to have as large a number of biological screens as possible. There are, however, many obstacles which seriously hamper this type of investigation. First, it is possible that a broad range of structurally diverse compounds contributes to the overall pharmacological activity of a plant and synergistic effects between the active principles may exist. This is particularly the case for medicinal plants possessing less specific activities. Secondly, there is an urgent need for more appropriate pharmacological models (Hamburger and Hostettmann 1991). Existing assays are quite often not reliably predictive for clinical efficiency.

Chemical screening of crude plant extracts which allows the localization and efficient targeted isolation of new types of constituents with potential activities, can be considered as an efficient complementary approach to bioassay-guided fractionation.

The addition of highly sensitive analytical techniques such as HPLC and related coupled instrumentation (LC/UV, LC/MS, LC/MS/MS and LC/NMR) enables rapid initial screening of crude plant extracts. These hyphenated techniques provide a great deal of preliminary information about the content and nature of constituents of the extracts. This is very useful when large numbers of samples have to be processed because unnecessary isolation of compounds is avoided. Once the novelty or utility of a given constituent is established, it is then important to process the plant extracts in the usual manner, to obtain samples for full structure elucidation and biological or pharmacological testing. In this chapter, it has been shown that in the search for xanthones with potential monoamine oxidase-inhibitory activity, LC/MS and LC/UV analyses have been of great value for the preliminary screening of numerous different species of Gentianaceae: this method has been used not only to determine the presence of xanthones but also to provide an indication of their structure. The use of LC/UV and LC/MS has also allowed the localization and targetted isolation of new types of constituents such as high molecular weight secoiridoids. Furthermore, by selective ion monitoring in LC/MS or even LC/MS-MS, it is possible to achieve the detection of specific target molecules – those, for example, which have already been found to exhibit a particular activity.

The recent introduction of LC/NMR as a complementary tool represents an important step forward for the on-line identification of plant metabolites. As shown, LC/NMR can provide data of crucial importance for a deeper structural investigation. LC/NMR has only been used very rarely until now, but as was the case for LC/UV and more recently for LC/MS, it will certainly become an important strategical analytical tool in the near future.

In order to rationalise the chemical screening approach and use it efficiently with a high sample throughput, the next challenge will be to find a way to centralise all the on-line data for rapid pattern recognition by reference to standard compounds. With such an analytical system for dereplication, phytochemists will be then able to concentrate their efforts on finding new biological targets. This aspect still remains the more difficult problem to solve when searching for new leads. Indeed both biological and chemical screenings provide important information on the plant constituents but these results will not be sufficient enough for the discovery of new potent drugs if no suitable pharmacological model exists. Thus, an important condition for success in the discovery of new bioactive plant constituents is the establishment of effective collaborations between botanists, phytochemists and pharmacologists in order to realise all the steps involved, starting with plant material in the field and concluding with viable pharmacologically-active compounds. Only in this manner can medicinal plants be investigated efficiently and new leads rapidly discovered.

ACKNOWLEDGEMENTS

Financial support was provided by the Swiss National Science Foundation. Thanks are due to Dr. Wolf Hiller, Varian AG, Darmstadt, for the LC/NMR measurements and to Dr. Winfried Wagner-Redeker, Spectronex AG, Basel, for the LC/ES-MS experiments.

REFERENCES

Albert, K. (1995) On-line use of NMR detection in separation chemistry. *J. Chromatogr. A*, **703**, 123–147.

Bhattacharya, S.K., Ghosal, S., Chaudhuri, R.K. and Sanyal, A.K. (1972) *Canscora decussata* (Gentianaceae) xanthones. III. Pharmacological studies. *J. Pharm. Sci.*, **61**, 1838–1840.

Blakley, C.R. and Vestal, M.L. (1983) Thermospray interface for liquid chromatography/mass spectrometry. *Anal. Chem.*, **55**, 750–754.

Bruins, A.P., Covey, T.R. and Henion, J.T. (1987) Ion spray for combined liquid chromatography mass spectrometry/atmospheric pressure chemical ionisation mass spectrometry. *Anal. Chem.*, **59**, 2642–2646.

Caprioli, R.M., Tan, F. and Cotrell, J.S. (1986) Continuous-flow sample probe for fast atom bombardment mass spectrometry. *Anal. Chem.*, **58**, 2949–2954.

Chace, D.H. and Callery, P.S. (1992) Diamine reagents in thermospray mass spectrometry. *Biol. Mass Spectrom.*, **21**, 125–132.

Dhasmana, H. and Garg, H.S. (1990) Xanthones of *Halenia elliptica*. *Phytochemistry*, **29**, 961–964.

Dreyer, D.L. and Bourell, J.H. (1981) Xanthones of the Gentianaceae. Part VI. Xanthones from *Frasera albomarginata* and *F. speciosa*. *Phytochemistry*, **20**, 493–495.

Garteiz, D.A. and Vestal, M.L. (1985) Thermospray LC/MS interface: principle and applications. *LC Mag.*, **3**, 334–346.

Hamburger, M., Hostettmann, M., Stoeckli-Evans, H., Solis, P.N., Gupta, M.P. and Hostettmann, K. (1990) A novel type of dimeric secoiridoid glycoside from *Lisianthius jefensis* Robyns et Elias. *Helv. Chim. Acta*, **73**, 1845–1852.

Hamburger, M. and Hostettmann, K. (1991) Bioactivity in plants: the link between phytochemistry and medicine. *Phytochemistry*, **30**, 3864–3874.

Hostettmann, K. and Jacot-Guillarmod, A. (1978) Contribution to the phytochemistry of the genus *Gentiana*. Part XXIII. Xanthones and flavone C-glucosides of the genus *Gentiana* (subgenus *Gentianella*). *Phytochemistry*, **17**, 2083–2086.

Hostettmann, K., Domon, B., Sehaufelberger, D. and Hostettmann, M. (1984) On-line high-performance liquid chromatography. Ultraviolet-visible spectroscopy of phenolic compounds in plant extracts using postcolumn derivatization. *J. Chromatogr.*, **283**, 137–147.

Hostettmann, K. and Hostettmann, M. (1989) Xanthones. In J.B. Harborne, (ed.), *Methods in Plant Biochemistry*, Academic Press, London, pp. 493–508.

Kingston, D.G.I. (1979) High performance liquid chromatography of natural products. *J. Nat. Prod.*, **42**, 237–260.

Niessen, W.M.A. and van der Greef, J. (1992) *Liquid Chromatography-Mass Spectrometry, Principles and Applications*, Marcel Dekker Inc., New York.

Rath, G., Touré, A., Wolfender, J.-L. and Hostettmann, K. (1995) Characterization of C-glycosylflavones from *Dissotis rotundifolia* T. by liquid chromatography-tandem mass spectrometry. *Chromatographia*, **41**, 332–342.

Recio-Iglesias, M.C., Marston, A. and Hostettmann, K. (1992) Xanthones and secoiridoid glucosides of *Halenia campanulata*. *Phytochemistry*, **31**, 1387–1389.

Rodriguez, S., Wolfender, J.-L., Odontuya, G., Purev, O. and Hostettmann, K. (1995) Xanthones, secoiridoids and flavonoids from *Halenia corniculata*. *Phytochemistry*, **40**, 1265–1272.

Rodriguez, S., Wolfender, J.-L., Odontuya, G., Purev, O. and Hostettmann, K. (1996) Corniculoside, a new biosidic ester secoiridoid from *Halenia corniculata*. *Helv. Chim. Acta*, **79**, 363–370.

Smallcombe, S.H., Patt, S.L. and Keiffer, P.A. (1995) WET solvent suppression and its application to LC NMR and high-resolution NMR spectroscopy. *J. Magn. Reson. A*, **Series A 117**, 295–303.

Suzuki, O., Katsumata, Y., Oya, M., Chari, V.M., Klapfenberger, R., Wagner, H. and Hostettmann, K. (1981) Inhibition of type A and type B monoamine oxydases by naturally occurring xanthones. *Planta Med.*, **42**, 17–21.

Watanabe, N., Niki, E. and Shimizu, S. (1979) An experiment on direct combination of high performance liquid chromatograpy with FT-NMR (LC-NMR). *Jeol News*, **15A** , 2–5.

Whitehouse, R.C., Dreyer, R.N., Yamashita, M. and Fenn, J. B. (1985) Electrospray interface for liquid chromatographs and mass spectrometers. *Anal. Chem.*, **57**, 675–679.

Wiloughby, R.C. and Browner, R.F. (1984) Monodisperse aerosol generation interface for combining liquid chromatography with mass spectrometry. *Anal. Chem.*, **56**, 2625–2631.

Wolfender, J.-L. and Hostettmann, K. (1993) Liquid chromatographic-UV detection and liquid chromatographic-thermospray mass spectrometric analysis of *Chironia* (Gentianaceae) species. A rapid method for the screening of polyphenols in crude plant extracts. *J. Chromatogr.*, **647**, 191–202.

Wolfender, J.-L. and Hostettmann, K. (1995) Applications of liquid chromatography-mass spectrometry to the investigation of medicinal plants. In J.T. Arnason, R. Mata and J. Romeo, (eds.), *Phytochemistry of Medicinal Plants*, Plenum Press, New York, pp. 189–215.

Wolfender, J.-L., Maillard, M. and Hostettmann, K. (1994) Thermospray liquid chromatography-mass spectrometry in phytochemical analysis. *Phytochem. Anal.*, **5**, 153–182.

Wolfender, J.-L., Rodriguez, S., Hostettmann, K. and Wagner-Redeker, W. (1995) Comparison of liquid chromatography/electrospray, atmospheric pressure chemical ionisation, thermospray and continuous-flow fast atom bombardment mass spectrometry for the determination of secondary metabolites in crude plant extracts. *J. Mass Spectrom. and Rapid. Commun. Mass Spectrom.*, S35–S46.

Wolfender, J.-L., Rodriguez, S., Hiller, W. and Hostettmann, K. (1997) LC/UV/MS and LC/NMR analyses of crude extracts of Gentianaceae species. *Phytochem. Anal.*, **8**, 97–104.

3. THE QUEST FOR NEW BIOLOGICALLY ACTIVE NATURAL PRODUCTS

ROBERT P. BORRIS[1]

Department of Natural Products Chemistry, Merck Research Laboratories, P. Roy Vagelos Research and Development Center, P. O. Box 2000, Rahway, New Jersey, 07065 U.S.A.

INTRODUCTION

Merck has maintained an active natural products research program continuously since the early days of antibiotic research in the 1940's. Over the years, the primary focus of this effort has been the discovery of new microbial metabolites with biological activity. While plants received some intermittent attention in the forties through seventies, this source of novel molecules was not addressed in a systematic manner until the 1980's. In the early eighties, Merck, with collaborators in the People's Republic of China, embarked on a ten year evaluation of traditional Chinese medicines. In the middle of that decade, the scope of these studies was broadened to include plants from other regions of the globe through an ongoing collaboration with the New York Botanical Garden. Then, in 1991, Merck entered into the much publicized agreement with INBio to study the flora (and parts of the fauna) of Costa Rica in greater depth. It is, then, clear that Merck scientists have been studying 'traditional medicines' and 'biodiversity' long before these buzzwords became fashionable.

Why is natural products an important field of research in a major pharmaceutical company? In order to address this question, it is essential to understand the basic mission of pharmaceutical corporations. These organizations are intended to discover and develop new drugs for the treatment of disease, *and* to manufacture and market these drugs in such a way as to maximize the financial return on the investments of their stockholders. Corporations such as Merck support a great deal of basic research, both in their own labs and in those of academic institutions, dealing with the fundamental understanding of wellness and disease. The support of this basic research is viewed as an investment necessary to keep our drug discovery efforts on the cutting edge of science, as it enables our scientists to select and develop the best and most appropriate targets for the diseases we are trying to treat. The main thrust of the research at a large pharmaceutical corporation is directed toward the discovery and development of new drugs. Natural products research is a part, and in truth a small part, of the discovery process. The screening of natural products is one of the earliest steps in drug discovery – Lead Identification.

[1]The opinions expressed in this paper are those of the author, and do not necessarily reflect the views of the management of Merck Research Laboratories or Merck and Co., Inc.

A lead compound, also frequently referred to as a chemical template, is a compound with many of the characteristics of a desired new drug which will be used as a model for chemical modification. It must be potent, but it need not possess the potency at the nanomolar or picomolar levels expected of a product candidate. The molecule must show specificity for the desired target, but it need not possess the exquisite biochemical specificity required for a new drug. Finally, it must be available in sufficient quantities to support the early stages of development such as biological characterization and toxicity studies while a total synthesis of the product candidate is completed. It is much easier to find a lead compound than to find a product candidate. The transition from looking for drugs to looking for leads has been made possible by the tremendous advances that have occurred in organic and medicinal chemistry in the last few decades. Medicinal chemists are *expected* to be able to improve the potency and/or specificity of a lead compound, perhaps by several orders of magnitude. It is thus the task of the medicinal chemist to take a lead compound and turn it into a drug.

The earlier question of why pharmaceutical corporations are involved in natural products research can now be addressed. Over the years, it appears that interest in natural products has been a cyclical phenomenon, with the last peak occurring about 20–25 years ago. At that time, the assays in use were primarily animal models of human diseases. By today's standards, they were slow, cumbersome, expensive, often difficult to interpret, and not very selective. These animal models have almost all been replaced in screening by the so-called "mode of action" assays in use today. In a mode of action assay, the effect of a compound on a discreet biochemical event is measured and compared with a similar event in the absence of test compound. This event may be the binding of a ligand to its receptor, or the activity of an enzyme, or the transcription of a gene in a cultured cell, or any of a number of other kinds of measurable biochemical phenomena. Mode of action assays are fast, quantitative, and easy to perform, making them ideal for miniaturization and automation. Once appropriately miniaturized, they are also relatively inexpensive as compared to the animal tests of a few decades ago. The mode of action assays, together with *in vitro* antibacterial, antifungal, antiviral and antiparasitic assays, have made it economically feasible to screen vast numbers of substances in a wide range of assays, and in a relatively short time.

The advances in assay technology have been a necessary factor in the resurgence of interest in natural products, but were not sufficient in themselves to achieve this end. The advances in chemistry and chemical instrumentation that have occurred concurrently are of equal importance. The advent and subsequent development of high performance liquid chromatography (HPLC), high performance centrifugal countercurrent chromatography (HPCCC), capillary zone electrophoresis (CZE), high resolution mass spectrometry (HRMS), high field and multidimensional nuclear magnetic resonance spectrometry (NMR) and X-ray crystallography have allowed the chemist to isolate submilligram quantities of new compounds, and confidently fully characterize them and identify their structures.

The newer methods available in biology, biochemistry and chemistry have produced the current peak in interest in natural products, but they have also resulted in the advent of other kinds of screening. Most major pharmaceutical corporations routinely screen their corporate compound collections in new assays as they become available.

This is a very cost effective utilization of existing resources. Corporations buy compound collections from other companies, perhaps not in the same industry. As the quantities of chemical needed are quite modest, the costs involved are not extreme. Of course, mergers and acquisitions can result in still larger libraries of defined chemicals. This kind of screening has the inherent advantage of explicitly knowing in advance which compounds are being evaluated in the assays, allowing the chemists to proceed rapidly with any lead that may be discovered. Modular and combinatorial synthesis schemes, in addition to rapidly optimizing leads, are logical extrapolations of this process, allowing thousands of chemicals to be screened as a relatively limited number of mixtures. Their low cost makes these kinds of programs attractive sources of molecular diversity. By comparison, natural products research is not low cost but it offers an extremely high level of molecular diversity. It is this diversity that is absolutely essential to the discovery of new drugs, and is the reason why corporations such as Merck are involved in the study of natural products.

While the natural products group at Merck is involved in the discovery of new lead compounds from plants, plants are not the main source of new natural products at Merck, or at any of the other major pharmaceutical corporations. Microbial metabolites have been, and continue to be, the mainstay of pharmaceutical natural products research.

MICROBIAL PRODUCTS RESEARCH

In a typical microbial products program, microbiologists isolate pure cultures of microorganisms from environmental samples which have been collected in the biosphere. Environmental samples can consist of almost anything, from the traditional soil samples to leaf litter, roots or branches of living or dead plants, animal droppings and insect carcasses. Each kind of environmental sample has its own characteristic spectrum of organisms which are likely to be found there. While it is not possible to say how many different organisms are present in a particular environmental sample, it is not unusual for a well-trained group of microbiologists to isolate over a hundred species from a single sample.

The isolated pure cultures are cultivated in a series of small scale fermentations, perhaps in a defined range of culture conditions, and run through a preliminary panel of screening assays. A major advantage of working with microorganisms is the ability to screen vast numbers of cultures on a small scale, before identifying the organisms, while retaining a high probability of success in reproducing the activity on a larger scale when necessary. Cultures which are inactive in all of the assays are usually destroyed, while those found to be active in one or more screening assay are regrown under conditions similar to the original small-scale fermentation. If the originally observed activity is reproduced, the culture extract is subjected to some form of dereplication.

Dereplication is the process of determining whether a particular active constituent has been encountered previously in a screening program. This is actually a formidable task, but a well conceived and well executed dereplication process is an essential part of any successful natural products program. It prevents a group from wasting resources by re-isolating known compounds, allowing it to focus on unique activities. In programs such as Merck's, in which the biological assays are frequently changing, one is confronted with the problem of rapidly recognizing not only the compounds known to be active in the current assays, but also those that have been isolated for earlier assays. There are, literally, thousands of compounds known to be active in one or more of the hundreds of assays used at various times in the natural products screen at Merck. Dereplication is a daunting but not impossible task. The use of HPLC-MS coupled with ultraviolet detection using a photodiode array detector greatly simplifies the task, especially when linked with computer databases and pattern matching algorithms.

If an extract survives dereplication, it becomes the focus of an intense interdisciplinary collaboration aimed at isolating and identifying the active constituent. Production of the compound is optimized by the microbiologist manipulating the media and growth conditions. The chemists perform a separation guided at each step with bioassay data. The assay biologist or biochemists evaluate samples from the chemists and microbiologists. The tightly collaborative effort allows the group to arrive at the pure active compound in minimal time. Using state of the art spectroscopic and chemical techniques, the natural products chemist determines the molecular structure of the active compound. Often, by the time the active molecule has been identified, the microbiologists have increased its titer in fermentations by a factor of ten or more, and the chemists have refined the steps in the isolation to allow direct implementation in a large-scale isolation in the pilot plant.

An identified, pure compound must be fully characterized in a biological and biochemical sense, also. It is important to know that it possesses the desired activity, but one must determine with what *other* molecular systems the molecule interacts. A potential lead compound must be evaluated against the entire spectrum of molecular targets available in the laboratory to see if it is specific for the desired target and what potential side effects or toxicity might be predicted. It is not an uncommon occurence for a new compound to interact with entire families of related targets, e.g. G-protein linked receptors or serine proteases. These promiscuous agents are usually of little interest.

A number of other discovery activities may be intertwined with the biological characterization. These include the isolation of minor constituents from the culture extract, microbial transformation of the active compound, partial synthesis or preparation of chemical derivatives, and directed biosynthesis experiments in which the microbiologist may feed the microbe an 'unnatural' precursor in an attempt to cause the biosynthesis of related molecules. All of these activities are aimed at the discovery of a family of related compounds to aid in the elucidation of a structure activity relationship for the molecular class. Eventually, the synthetic chemists refine the SAR to select the best compound in the class for total synthesis. With this biochemical knowledge in hand, the compound progresses into pharmacologic evaluation. Does it actually have the predicted activity in an intact animal? The transition from *in vitro* to *in vivo* studies often requires a further refinement in the structure activity relationship.

When the best compound in the series is identified it is subjected to a rigorous safety assessment program, involving several species of animal and high doses of compound. The observation of *any* significant toxicity in *any* species of test animal, even at very high doses of compound, can result in the termination of active interest in a compound. When this happens, often one or more related compounds, the backup candidates, will be similarly evaluated to determine whether the observed toxicity is due to an idiosyncrasy of the initial safety assessment candidate, or whether it is a more general phenomenon occurring throughout the class of compounds. The structure activity relationship may again require refinement in light of the safety assessment data as the synthetic chemists attempt to reduce or eliminate unwanted biological activities. Obviously the most active compound from the original discovery process is frequently not the one developed into a product. The synthetic medicinal chemist's ability to subtly manipulate the chemical and biological properties of a molecule is absolutely critical to this kind of drug discovery program, and its importance cannot be overstressed. If the chemists are unable to reduce or eliminate the observed toxicity of a safety assessment candidate, it is likely that development of the entire class of compound will be abandoned.

In the best case scenario, the active compound will pass through safety assessment without any indication of toxicity. An Investigational New Drug Application (INDA) is then filed and (hopefully) approved, and the compound will make the transition from preclinical studies in animals to clinical trials in humans. Assuming that safety and efficacy are adequately observed and documented in Phase I–III trials, a New Drug Application (NDA) will be filed with the regulatory authorities (e.g., the U.S. Food and Drug Administration). After a necessarily detailed and critical evaluation of the NDA, and possibly more clinical trials required by the regulatory authority to clarify relevant issues in the application, the drug will receive approval for marketing in a particular country. In recent years, the regulatory authorities have begun to require in some cases that additional clinical studies be performed *after* regulatory approval. The entire process of clinical trials and regulatory approval can be summarized for our purposes in a few lines of text, but this is a long and costly process, estimated to take 10 years to complete at an average pretax cost estimated by the Office of Technology Assessment of the U.S. Congress at $359 million (U.S. Congress, Office of Technology Assessment, 1993). It is only *after* the completion of this process and the massive investment in research and development that a product can be brought to market and begin generating revenue.

Working with microorganisms in a natural products discovery program offers a number of significant advantages. The microbial kingdom is incredibly diverse in both a taxonomic sense and a biochemical sense. Importantly, much of this diversity is accessible even without traveling to distant locations. The accessibility of a wide range of organisms makes a high screening throughput both practical and profitable. High throughput would be of little value if we were not confident of the ability to successfully reproducing the activity in subsequent regrowths. From the perspective of development, the ability to produce large quantities of microbial products by fermentation is well established. Supply is unlikely to be a persistent problem in the development of a microbial product. Finally, there is real potential for eventually making a profit, as exemplified by the many antibiotics derived from natural products leads which are currently on the market and such compounds as Merck's cholesterol lowering agent, lovastatin (Alberts *et al.* 1980), and our broad spectrum antiparasitic drugs, the avermectins (Miller *et al.* 1979), which are both expected to achieve $1 billion in sales this year.

PHYTOCHEMICAL RESEARCH

If microbial products are so good, why should we study plants? Simply stated, plants are complementary to microorganisms in a chemical sense. A careful survey of the natural products literature suggests that the natural products researcher is unlikely to encounter the same kinds of secondary metabolites in plants that will be produced by microorganisms. In order to maximize the scope of the chemical diversity evaluated in a screening program, it is necessary to screen both plant and microbial products. Like microorganisms, plants are a biologically and chemically diverse resource, with 250–500,000 species estimated to be in existence on the planet. The rapid rate of species extinction which we are currently witnessing, due in part to habitat destruction, not only lends a certain urgency to the quest for plant-derived drugs, but also calls into question the ability to acquire large amounts of a plant from natural populations to support a development program. As chemical diversity is of paramount importance to our goals in drug discovery, we are faced with what amounts to a now or never situation.

Long before ever collecting the first sample for a phytochemistry program, one is faced with a number of issues which must be addressed according to the objectives of the screening program. An early, multifaceted key issue is the selection of species for investigation. A decision must be made regarding where samples will be collected, considering not only the diversity of the flora in a particular country, but also the species density. Can one practically collect a large number of diverse species without spending an excessive amount of time and money on travel? Assuming that there are a large number of species in a limited geographic area, is there enough of a local infrastructure to allow successful collection? There is little point in gaining access to a dense, diverse flora if it is not possible to get screening samples, in the range of a few hundred grams to a few kilos, collected, dried and delivered to the extraction laboratory before significant deterioration occurs. It must then be determined who will actually select the species which will be studied, who will collect them in their natural habitat and who will rigorously identify them. Are well-trained local botanists available and willing to collaborate, or must you bring in your own staff? Is there a local or national herbarium in the source country to use in identifying new collections and in which you must deposit specimens? The competence of the project's botanical staff is of critical importance to its success.

Once these issues have been fully addressed, the scientific aspects of species selection must be considered. What selection strategy best fits the overall goals of the screening program? There are probably as many different approaches to the selection of plant species for study as there are practitioners in the field.

The most general approach to the selection of plant materials is random or empirical selection. This is, in essence, the 'walk in the woods approach', collecting whatever is readily available. There is little intellectual input required, as selection is a chance occurrence. Assuming that one is working with a diverse flora, it is possible to generate a high degree of chemical diversity in the extracts produced. Empirical selection works best in situations where a program can afford to produce a very high volume of samples and has sufficient assay capabilities to screen that high volume.

The most highly publicized method of plant selection is ethnobotany or ethnomedicine. Whether one takes the approach of learning directly from the indigenous populations or the approach of using the published uses of plants to predict which species will be of interest, this method requires a significant investment of intellectual input to be a success. In my opinion, it is best suited to programs having a narrow scope of disease targets which are readily discernable to the naked eye. As the range of targets increases, this approach becomes more difficult to apply, limiting its utility in a broad-based screening program.

If it is known at the outset what kind of chemicals will be of interest to a program, it is possible to develop a program based on chemical taxonomy, focusing on one or more particular groups of compounds. Chemotaxonomic studies also require a significant intellectual investment at the outset, and are not really amenable to broadbased screening as the concept, as typically applied, minimizes diversity to maximize the depth of coverage of a narrow group of chemicals.

A potentially more efficient alternative to empirical selection is the more focused, biodiversity-driven selection based on botanical taxonomy. This method attempts to create a *representative* cross section through the entire plant kingdom. The exact nature of this cross section depends on the taxonomic system employed and the taxonomic level at which it attempts to be representative. This method assumes that chemical diversity will parallel botanical diversity and is intended to generate a representative sampling of chemotypes available in the plant kingdom. Once hits are encountered it is straightforward to expand collections of related taxa. The goal of this selection method is to attain a level of chemical diversity comparable to that attainable in random selection, but in significantly fewer samples.

Of course, it is possible and perhaps desirable to merge more than one of these methods into an overall screening philosophy, or to use different methods for different aspects of a program.

Regardless of the approach taken to selection and the details of the collection process, a phytochemistry program begins with samples being collected from some natural population. The plant samples are rigorously documented with appropriate voucher specimens, photographs and written notes which allow for proper botanical identification. The plant samples are processed, perhaps by drying and milling, and then extracted to remove the compounds of interest from the biomass, according to the protocols of the program. After appropriate processing, the extracts are then screened in a primary screening panel of assays. Throughput, a key issue in any screening program, is limited by a host of factors including the logistics of getting samples to the lab, the tedious process of identifying the plant specimens, and the cost of collecting samples in remote locations. In a climate of competing programs and limited resources, it is unlikely that a throughput of more than a few thousand samples a year will be attained.

Any activities observed in the initial screening panel are then confirmed, frequently using a second portion of the same extract used in primary screening, and confirmed actives are subjected to dereplication. The dereplication process is similar to that used for microbial products except that knowledge of the identity of the plant also enables one to use the extensive phytochemical literature as an aid to rapidly identifying known compounds.

As was the case for microbial products, the isolation of suspected new active compounds from extracts that show activity in screening involves a tight collaboration between the chemist performing the isolation and the biologist or biochemist performing the assay. Experience indicates that plant extracts are usually more complex than extracts of fermentation broths, but modern isolation methodologies routinely allow the purification of the active constituent in an acceptable timeframe. Recollection or resupply is a much more complicated procedure for plants. A fermentation can be repeated to acquire more of a microbial product in a matter of days, or at most weeks, but plants must be collected at the same stage of their life cycle, in the same season, and preferably in the same location to be completely confident of getting comparable material. Even when these variables are all carefully controlled, success is not assured. Management rarely views the introduction of a delay of six months to two years into a development timeline in a favorable manner.

Once a pure, active phytochemical is in hand, research activities surrounding it are essentially the same as those for a microbial product. The compound is identified and, along with some related compounds, is characterized biologically. The best member of the group is taken forward through pharmacologic evaluation, safety assessment and on into clinical trials. If all goes extraordinarily well, after ten years and an investment of $359 million, a new drug based on this lead may reach the market.

From this discussion, it should be evident that this is a modular approach to screening, which can be expanded to include many other kinds of organisms, such as marine invertebrates, marine algae, insects, protozoa, or other kinds of plants and animals. The limiting factors include the actual goals of the program, the imagination of the people involved, and, of course, one's resources.

Phytochemistry is conducted today in essentially the same manner as it has been for the last several decades. Plant samples are collected from some wild or cultivated population, extracted and fractionated to yield pure compounds. In my opinion, this will change dramatically in the next five to ten years, as collection of specimens becomes more difficult and expensive. Production of secondary metabolites by plant tissue culture may allow the discovery of interesting compounds produced by natural populations in too small a quantity to allow isolation and identification, or not produced at all in the wild. This area of endeavor is the next frontier. The production of secondary metabolites by tissue culture on an industrial scale, making hundreds or thousands of kilograms of product per year, could replace large-scale collection from wild populations or mass cultivation with a more readily controlled and cost effective alternative that is competitive with microbial fermentation. Biotransformations using cultured plant cells may prove complementary to those using microorganisms. Transformations using cell-free systems or isolated enzymes are a logical extension of this technology.

Merck currently maintains two major collaborations for the acquisition of plant samples for screening, with INBio and the New York Botanical Garden. The INBio collaboration has garnered significant attention in the press since its signing in September 1991. Under this agreement, INBio provides Merck with a relatively small number of plant, insect and environmental samples for use in its screening programs. Merck provides a number of different forms of compensation for these samples, and in a unique feature of this agreement, 10% of the total budget for the collaboration is donated to support the Costa Rican system of National Parks and Protected Areas – a direct payment to promote conservation. Not only has an upfront payment been made to cover the expenses of doing this research, but Merck has also established and equipped a laboratory at INBio in which the work is being performed. Costa Rican scientists have been trained in extraction methods and in advanced laboratory techniques, both at INBio and in the Department of Natural Products Chemistry at Merck in New Jersey. Of course, when publications result from this work, they are co-authored by scientists from both institutions. Similarly, patents resulting from this work will include inventors from both institutions as appropriate under prevailing patent law. If a marketable product results from a discovery made under this collaborative agreement, Merck will pay royalties to INBio, a portion of which will be used to support Costa Rica's system of National Parks and Protected Areas.

The agreement with the Institute of Economic Botany of the New York Botanical Garden, while less publicized, is very similar in content. Under this agreement, which was first established in 1986, the New York Botanical Garden provides Merck with plant samples that are collected by its botanists in habitats around the world. In return, Merck has established and equipped a laboratory at the New York Botanical Garden, trained NYBG scientists in various aspects of plant processing and extraction, and provides payments to support the research. Publications that result from the work are co-authored by scientists from Merck and the New York Botanical Garden, as well as collaborators in the source countries. Patents resulting from this work will include inventors from both institutions as appropriate under patent law. Merck has also agreed to pay royalties, when a marketable product results from this research, which will be split between the New York Botanical Garden and its collaborators in the source countries.

These two collaborations are examples of how multinational corporations can approach the sourcing of natural products samples. Both explicitly recognize the scientific contributions of collaborators in the developed and developing countries. Both compensate source countries with royalties on the profits from drugs resulting from this research. Each is tailored to the specific needs of our collaborating institutions. This is an important point, as each collaborator has a different set of interests, needs and problems. Hopefully, both of these collaborations will yield the kind of major discovery that will justify the effort and expense that have been made by all parties involved.

EXAMPLES

As mentioned above, the main thrust of research at a pharmaceutical company is the discovery of really useful compounds – new drugs. While it would be pleasant to think that mankind is gradually eliminating the diseases which plague its existence, in truth there is a constant stream of new disease targets on which to work. These new targets stem from our ever increasing understanding of the underlying physiology of individual diseases, and the recognition of "new" diseases. The discovery of AIDS in the late 1970's and early 1980's is an example of such a "new" disease which previously existed unrecognized. Adding to the need for new drugs is the emergence of resistance to current therapies. Infection with drug resistant microorganisms now poses a serious threat to the chemotherapy of infectious diseases. The following are examples of some disease targets which we consider to be of major importance. This list is not exhaustive. The phytochemicals which are described are ones which, although generating considerable excitement at the time of discovery, did not merit further development into drugs. An explanation for these decisions is beyond the scope of this paper. While it would be a pleasure to say that we have found phytochemicals with activity against each target we have studied, this is not the case.

Resistant Microorganisms

Methicillin resistant strains of *Staphylococcus aureus* (MRSA) are now the number one cause of nosocomial infections in hospitals in the United States. The number two cause is vancomycin resistant *Enterococcus* (VRE). In New Jersey, over 80% of drug resistant bloodstream isolates reported from 1992 to 1994 were methicillin resistant *Staphylococcus aureus* while 10% were vancomycin resistant *Enterococcus* (Anon 1995). As there are few, if any, antibiotics to effectively treat these infections, a need clearly exists for new antibiotics with activity against these organisms. Similarly, tuberculosis, a disease thought under control for many years, is developing resistance to many of its standard therapies. Currently, resistance to isoniazid is observed in 13–26% of patients in the U.S., and multidrug resistance, minimally defined as resistance to isoniazid and rifampicin, is found in 3–19% of patients (Collazos *et al.* 1995). The World Health Organization predicts that by the year 2005 tuberculosis will kill 4 million people per year, up from the current 2.5 million (Nowak 1995).

Using a disk diffusion assay against a wide range of microorganisms, including several drug resistant pathogens, a methanol extract of the stems of *Minquartia guianensis* Aubl. (Olacaceae) collected in Costa Rica was found to have significant activity against MRSA. Activity was limited to the non-polar fractions via solvent partitioning. Column chromatography followed by preparative thin layer chromatography yielded the pure active compound, minquartynoic acid (Figure 1). This known compound was shown to also have antifungal activity and cytotoxicity, and this lack of any specificity suggests that it is a generally toxic compound (Borris, unpublished results).

Fungal Infections

In the decade from 1980 to 1990, the incidence of nosocomial fungal infections doubled from 6% to 12%. *Candida* is now the fourth most common nosocomial pathogen (8%). Forty percent of hospital deaths can be attributed to nosocomial fungal infections with 30–40% of these being due to non-*albicans Candida.* The spectrum of recognized fungal pathogens now includes over 50 species, including such major

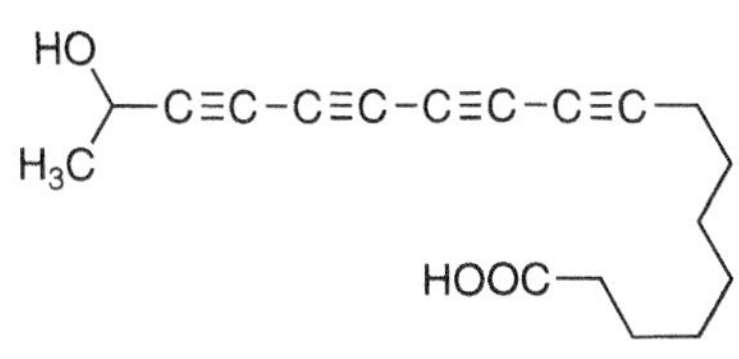

Figure 1 Minquartynoic Acid

life-threatening pathogens as *Candida*, *Aspergillus* and *Cryptococcus*. Over 90% of AIDS patients acquire fungal infections, and many die from them. To date, there remains a major need for a safe, non-toxic, systemic antifungal agent (Sternberg 1994). Our phytochemistry program has not yet found one.

Parasitic Infections

Throughout much of the world, parasitic diseases remain one of the most prevalent causes of morbidity and mortality. The losses due to protozoal diseases such as Chagas disease (*Trypanosoma* spp., 18 million people infected), leishmaniasis (*Leishmania* spp., 12 million infected) and malaria (*Plasmodium* spp., 270 million infected) are staggering. Helminth infections are comparably prevalent, with river blindness (*Onchocerca* spp., 20 million infected), lymphatic filariasis (*Wuchereria* spp., 120 million infected) and schistosomiasis (*Schistosoma* spp., 200 million infected) as examples (Kolberg 1994).

For several years, Merck scientists have been using *Caenorhabditis elegans*, a free living nematode, as a screening tool for parasitic helminths. *C. elegans* can be conveniently cultured and handled in the laboratory, and does not require a mammalian host. When it is exposed to an anthelminthic agent, *C. elegans* loses motility, stiffening to a rod before death. It is thus straightforward to count the number of motile organisms and the number of non-motile organisms in the field of view of a microscope to determine the percent of efficacy. Using this method, the methanol extract of the roots of *Curcuma comosa* Roxb. (Zingiberaceae) from Thailand were found to possess significant nematocidal activity. Solvent partition experiments showed the activity to be present in the most non-polar, hexane soluble fractions. Column chromatography followed by repetitive preparative thin layer chromatography afforded a series of active compounds (Figure 2) (Jurgens *et al.* 1994). The presence of a family of related compounds in a plant extract

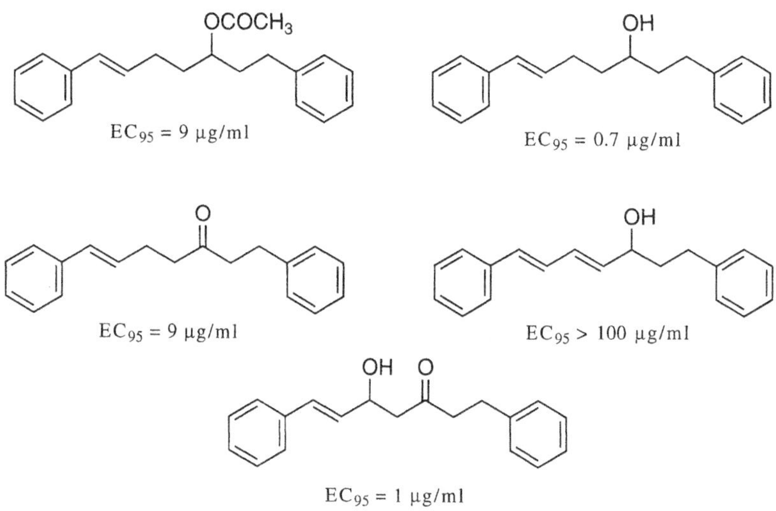

Figure 2 Nematocidal diphenylheptanoids from *Curcuma* sp.

is not at all extraordinary. From the perspective of drug discovery, while the presence of a family of related compounds with similar chemical properties may make the isolation of each compound somewhat more challenging, it also offers, as in this case, the beginnings of an understanding of the structure–activity relationships of the series. This is very valuable information for the medicinal chemist.

HIV and AIDS

Through 30 September 1996, 566,002 cases of AIDS had been reported to the U.S. Centers for Disease Control. Over 353,000 deaths had been reported from this disease. Worldwide, WHO estimates 7,700,000 cases of AIDS (Anon 1996). AIDS is the number one killer in the 25–44 year old age group in the U.S. Globally, the picture is even more dismal. WHO estimates 18.5 million people are infected with the virus, and of these, two-thirds to three-fourths are in sub-Saharan Africa. WHO found the highest proportional rate of HIV infection in Botswana, where 18% of the sexually active adult population is infected. Zambia (17%), Zimbabwe (17%), and Uganda (15%) had similarly high rates of infection. In the Americas, the highest rate of infection was found in Haiti (4.4%), followed by the Bahamas (3.9%) and Barbados (2.8%). It is estimated that about 0.5% of the population of the United States is infected with HIV (Anon 1996). While the reverse transcriptase inhibitors and protease inhibitors already on the market are showing tremendous promise in the treatment and management of AIDS, they are not cures, and until a cure is found and made readily available throughout the world HIV/AIDS will continue to be an epidemic.

Central Nervous System

The central nervous system is involved in a wide range of diseases which are often poorly recognized and treated. Mood disorders afflict over 5% of the population of the United States at any one time, while depression afflicts 2.2% of the population (Judd *et al.* 1994). Alzheimer's disease continues to be an increasingly common problem as our population ages. Pain, while perhaps more properly considered a symptom rather than a disease in itself, has certainly touched the lives of everyone. A clear need exists for a safe, non-narcotic analgesic for the treatment of moderate to severe pain.

Migraine disease is a syndrome characterized by intense vascular headaches. While serotonin (5-HT) has numerous receptors throughout the body, the $5HT_{1D}$ receptor subtype has been implicated in migraine disease, and an agonist selectively acting at that receptor subtype was considered to be of potential therapeutic utility in the management of that disease. Screening of the Merck plant extract library in a $5HT_{1D}$ receptor binding assay first afforded the methanol extract of *Ochrosia oppositifolia* (H.B.K.) Baill. (Apocynaceae) as a hit. A straightforward isolation involving solvent partition, acid/base extraction and preparative thin layer chromatography afforded the known indole alkaloid, isoreserpiline (Figure 3, **A**), as the main active constituent. With this compound in hand, the scope of the search was narrowed to include only compounds not containing the indole nucleus. A methanol extract of the roots of *Aconitum columbianum* Nutt. (Ranunculaceae), collected in Utah, was found to be active in the binding assay. As this taxon was considered unlikely to contain indole alkaloids,

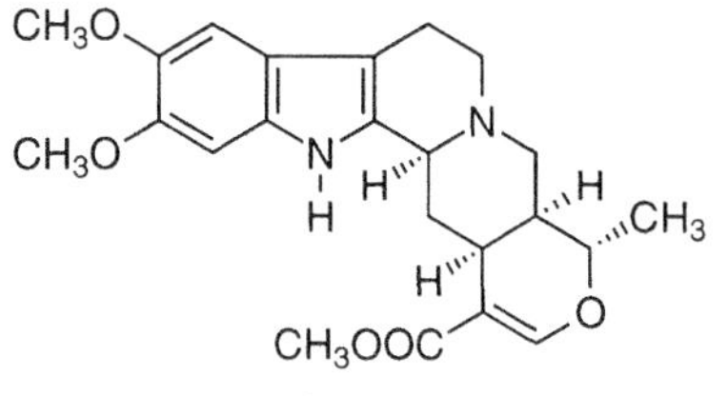

A

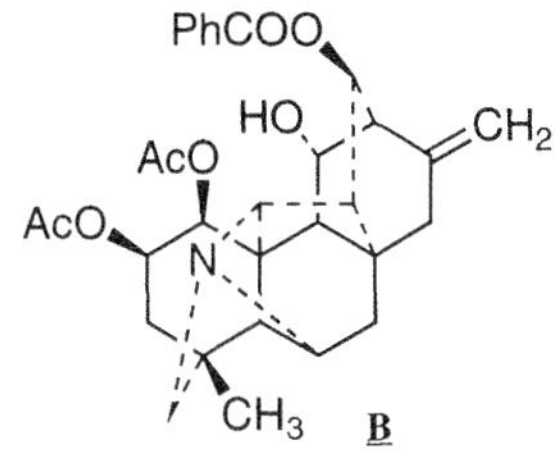

B

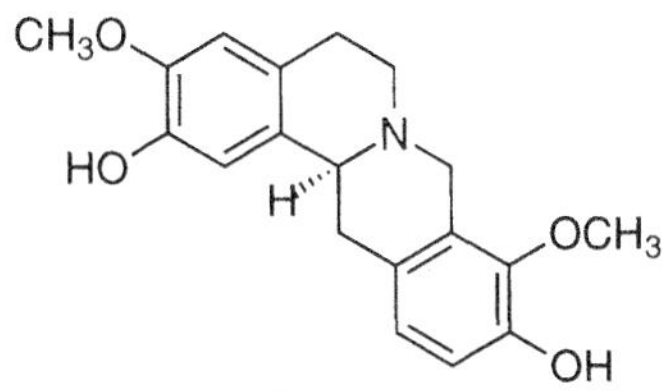

C

Figure 3 Serotonin binding antagonists: A. isoreserpiline; B. paniculatine; C. stepholidine

fractionation was commenced. Another straightforward fractionation led to the isolation of a single active component which was identified as the known hetisine alkaloid, paniculatine (Figure 3, B). As the *Aconitum* alkaloids have a reputation for severe toxicity, this discovery was considered more of a curiosity than a lead, so screening continued and eventually found the methanol extract of the roots of *Stephania pierrei* Diels (Menispermaceae) as another active which was unlikely to contain indole alkaloids. Yet another straightforward isolation yielded the known tetrahydroisoquinoline alkaloid, stepholidine (Figure 3, C), as the main active constituent (Borris, unpublished results).

Bradykinin, a nine residue polypeptide, is a potent vasodilator and producer of pain. As the receptor for this molecule has been isolated and characterized, a receptor binding assay was developed to find antagonists of bradykinin binding which should be analgesics. A methanol extract of the bark of *Ochrosia oppositifolia* was found to be a potent inhibitor of bradykinin binding. While this was the same extract that had previously been found to be active in the $5HT_{1D}$ binding assay, isoreserpiline was not active in the bradykinin binding assay. Solvent partitioning of this extract, followed by a typical acid/base extraction scheme, indicated that the activity was found only in the alkaloidal fraction. Column chromatography followed by repetitive preparative thin layer chromatography afforded a single, pure, active component which was identified as the known dimeric indole alkaloid, ochrolifuanine A (Figure 4), which was a micromolar inhibitor of bradykinin binding (Borris, unpublished results).

Cancer

The American Cancer Society has stated that one third of the population of the United States will develop some form of cancer during their lives. Twenty percent of that

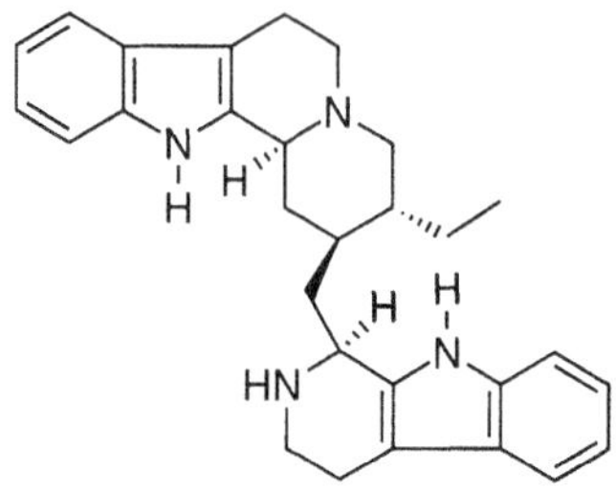

Figure 4 Ochrolifuanine A

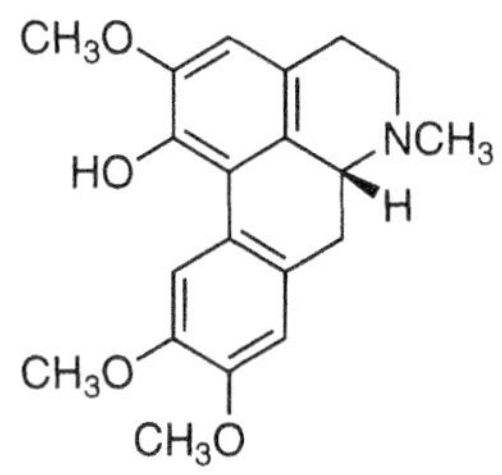

Figure 5 (-)-Thaliporphine

population will die from the disease, and despite the discovery and development of new drugs such as taxol and the camptothecin analogs, the rate at which Americans are dying from cancer continues to increase. In 1992, approximately 1,170,000 new cases of cancer were reported in the U.S. while 526,000 people died from the disease (Mendelsohn 1994). These numbers are overwhelming and unmistakably point to a glaring need for new and effective therapeutic entities.

Inflammatory Diseases

Rheumatoid arthritis is a chronic debilitating disease afflicting about 1% of the population (Lipsky 1994). Interleukin-1β, a major determinant in the etiology of acute and chronic inflammatory disease (Black *et al.* 1989; Dinarello 1992; Thornberry *et al.* 1992), is synthesized by monocytes as an inactive 31 kDa precursor which is then cleaved to the active form by interleukin-1β converting enzyme (ICE), a unique heterodimeric cysteine-containing protease with high substrate specificity. Inhibition of this enzyme was considered an attractive target for potential antiinflammatory drugs.

Screening of the Merck plant extract library showed that an extract of the stems and leaves of *Croton urucurana* (H.B.K.) Baill. (Euphorbiaceae) had significant ICE inhibitory activity. Fractionation, using a straightforward series of partition and chromatographic steps, rapidly afforded the single active constituent which was identfied as the known aporphine alkaloid, (-)-thaliporphine (Figure 5) (Salvatore *et al.* 1994).

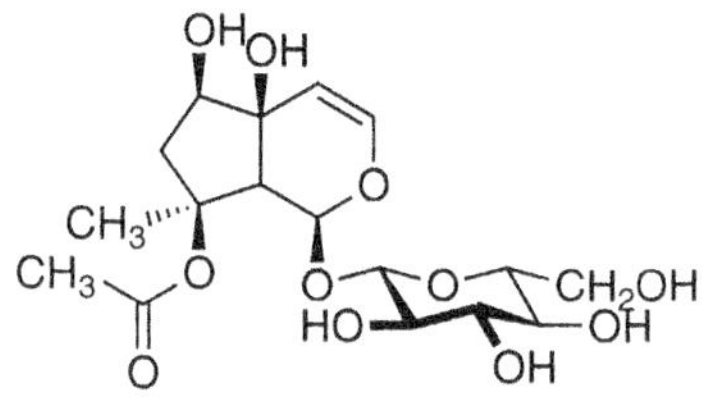

Figure 6 8-*O*-Acetylharpagide

Agricultural Production

The potential utility of natural products is certainly not limited to medicines for humans. A niche also exists for the discovery of chemicals for use in agriculture. In 1994, the total global agricultural output capability was estimated at US$579 billion. Of this, the potential loss due to disease and pests was estimated at $404 billion. The use of agricultural chemicals prevented $160 billion of this loss, reducing the actual losses in agricultural production to $244.1 billion (42.1%). With many parts of the world plagued by chronic hunger, these losses are simply not acceptable. These agricultural losses can be broken down by causative agents. The potential loss due to weeds is estimated at $171.4 billion, but $95 billion (55%) of that potential loss is prevented by use of chemicals. Animal pests account for $131.4 billion in potential losses, $41.1 billion (31.3%) of which is prevented through chemical intervention. Mankind is doing an even poorer job of treating plant diseases which account for $101.3 billion in potential losses. Only $24.3 billion (24%) of the potential losses due to plant disease are prevented by use of chemicals (Oerke *et al.* 1994). A need exists for safe, effective, selective herbicides, fungicides, insecticides and agents to deter herbivory.

As part of Merck's research in this area, it was felt that interaction at the ecdysone receptor may be a significant target for the development of selective insecticides. Non-steroidal ecdysone receptor agonists, compounds which could cause fatal precocious molting in insects, should be relatively nontoxic to mammals which lack that receptor. Screening of the Merck plant extract library in a transactivation assay using β-galactosidase as the reporter gene regulated by ecdysone response elements showed that a methanol extract of the entire plant of *Ajuga reptans* L. (Lamiaceae) collected from a cultivated population in New York had the desired activity. Isolation via column chromatography and preparative HPLC afforded the known iridoid glycoside, 8-*O*-acetylharpagide (Figure 6) as the active compound. This iridoid, although not a steroid, activates transcription of an ecdysteroid inducible promoter in the transactivation assay, competes with known ecdysteroids for binding to a *Drosophila* ecdysteroid receptor preparation, and causes differentiation in ecdysteroid responsive cells, analogous to that seen in the presence of ecdysteroids (Elbrecht *et al.* 1996).

CONCLUSIONS

Phytochemistry continues to offer the pharmaceutical and chemical industry a source of molecular diversity which is complementary to other forms of natural products.

While research in this field is still conducted largely in the same way that it has been for many years, the incremental advances in chemical and biological methodologies have greatly expanded its utility. Still, the ability of phytochemistry to compete with other sources of molecular diversity is limited by the cost of acquiring and processing relatively large samples in the field. As access to natural populations diminishes and/or grows more expensive, the focus of phytochemical research will likely shift from the study of metabolites in natural populations to the study of metabolism at a molecular level in cultured cells and tissues.

ACKNOWLEDGEMENTS

Natural products research is rarely an individual task, especially in an industrial environment. I would like to thank the following colleagues for their active participation in the projects used as examples in this paper: Dr. Michael Balick (NYBG), Ms. Jane Bearinger, Dr. Hans T. Beck (NYBG), Ms. Aurora Bernard, Ms. Elizabeth Birzin, Ms. Yuli Chen, Dr. Alex Elbrecht, Ms. Easter Frazier, Dr. Otto Hensens, Ms. Amy Hsu, Dr. Quirico Jimenez (INBio), Ms. E. Tracey Jones, Dr. Tannis Jurgens, Dr. Steven King (NYBG), Mr. Greg Koch, Ms. Nailin Lee, Dr. Jerrold Liesch, Dr. Russell Lingham, Dr. Weerachai Nanakorn (Royal Forest Department, Bangkok), Dr. Lisa O'Rourke (NYBG), Dr. C. Ian Ragan, Dr. Richard Ransom, Dr. Michael Salvatore, Dr. James Schaeffer, Dr. C. F. Shen (NYBG), Dr. Ana Sittenfeld (INBio), Mr. Jack Smith, Dr. Giselle Tamayo (INBio), Dr. Karl Vincent (NYBG), Dr. Nelson Zamora (INBio) and Ms. Debra Zink. Additionally, I'd like to thank Dr. Georg Albers-Schönberg, Dr. James V. Heck, Dr. Mervyn J. Turner and my colleagues in Natural Products Chemistry and Natural Products Drug Discovery for their help and continued support for the phytochemistry program at Merck.

REFERENCES

Alberts, A.W., Chen, J., Kuron, G., Hunt, V., Huff, J., Hoffman, C., Rothrock, J., Lopez, M., Joshua, H., Harris, E., Patchett, A.A., Monaghan, R., Currie, S., Stapley, E.O., Albers-Schonberg, G., Hensens, O.D., Hirschfield, J.M., Hoogsteen, K., Liesch, J. and Springer, J.P. (1980) Mevinolin, a highly potent competitive inhibitor of HMG-CoA reductase and cholesterol lowering agent. *Proc. Nat. Acad. Sci. (USA)*, **77**, 3957–3961.

Anon. (1995) Statewide surveillance for antibiotic-resistant bacteria – New Jersey, 1992–1994. *Morbidity and Mortality Weekly Report*, **44**, 504–506.

Anon. (1996a) Case Watch. *AIDS Clinical Care*, **8**, 1.

Anon. (1996b) WHO reports 18.5 million have HIV. *AIDS Weekly Plus*, Jan. 1, 1996, 10–11.

Black, R.A., Kronheim, S.R. and Sleath, P.R. (1989) Activation of interleukin-1β by a coinduced protease. *FEBS Letters*, **247**, 386–390.

Collazos, J., Mayo, J. and Martinez, E. (1995) The chemotherapy of tuberculosis – from the past to the future. *Respiratory Medicine*, **89**, 463–469.

Dinarello, C.D. (1992) Role of Interleukin-1 in infectious diseases. *Immunol. Rev.*, **127**, 119–146.

Elbrecht, A., Chen, Y., Jurgens, T., Hensens, O.D., Zink, D.L., Beck, H.T., Balick, M.J. and Borris, R.P. (1996) 8-*O*-Acetylharpagide is a nonsteroidal ecdysteroid agonist. *Insect Biochemistry and Molecular Biology*, **26**, 519–523.

Judd, L.L., Britton, K.T. and Braff, D.L. (1994) Mental disorders. In K.J. Isselbacher *et al.* (eds.), *Harrison's Principles of Internal Medicine, 13th. Ed.*, McGraw Hill Inc., New York, pp. 2400–2419.

Jurgens, T.M., Frazier, E.G., Schaeffer, J.M., Jones, T.E., Zink, D.L., Nanakorn, W., Beck, H.T. and Balick, M.J. (1994) Novel nematocidal agents from *Curcuma comosa*. *J. Nat. Prod.*, **57**, 230–235.

Kolberg, R. (1994) Finding 'sustainable' ways to prevent parasitic diseases. *Science*, **264**, 1859–1861.

Lipsky, P.E. (1994) Rheumatoid Arthritis. In K.J. Isselbacher *et al.* (eds.), *Harrison's Principles of Internal Medicine, 13th. Ed.*, McGraw Hill Inc., New York, pp. 1648–1655.

Mendelsohn, J. (1994) Principles of neoplasia. In K.J. Isselbacher *et al.* (eds.), *Harrison's Principles of Internal Medicine, 13th. Ed.*, McGraw Hill Inc., New York, pp. 1814–1826.

Miller, T.W., Chaiet, L., Cole, D.J., Cole, L.J., Flor, J.E., Goegelman, R.T., Gullo, V.P., Joshua, H., Kempf, A.J., Krellwitz, W.R., Monaghan, R.L., Ormond, R.E., Wilson, K.E., Albers-Schonberg, G. and Putter, I. (1979) Avermectins, a new family of potent anthelminthic agents: Isolation and chromatographic properties. *Antimicrobial Agents and Chemotherapy*, **15**, 368–371.

Nowak, R. (1995) WHO calls for action against TB. *Science*, **267**, 1763.

Oerke, E.C., Dehne, H.W., Schönbeck, F. and Weber, A. (1994) *Crop production and crop protection – estimated losses in major food and cash crops*. European Crop Protection Association, Brussels, p. 9.

Salvatore, M.J., Hensens, O.D., Zink, D.L., Liesch, J., Dufresne, C., Ondeyka, J.G., Jurgens, T.M., Borris, R.P., Raghoobar, S., McCauley, E., Kong, L., Gartner, S.E., Koch, G.E., Pelaez, F., Diez, M.T., Cascales, C., Martin, I., Polishook, J.D., Balick, M.J., Beck, H.T., King, S.R., Hsu, A. and Lingham, R.B. (1994) L-741,494, a fungal metabolite that is an inhibitor of interleukin 1β converting enzyme. *J. Nat. Prod.*, **57**, 755–760.

Sternberg, S. (1994) The Emerging Fungal Threat. *Science*, **266**, 1632–1634.

Thornberry, N.A., Bull, H.G., Calaycay, J.R., Chapman, K.T., Howard, A.D., Kostura, M.J., Miller, D.K., Molineaux, S.M., Weidner, J.R., Aunins, J., Elliston, K.O., Ayala, J.M., Casano, F.J., Chin, J., Ding, G.J. F., Egger, L.A., Gaffney, E.P. Limjuco, G., Palyha, O.C., Raju, S.M., Rolando, A.M., Salley, J.P., Yamin, T.T., Lee, T.D., Shively, J.E., MacCoss, M., Mumford, R.A., Schmidt, J.A. and Tocci, M.J. (1992) A novel heterodimeric cysteine protease is required for interleukin 1β processing in monocytes. *Nature*, **356**, 768–774.

U.S. Congress, Office of Technology Assessment. (1993) *Pharmaceutical R & D Costs, Risks and Rewards*, Washington DC: U.S. Government Printing Office (OTA-H-522).

4. PHARMACOCHEMISTRY OF NEW COMPOUNDS FROM SOUTH AMERICAN PLANTS

R. ANTON[1], A. LOBSTEIN[1], M. SAUVAIN[2] and B. WENIGER[3]

[1]*Laboratoire de Pharmacognosie, Faculté de Pharmacie, B.P. 24, 67401 Illkirch Cédex, France (Groupement de Recherches CNRS G 1206)*
[2]*ORSTOM UR 45, Faculté de Pharmacie, 51 Rue Cognacq-Jay, 51096 Reims Cédex, France*
[3]*ORSTOM, Universidad del Valle, Cali, Colombia*

DRUGS CONTAINING TRITERPENE SAPONINS: NEW DEVELOPMENTS AND BIOLOGICAL EFFECTS

The number of publications containing chemical and pharmacotoxicological data is constantly on the increase and the research emphasis in Pharmacognosy is increasingly directed towards pharmacochemistry of natural substances. Among these natural compounds, saponins seem to play a more and more important role because these biologically active compounds benefit from new structural elucidation techniques. Their structural elucidation is in fact very complex because we need information about:

- the nature of aglycones
- the nature of sugars
- the number of sugars in the glycosidic chain
- the linkage position of the glycosidic chain on the aglycone
- the nature of linkage between aglycones and sugars
- the sequence of sugars in the glycosidic chain
- the nature of the linkages between the different sugars
- the configuration of the sugars.

This elucidation involves very sophisticated NMR techniques, each of which gives some complementary information, especially concerning the nature, the sequence and the linkage of sugars. In reality, numerous possibilities of structures are available and some saponins have been found to contain up to 12 sugar residues. The numerous sugar residues make these molecules highly polar and hence difficult to isolate and purify.

A great number of new molecules has been described in the literature in relation to biological effects (Hostettmann and Marston 1995). These activities can be:

- inherent to the whole molecule. In this case, no structure–activity relationship is presented.
- associated with the aglycone. Numerous hemisyntheses show that a modification of certain functions may increase the solubility and bioavailability of the molecules. In this case, the glycosidic chains have relatively little influence on the type of activity and play mainly a role of vectorisation.

- linked with the presence of the aglycone but also with the nature of the glycosidic chains, in relation to the:
  - number of glycosidic moieties (antifungal activity)
  - ramification of the glycosidic chain (anti-inflammatory activity)
  - number of sugars (antiparasitic activity)
  - nature of the sugars (spermicide)
  - sequence of the sugars (cytotoxicity).

However, it is very difficult to determine structure–activity relationships because of:

- the numerous biological tests which exist and are not systematically used
- the impossibility of having at our disposal all the structures derived from that of the basic molecule.

Some examples of our results concerning new structures with biological activities isolated from South American species are presented here.

Mimosa tenuiflora (Willd.) Poiret

The bark of *Mimosa tenuiflora* (Mimosaceae), an American shrub, is used in Mexican traditional medicine for the treatment of burns and the prevention of inflammation. This material was used to alleviate the suffering of burn victims following the San Juanica natural gas explosion in 1982 and the 1985 earthquake. The results of this treatment have been observed by a great number of journalists and scientists worldwide. Analgesic effects lasting 2 or 3 hours and a complete reconstitution of the epidermis within a few weeks are observed by the use of a powder of this bark.

Chemical investigations were developed (Jiang *et al.* 1991a; Anton *et al.* 1993), which allowed the isolation and identification of 3 new triterpenoid saponins: mimonosides A, B and C (Figure 1), obtained for the first time from the genus *Mimosa*. The two main

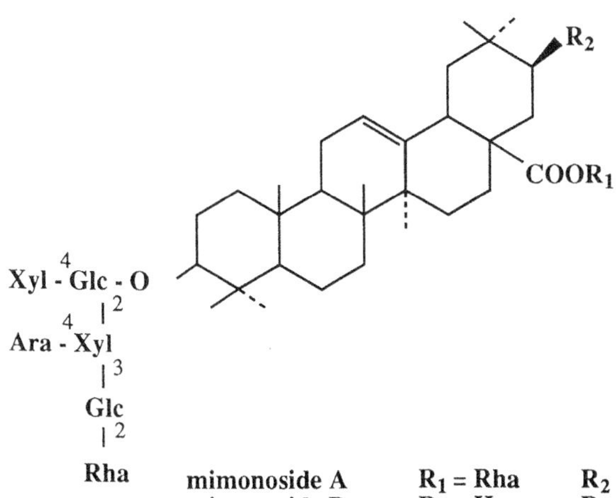

mimonoside A	R_1 = Rha	R_2 = H
mimonoside B	R_1 = H	R_2 = H
mimonoside C	R_1 = Rha	R_2 = OH

Figure 1 Saponins from *Mimosa tenuiflora*

saponins, mimonosides A and B, are oleanolic acid derivatives and the minor one is a machaerenic acid glycoside. The structural elucidations were carried out by chemical methods (acid and base hydrolysis and peracetylation) and spectroscopic techniques. Sequencing of the sugar chain was done on the basis of NMR of the peracetylated material. Sequencing of the sugar chain requires the observation of the correlation between elements of different sugar residues. This is done using the HMBC experiment which correlates protons with carbon atoms separated by 2 or 3 bonds. This sequencing was also obtained and confirmed by measuring inter-residue Overhauser effect (ROESY) (Jiang *et al.* 1991b,c).

The biological effects of these mimonosides have shown significant cytotrophic properties on murine (LMTK) and human cultured fibroblasts (Jiang *et al.* 1991). For instance, on the third day of culture, the reference cells appear as non-confluent with a small nucleus and large cytoplasm. The same fibroblasts after 9 days of culture show a much higher cell density. On the contrary, these same cells treated with saponins and observed after the 9th day of culture show a much higher cell density compared to the control, with fusiform shapes, and organised in parallel stripes. No degenerative effect is detectable. By measuring the amount of protein in the control and treated fibroblasts, it was found that the treated culture (30 μM mimonoside A or B) contained twice as many proteins (control: 12 μg/well; test: 25 μg/well). We also observed that these saponosides do not develop any cytotoxic effect on human or murine tumor cells ($Molt_4$ or RDM_4), even at very high concentrations (10^{-3} M). Moreover, these substances stimulate murine lymphocytes (splenocytes and thymocytes) in *in vitro* cultures. Even with saponins, a synergic effect in the presence of classical mitogenic agents can be observed in *in vitro* tests like concanavalin A (4 μg/ml), lipopolysaccharide, phytohemagglutinin, increasing significantly the activation of the lymphocytes (Jiang 1992).

Topical therapy using powder of the bark on second degree burns has been clinically experimented. A multicenter study on 110 patients in different French hospitals was investigated. The sterile powder is simple to use and applied on the burn. Parameters such as the change in pain on palpation or in spontaneous pain and the quality of the healing process have been successfully evaluated. An excellent tolerance and a good quality of the cicatrization were generally observed.

To sum up, ethnopharmacological observation could be confirmed with this example. Microscopic observation showed that vegetable fibers and starch abound in this bark. Other phytochemical investigations show a great amount of condensed tannins. It is well-known that such vegetable fibers, starch and tannins can ameliorate the treatment of burns and skin regeneration. The significant mitogenic effect and cytotrophic properties on human fibroblast cells of these new saponins can contribute to explain the traditional use. Unfortunately, these *Mimosa* species like other members of the Mimosaceae contain psychotropic alkaloids such as tryptamine derivatives which may limit its development in France because of legal requirements. Nevertheless, the case of *M. tenuiflora* is a good example of an *a posteriori* scientific validation of a traditional remedy (Anton *et al.* 1993).

Mimosa pigra L.

This species of the Mimosaceae is a shrub widespread in the southern hemisphere. No traditional use and no clinical observations have been reported for this plant. It was therefore interesting to compare the chemical composition with *M. tenuiflora*. A series of saponins, mimonosides D to L (Figures 2 and 3) of the same type as those of *M. tenuiflora* was isolated by complementary chromatographic methods (flash, high speed counter-current chromatography etc.) (Englert *et al.* 1995a). They derived from machaerenic and 2-hydroxymachaerenic acids with the same sugars as *M. tenuiflora* but with different sequences and different total sugar numbers. The substitution at position C-21 was either a *Z*/*E*-methoxycinnamic unit or an *E*-cinnamic acid unit.

These saponins have no positive effect on fibroblasts but show growth inhibition activity on $Molt_4$ tumour cells. The methoxy groups seem to be responsible for the growth inhibition effect on tumoral $Molt_4$ cells: a decreasing number of methoxy groups leads to the decrease in the growth inhibition effect, probably in relation to a modification of the degree of lipophilicity of the molecule (Englert *et al.* 1995b).

Colubrina elliptica (Sw.) Briz & Stern.

Some species of the genus *Colubrina* (Rhamnaceae) are traditionally used in the Caribbean for the treatment of skin diseases. Four new saponins were isolated from

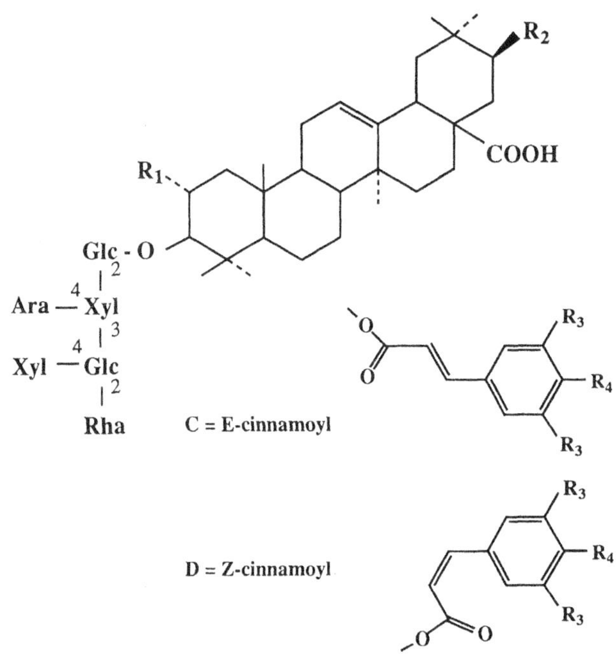

mimonoside D	$R_1 = H$	$R_2 = C/D$	$R_3 = H$	$R_4 = OCH_3$
mimonoside E	$R_1 = H$	$R_2 = C/D$	$R_3 = OCH_3$	$R_4 = OCH_3$
mimonoside F	$R_1 = OH$	$R_2 = C$	$R_3 = H$	$R_4 = H$
mimonoside G	$R_1 = OH$	$R_2 = C/D$	$R_3 = H$	$R_4 = OCH_3$
mimonoside H	$R_1 = OH$	$R_2 = C/D$	$R_3 = OCH_3$	$R_4 = OCH_3$

Figure 2 Saponins from *Mimosa pigra*

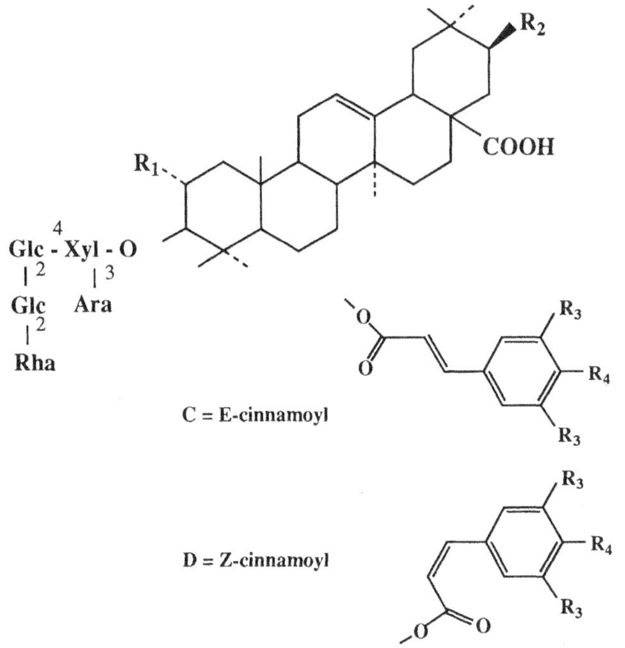

mimonoside J	$R_1 = H$	$R_2 = C$	$R_3 = H$	$R_4 = H$
mimonoside K	$R_1 = H$	$R_2 = C$	$R_3 = H$	$R_4 = OCH_3$
mimonoside L	$R_1 = H$	$R_2 = D$	$R_3 = H$	$R_4 = OCH_3$

Figure 3 Saponins from *Mimosa pigra*

Colubrina elliptica : mabiosides B to E (Figure 4). Because of their easy modification under acidic conditions, the structure investigation was made using only non degradative methods. Only mabioside B contains a seco ring D dammarane triterpenoid aglycone (Oulad-Ali *et al.* 1993, 1994).

To assess the biological effects of the isolated saponins, their activities were investigated in different cell culture systems. Cell growth inhibition and cytotoxic activity were evaluated on cultured tumoral $Molt_4$ lymphoma cells and on non-tumoral 3T3 and L929 fibroblasts. Their possible immunomodulatory effects on cultured murine spleen lymphocytes were also tested. The crude ethanolic bark extract of *C. elliptica* was strongly cytotoxic to the $Molt_4$ cells and also on the adherent fibroblasts. Mabiosides B and C were only weakly cytotoxic.

Considering the mabiosides B, C and D, the glycosidic chain in position 3 is the same and despite a difference in skeleton concerning the cycle D, both the cytotoxicity and the effect on fibroblasts and splenocytes remain low. On the contrary, for mabioside E, some differences are related not only to cycle E but also to the glycosidic chain, with an additional glucose and the presence of a sulphated group. Thus, the biological activity becomes very important (Oulad-Ali 1994).

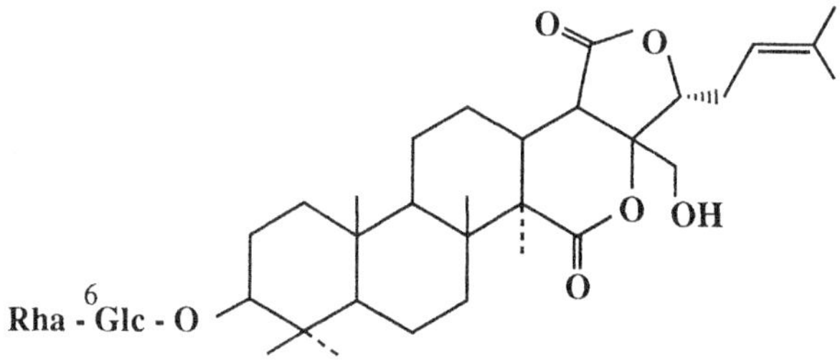

mabioside B

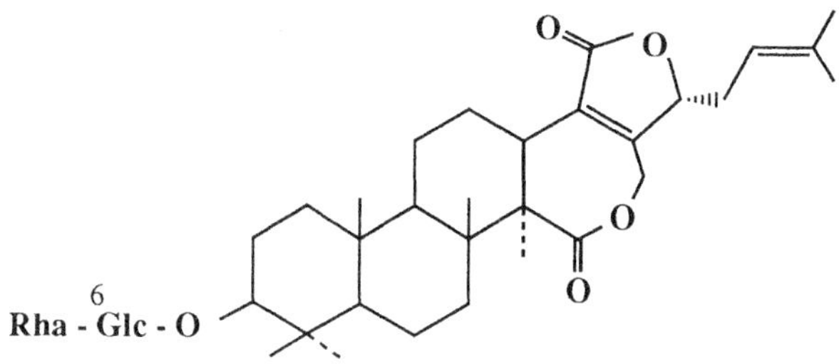

mabioside C

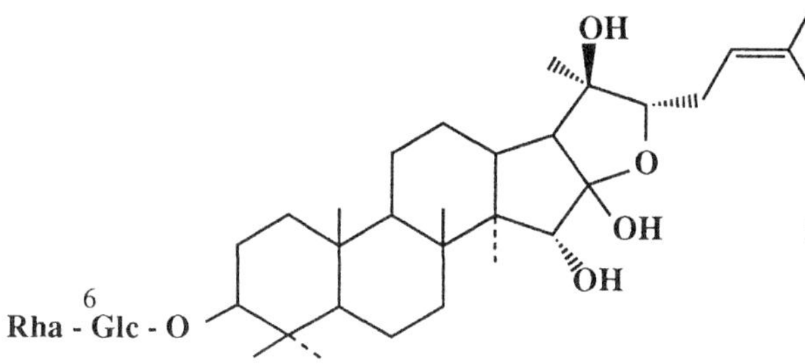

mabioside D

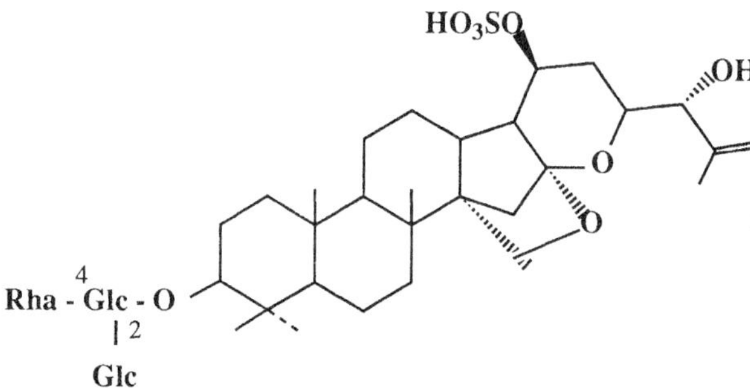

mabioside E

Figure 4 Saponins from *Colubrina elliptica*

Sideroxylon cubense (Greiseb.) T. Pennington

Analogous biological considerations can be observed with the roots of another Caribbean plant, *Sideroxylon cubense* (Sapotaceae), used externally for the treatment of fractures. Sideroxyloside A (Figure 5) and a glucoside of protobassic acid, which differs only by the absence of the glycosidic chain linked to the carboxyl function at C-28, were isolated (Jiang *et al.* 1994).

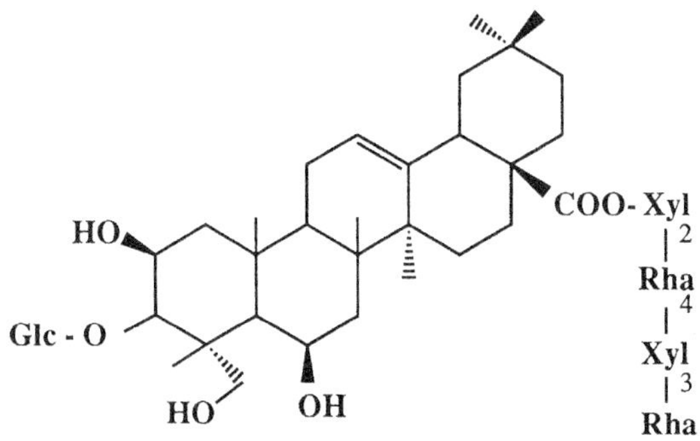

sideroxyloside A

Figure 5 Saponin from *Sideroxylon cubense*

Sideroxyloside A shows a cytotoxic effect on $Molt_4$ cells. On the contrary, the same molecule without the glycosidic chain at C-28 is inactive and stimulates the development of fibroblast cells. A parallel observation concerns two other reference molecules: chrysanthelline, a pentasaccharide of echinocystic acid, and echinocystic acid 3-*O*-glucoside. Chrysanthelline and sideroxyloside A have the same sugar chains at C-28 and are very cytotoxic. The same substances without this sugar moiety are devoid of cytotoxicity on $Molt_4$.

Sideroxylon foetidissimum (Jacq.)

This is an additional member of the Sapotaceae and is another endemic shrub of Caribbean. From its root have been isolated two new triterpenoid saponins named sideroxylosides B and C (Figure 6). On the basis of spectroscopic (^1H, ^{13}C, 2D homonuclear and heteronuclear NMR experiments) and chemical (alkaline and acid hydrolysis) methods, their structures were established as protobassic acid derivatives. Unfortunately, no biological evaluation was possible due to the very small quantity of substances available (Nicolas *et al.* 1995).

At the same time, our interest in the genus *Sideroxylon* has extended to another species, *Sideroxylon spinosum* L., also named *Argania spinosa* L. (Sapotaceae), which is not an American plant but a strictly endemic tree growing in the area of Agadir in south Morocco. New saponins, bayogenin derivatives, have been isolated from the root bark, namely arganines G, H and I (Figure 7) (Oulad-Ali *et al.* 1996). In the case of arganines G and H, the glycosidic side chain is very simple, contrasting with arganine I, with three sugars, which stimulate fibroblast growth.

In conclusion, our results show that the glycoside chains of the triterpene saponins play an important role in cellular recognition. Numerous works mention that the interaction of the cells are linked with the particular glycosides present on the surface of these cells and depend on the nature of the tissue and the animal species. Drugs that could interfere with these glycosides could stop certain pathologies like infections,

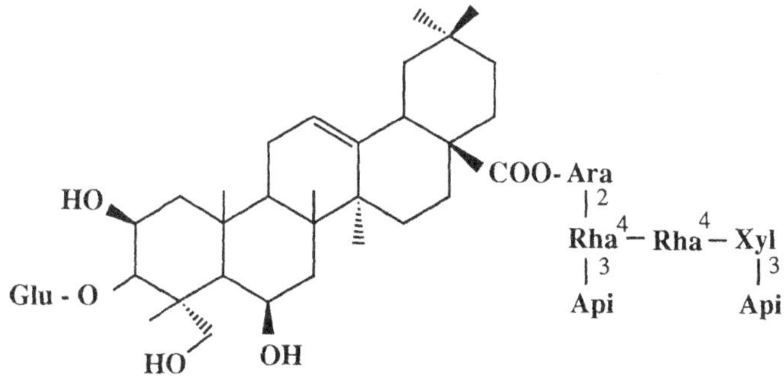

sideroxyloside B

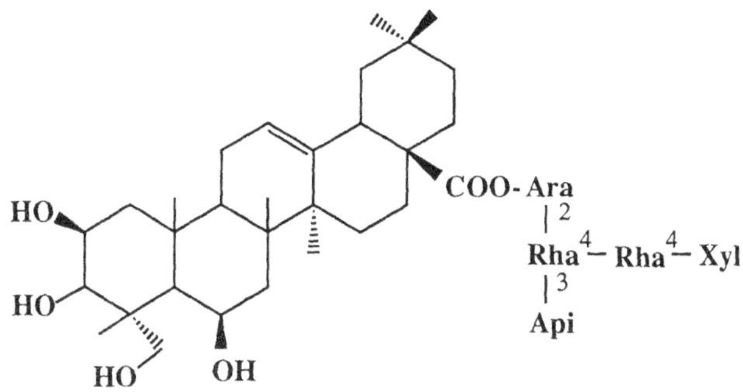

sideroxyloside C

Figure 6 Saponins from *Sideroxylon foetidissimum*

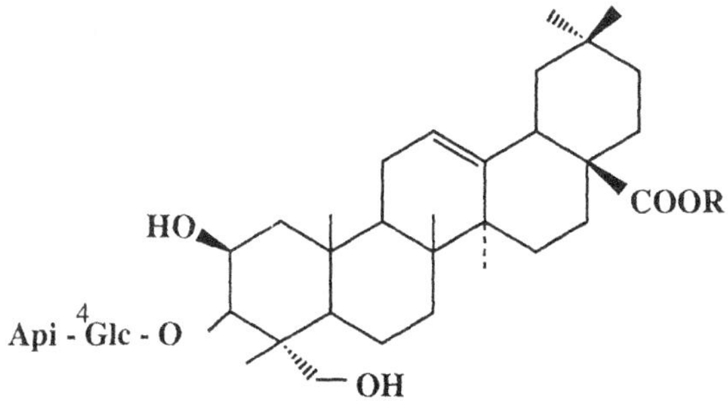

arganine G R = Glc
arganine H R = Ara
arganine J R = Ara - Rha - Xyl - Api (2, 4, 3)

Figure 7 Saponins from *Argania spinosa*

inflammations etc. and perhaps prevent the appearance of cancer. In this direction, saponins are interesting molecules because the aglycone probably plays an important role in modifying the membrane fluidity. The glycoside chain presents the potential for a considerable variability, and a far higher complexity than the linkage of amino-acids in the proteins. This could explain numerous biological effects described in the literature for this chemical class of natural compounds.

ALKALOIDS

The TRAMIL Research Program

The Caribbean Basin represents a privileged geographical region in terms of its biological and cultural diversity. It is characterized by a unique concentration of animal and plant species and its tropical moist forests are classified among those of largest habitat diversity worldwide. Within this biological and cultural diversity, medicinal plants are playing an important role in the traditional health systems in this region (Weniger *et al.* 1993a).

Recognizing this value, the TRAMIL research program has been working since 1984. The main objective of the program is the validation and revaluation of useful aspects of the local popular therapeutic traditions, through bibliographic and scientific studies conducted by Latin-American and European collaborators integrated in the TRAMIL Network (Weniger 1991). This continuous scientific evaluation makes it possible to classify the selected medicinal plants into 3 categories:

- potential toxic plants for which the traditional use has to be discontinued
- plants which require further research (lacking or insufficient scientific information)
- recommended plants for their proven efficacy or known for their innocuity

TRAMIL also supports dissemination projects at the community level, bringing back the evaluated information regarding treatment of certain ailments which may be cured by medicinal plants, at minimal cost and in harmony with the popular tradition (Robineau 1991).

The TRAMIL Program was initiated by a multidisciplinary group of scientists who began working on health care in Haiti and the Dominican Republic in 1984 (Weniger *et al.* 1986a,b). The Program has now expanded to the whole Caribbean Basin and the results of this research, the first edition of the Caribbean Pharmacopoeia, in Spanish, has been published recently (Carballo 1997). French and English versions are due in the near future. The TRAMIL program is supported by DED (German Cooperation), ACCT (Cultural and Technical Cooperation Agency), Fondation de France, French Ministry of Cooperation, ORSTOM and UNESCO.

Some Results

The Rubiaceae family is a large botanical family of more than 400 genera, widely distributed, in particular, in tropical and sub-tropical areas. The family is usually divided into five sub-families: *Hillioideae*, *Henriquezieae*, *Guettardoideae*, *Cinchonoideae* and *Rubioideae*.

While many phytochemical and biological studies have been performed on *Cinchonoideae* and *Rubioideae* species, little is known about the *Guettardoideae* sub-family. This sub-family is well represented in the Caribbean region, and we put our efforts into the study of 4 *Guettardoideae* species from this area.

Neolaugeria resinosa (Vahl) Nicols

This species is a small tree distributed through the Caribbean area and known traditionally for its toxicity. From the root bark of a sample collected in Puerto-Rico three new alkaloids with a novel oxindolic skeleton were isolated: neolaugerine, isoneolaugerine and 17-hydroxy-isoneolaugerine (Figure 8). Structural elucidation was performed using IR, UV, MS and ^{1}H and ^{13}C-NMR mono- and bidimensional spectroscopy.

The three alkaloids were evaluated for their cytotoxic activity against one murine non-cancerous cell line (3T3), 2 murine cancer cell lines (FAZA and B16) and 4 human cancer cell lines ($Molt_4$, U937, Hep G2 and KB). All of them showed weak activity on Molt4 cell lines, but strong growth inhibition on 3T3 and Hep G2 cell lines ($EC_{50} < 1$ μg/ml after 48 hours of contact). Neolaugerine showed also significant receptor binding with muscarinic receptors ($IC_{50} = 3$ μM) and sigma receptors ($IC_{50} = 3$ μM) (Weniger *et al.* 1993b).

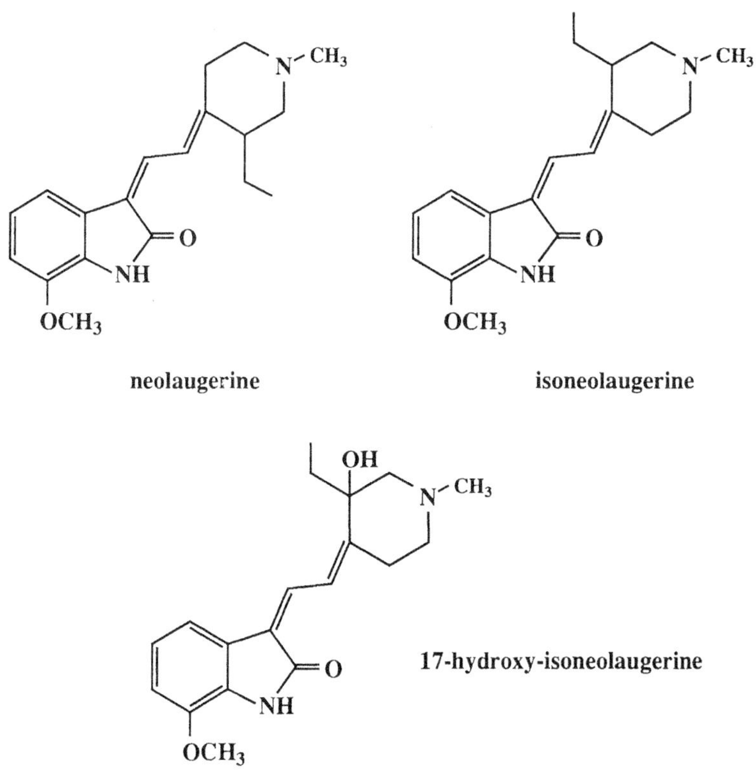

Figure 8 Oxindole alkaloids from *Neolaugeria resinosa*

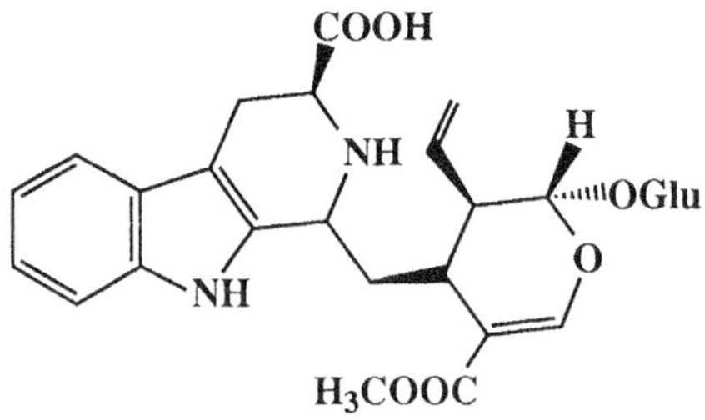

3α, 5α-tetrahydrodeoxy-cordifoline

Figure 9 Glucoalkaloid from *Guettarda ovalifolia*

Guettarda ovalifolia Urb.

This small tree is found only in Puerto-Rico and in the Dominican Republic. From a root sample collected in Puerto-Rico a glucoalkaloid, 3*α*,5*α*-tetrahydrodeoxy-cordifoline (Figure 9) was isolated, using flash chromatography and purification on Sephadex LH20. This molecule is considered to be a biogenetic precursor of the 5-carboxylated indole alkaloids but, surprisingly, no other indole alkaloids could be found in the aerial and subterranean parts of the plant. Structure elucidation was performed using IR, UV, MS and ^{1}H and ^{13}C-NMR mono- and bidimensional spectroscopy .

3*α*,5*α*-tetrahydrodeoxy-cordifoline did not show any activity on the cell lines mentioned above, except for a slight growth inhibition of the $Molt_4$ cell line (ED_{50} = 100 µg/ml after 48 hours of contact) (Jiang *et al.* 1994).

Antirhea portoricensis (Britton & Wilson) Standl.

This small West Indian tree is found from Bahamas to Trinidad. Using chromatographic methods (flash chromatography and gel filtration) we isolated 6 alkaloids: yohimbol, *epi*-yohimbol, antirhine, iso-antirhine, 20-*epi*-antirhine and 19(*S*)-hydroxy-dihydro-corynantheol (Figure 10) from a root sample collected in Puerto-Rico. This is the first time that the latter two have been isolated from plant material. Structural elucidation was performed using IR, UV, MS and ^{1}H and ^{13}C-NMR mono- and bidimensional spectroscopy .

While yohimbol and *epi*-yohimbol showed growth inhibition only on the KB cell line, antirhine and iso-antirhine were active on all of the cell lines mentioned above, and particulary on the $Molt_4$ cell line with EC_{50} = 3 µg/ml and EC_{50} = 13 µg/ml, respectively, after 48 hours of contact (Weniger *et al.* 1994).

Antirhea lucida (Sw.) Hook.

This is also a small West Indian tree found from the Bahamas to Trinidad. A new indole alkaloid, *N,N*-methyl-3′-indolylmethyl-5-methoxytryptamine, as well as the known gramine, *N,N*-dimethyltryptamine and 1,2,3,4-tetrahydro-6-methoxy-2-methyl-*β*-carboline (Figure 11) were isolated from a sample of root material collected in the

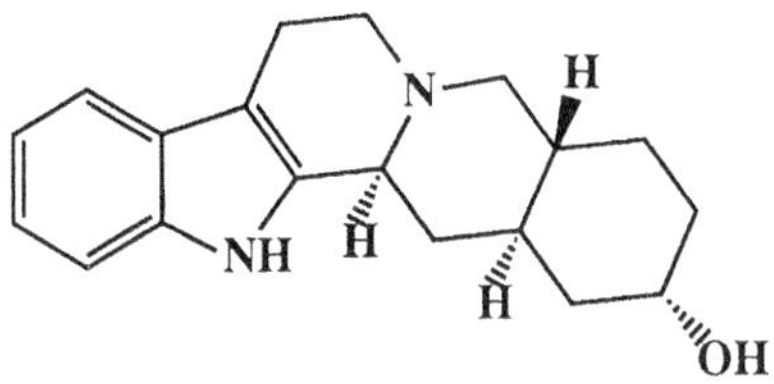

yohimbol

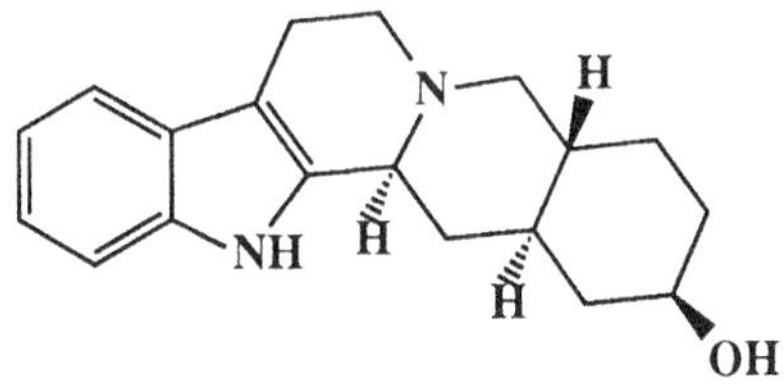

epi-yohimbol

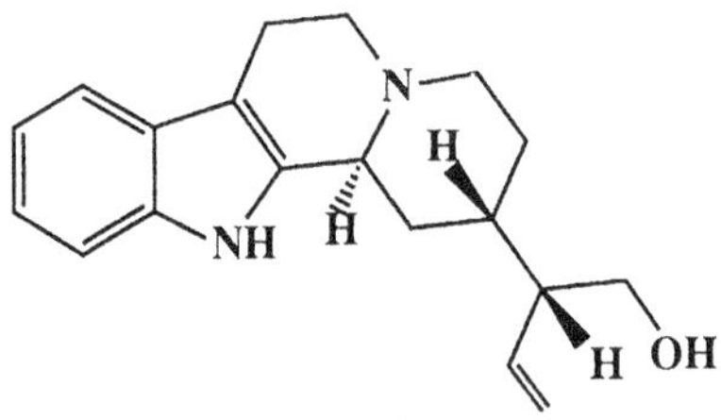

antirhine

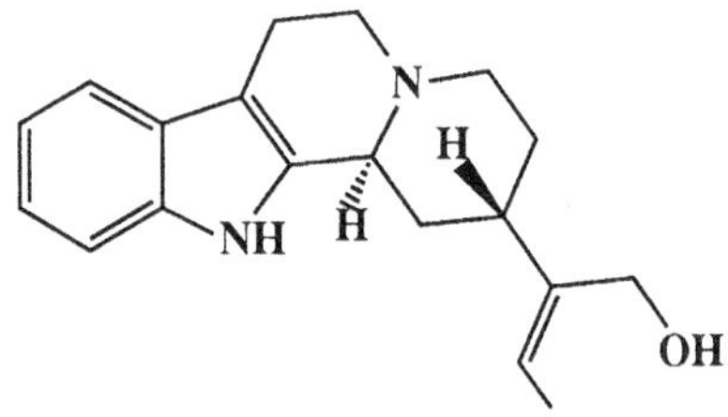

iso-antirhine

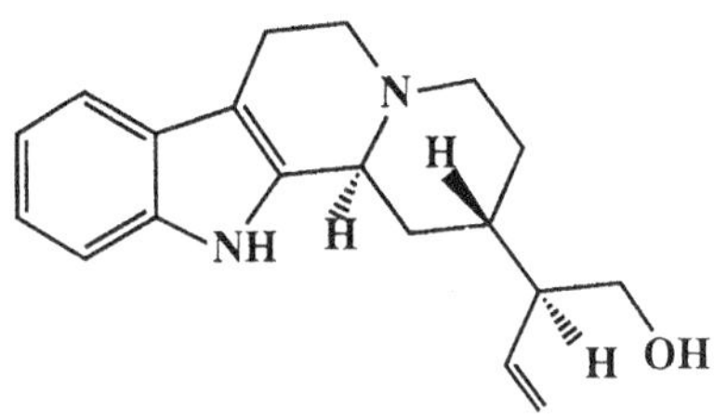

20-epi-antirhine

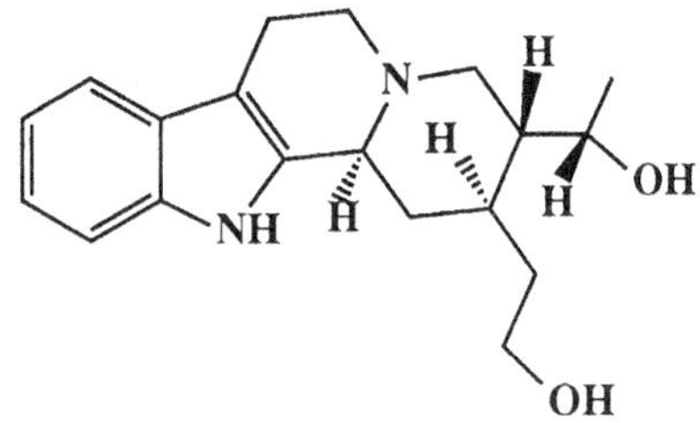

19(S)-hydroxydihydro-corynantheol

Figure 10 Alkaloids from *Antirhea portoricensis*

Dominican Republic using vacuum chromatography and TLC. Structural elucidation was performed using IR, UV, MS and ^{1}H and ^{13}C-NMR mono- and bidimensional spectroscopy (Weniger *et al.* 1995).

NATURAL COMPOUNDS AS ANTIMALARIALS

The treatment of tropical diseases continues to overwhelm the health care services of developing countries. Although diseases such as malaria affect millions of people of these countries, they are considered as non-profitable by the pharmaceutical industries. However, the discovery of antimalarial natural products is an old story for pharmacognosy. The recent success of artemisinin (from *Artemisia annua*), an unusual sesquiterpene lactone containing an endoperoxide moiety, used for the treatment of cerebral malaria and the renewal of the therapeutic value of quinine against chloroquine resistant

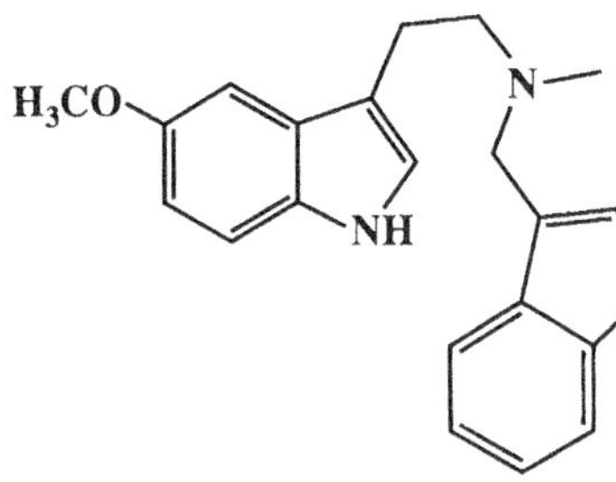

N,N-methyl-3'-indolylmethyl-5-methoxytryptamine

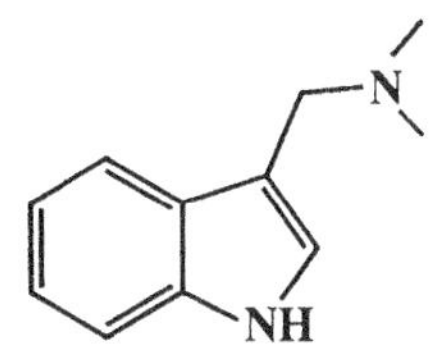

gramine

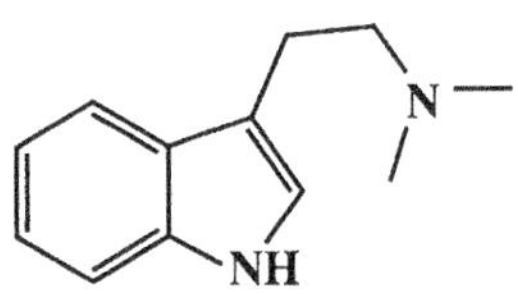

N,N-dimethyltryptamine

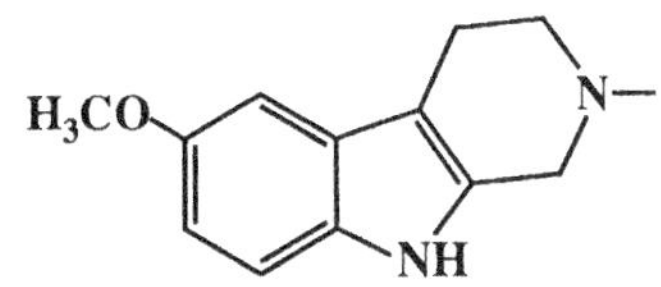

1,2,3,4-tetrahydro-6-methoxy-2-methyl-β-carboline

Figure 11 Indole alkaloids from *Antirhea lucida* (tryptamine derivatives)

strains of *Plasmodium* show that natural products from higher plants have potential as antiprotozoal agents.

The French Scientific Research Institute for Development in Cooperation (ORSTOM) has initiated and developed original investigations of alternative compounds for treatment of malaria, as templates for new lead discoveries, jointly with South American scientific institutions. This project began 14 years ago in French Guyana in collaboration with the Pasteur Institute of that region and then was extended to Bolivia with the collaboration of the University of La Paz, and recently to Colombia with the University del Valle of Cali. It is conducted through a close cooperation with the CNRS, the Universities of Reims, Paris XI, Strasbourg and also Barcelona (Spain).

Field Work

A survey of ethnobotanical knowledge in traditional pharmacopoeia was carried out in the areas where malaria is endemic. Then the ORSTOM group and their associates screened more than 300 antimalarial plants from 9 pharmacopoeias of French Guyana (Wayãpi, Palikur, Creole), Surinam (Saramaka) and Bolivia (Chimane, Chacobos, Alteños, Moseten, Tacana). Each pharmacopoeia of a determined group contains approximately 200 to 300 different plants of which 10 percent are antiprotozoal remedies.

Pharmacological Methods

The classical tests employed measure the specific antimalarial effect of plant extracts on protozoal cells and infected mice and the non-specific effect on host cells and toxicity on animals.

In vitro Testing against Plasmodium falciparum

Culture of chloroquine sensitive and chloroquine resistant strains of *Plasmodium falciparum* was carried out according to Trager and Jensen (1976) on red blood cells suspended in glucose-enriched RPMI 1640 medium supplemented with HEPES and 10% human serum at 37°C. The experiments were conducted in 96 flat bottom well plates. Each well containing 0.2 ml of a ring stage culture (1% hematocrit, 2% parasitaemia) was exposed to increasing concentrations of the extract. Control wells received the solvent, 0.9% saline solution or ethanol. After 48 hours under culture conditions, the supernatant was removed and parasitaemia was determined by examining Giemsa-stained thin blood smears made from remaining cells for each well.

The growth inhibition percentage of the parasite was calculated by the following formula:

$$\frac{(\text{parasitaemia in control} - \text{parasitaemia with drugs})}{\text{parasitaemia in control}} \times 100$$

Cytotoxicity

IC_{50} values against KB human buccal carcinoma cells were determined by standard methods established at the National Cancer Institute and currently used in ORSTOM laboratories.

In vivo Testing against P. berghei and/or P. vinckei petteri

The *in vivo* antimalarial activity of plant extracts was determined by the classical 4-days suppressive test (Peters 1980) against the rodent malaria *Plasmodium berghei* and *P. vinckei petteri*. Swiss male mice, mean body weight 20 ± 2 g, were bred in the department of animal breeding and kept in groups of five in plastic cages. The mice were fed on standard mice cubes and drinking water *ad libitum*. They were infected with 10^8 parasitized cells in 0.9% saline on day 0. Groups of 5 mice were treated intraperitonealy from day 0 to day 3 with increasing doses of plant extracts to 1 g/kg. The suppressive effect was estimated on day 4, by examining Giemsa-stained thin blood smears made from the tail of the treated mice and by comparison with a control group of mice treated with the solvent of the extract (saline, dimethyl sulfoxide or Tween). The stained thin blood smears were examined under × 1000 magnification, and the percentage of parasitized red blood cells was counted on at least 9000 red blood cells observed for each concentration. The percentage of growth inhibition of the parasite was calculated using the same procedure as for *in vitro* determination.

Measurement of the Acute Toxicity of Antimalarial Compounds

Volumes of increasing doses of compounds were randomly administered by subcutaneous injections to groups of 10 Swiss mice. The mortality was noted after 72 hours. The LD_{50} of compounds was determined from the graph obtained by plotting the doses (logarithmic scale) against the percentage of mortality (probit scale).

Antimalarial Natural Products

Chloroquine

The *in vitro* activity in our experimental models for chloroquine (Figure 12) chosen as reference compound, is an IC_{50} of 0.02 μg/ml on a sensitive strain and an IC_{50} of 0.08 μg/ml on a resistant strain of *P. falciparum*. *In vivo* the effective dose or ED_{50} was 0.5 mg/kg/day on *P. berghei* infected mice, with a LD_{50} of 79 mg/kg. In this case, the therapeutic index has a value of 158.

Sergeolide

Sergeolide (Figure 13), a quassinoid from *Picrolemma pseudocoffea* Ducke (Simaroubaceae) showed very strong antiplasmodial activities *in vitro* as well as *in vivo*. A low concentration (0.006 μg/ml) was able to fully inhibit the *in vitro* growth of chloroquine-sensitive

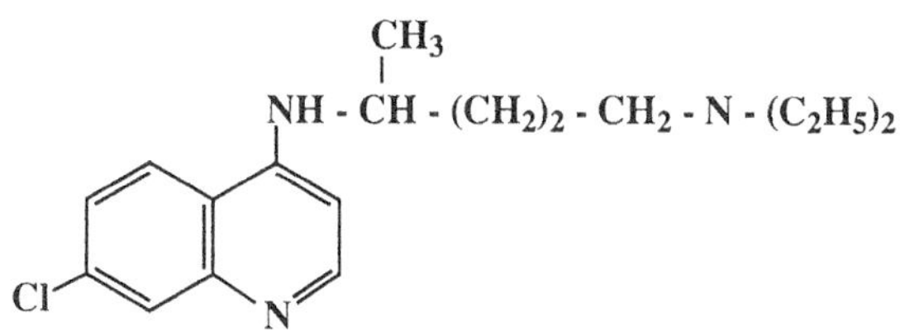

chloroquine

Figure 12 Antimalarial reference compound

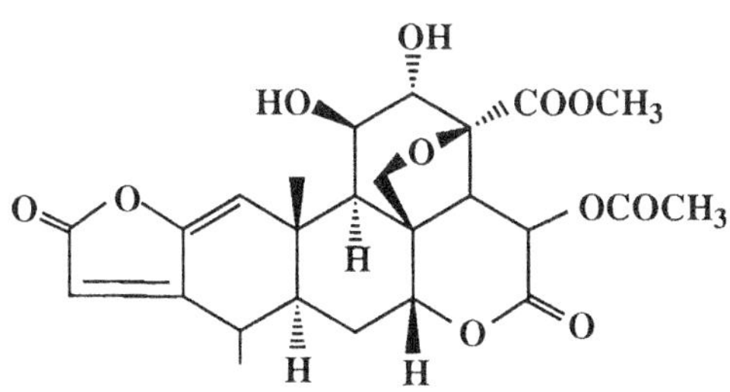

sergeolide

Figure 13 Quassinoid from *Picrolemna pseudocoffea*

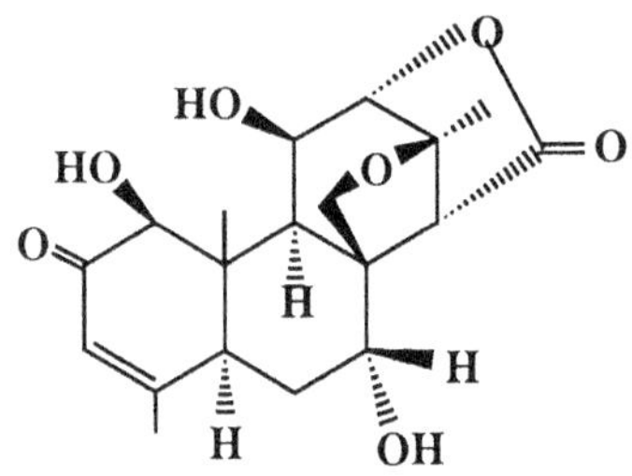

cedronin

Figure 14 Quassinoid from *Simaba cedron*

and resistant strains of *P. falciparum*. Small amounts (0.26 mg/kg/day) markedly reduced the virulence of experimentally induced *P. berghei* infection in mice. However sergeolide, because of its high toxicity (LD_{50} = 1.8 mg/kg) does not seem to be useful for malaria curative treatment. This study by C. Moretti from ORSTOM and the Pasteur Institute group in French Guyana was one of the first to show the pharmacological interest of quassinoids against malaria (Fandeur *et al.* 1985).

Cedronin

Cedronin (Figure 14) was isolated from *Simaba cedron* Planchon (Simaroubaceae), a species popularly believed in South America to have antimalarial properties. It was examined for *in vitro* and *in vivo* antimalarial activities and for cytotoxicity against KB cells. Experimental results showed that cedronin was active against chloroquine-sensitive and resistant strains of *P. falciparum*, with an IC_{50} of 0.25 µg/ml (0.65 mmol/ml). It was also found to be active *in vivo* against *P. vinkei* with an IC_{50} of 1.8 mg/kg (4.7 nM/kg) in the classical four days test. Cedronin belongs to the small group of quassinoids with a C_{19} basic skeleton and shows a rather low cytotoxicity against KB cells (IC_{50} = 4 µg/ml) as compared to the C_{20} biologically active quassinoids (Moretti *et al.* 1994).

n-Hentriacontanol

From the petroleum extracts of *Cuatresia forsteriana* (Solanaceae), we have isolated n-hentriacontanol as the active compound. This fatty alcohol markedly reduced the virulence of experimentally induced *P. vinckei* infection (ED_{50} = 3.1 mg/kg/day and LD_{50} = 412 mg/kg). The therapeutical index of 130 is equivalent to that of chloroquine in this model (Deharo *et al.* 1992).

The mode of action of this simple C_{31} linear alcohol and other fatty molecules was investigated. The C_{18} oleic, elaidic and linoleic fatty acids were active *in vitro* on *P. falciparum* whereas n-hentriacontanol did not show any inhibition of this parasite. In order to understand the mechanism of their antimalarial action, several tests were performed. No hemolysis of infected cells could be observed up to 500 µg/ml. No effect on the lipid peroxidation, on the ATP level, on the transport through the parasite induced permeability pathways or on the phagocytosis of the infected cells could be observed. The cytotoxicity effect of the fatty acids was very rapid: full inhibition of nucleic acids and protein synthesis was observed in less than 30 min. This inhibition was not relieved by the addition of deferrioxamine or $FeCl_3$, indicating that fatty acids do not act by facilitating the transport of iron. Inhibition was not relieved in the presence of orotic acid or its methyl ester, indicating that fatty acids do not act at the mitochondrial level of pyrimidine synthesis. Inhibition was not relieved when cells were incubated in a high potassium medium, implying that fatty acids did not act by depleting the parasite of this physiologically important cation. Vitamin E had no protective effects and no evidence for lipid peroxidation could be obtained, suggesting that the inhibitory effect of fatty acids was not due to lipid peroxidation. These fatty molecules have probably a specific target inside the parasite (Krugliak *et al.* 1995).

Tubulosine

Pogonopus tubulosus (D.C.) Schumann (Rubiaceae), a tree growing in the southern South American subtropical rainforest, provides one of the numerous drugs called "falsa quina" in South America which is used against malaria. The bark of this medicinal plant yielded three alkaloids: tubulosine, psychotrine and cephaeline. Tubulosine (Figure 15) showed an interesting activity *in vitro* with an IC_{50} of 0.006 µg/ml against the sensitive strain of *P. falciparum* and an IC_{50} of 0.011 µg/ml against the resistant strain of *P. falciparum*. This compound had good *in vivo* antimalarial activity with an ED_{50} of 0.05 mg/kg/day on the *P. vinckei petteri* strain and an ED_{50} of 0.45 mg/kg/day on the *P. berghei* model at a lower concentration than its lethal dose. Psychotrine was less active (Sauvain *et al.* 1996).

The results presented here lend support to the traditional common use of this plant as antimalarial. Unfortunately, however, tubulosine, the principal antimalarial compound isolated from this species presents little interest in the curative treatment of malaria because of its toxicity.

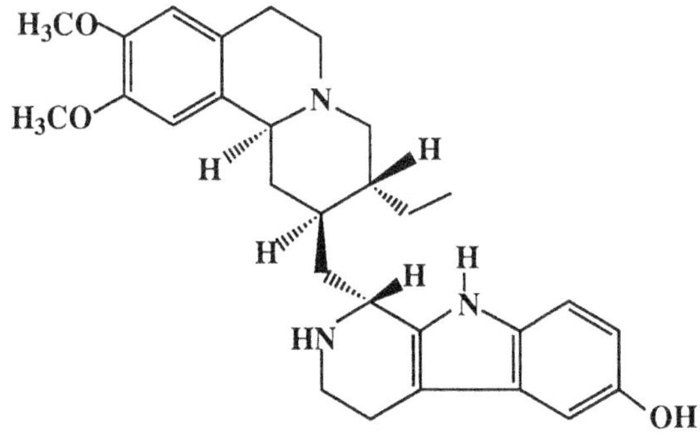

tubulosine

Figure 15 Alkaloid from *Pogonopus tubulosus*

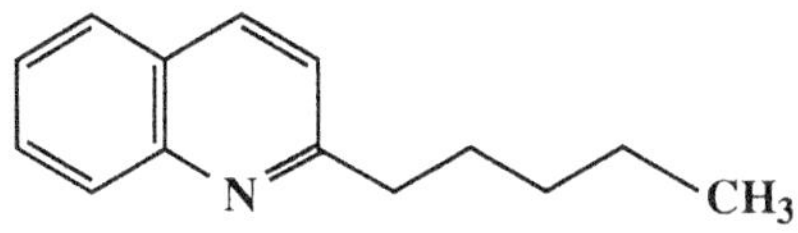

2n-pentylquinoline

Figure 16 Quinoline from *Galipea longiflora*

2-Substituted Quinolines

A. Fournet and the group of the Faculty of Pharmacy at the University of Paris XI have evaluated the *in vivo* antiplasmodial activity of six 2-substituted quinolines and a total alkaloid extract of *Galipea longiflora* Kr. (Rutaceae). Mice infected with *P. vinckei petteri* were treated orally at a single dose of 50 mg/kg of the quinolines or of the extract. 2-n-pentylquinoline (Figure 16) showed activity against *P. vinckei petteri.* This result seems to confirm the antimalarial efficacy of an infusion of the stem bark of *G. longiflora* used by the Bolivian natives against fever (Gantier *et al.* 1996).

More recently, some promising results *in vitro* have been obtained with two Colombian plants, *Guarea trichiloides* L. (Meliaceae) and *Guatteria amplifolia* Tr. & Pl. (Annonaceae). The methanolic extract of the fruit of *Guarea trichiloides* showed good leishmanicidal and trypanocidal activity at 200 µg/ml. The same extract showed also antimalarial activity, with $IC_{50} = 4$ µg/ml against a sensitive strain of *Plasmodium falciparum.* The methanolic extract of the aerial parts of *Guatteria amplifolia* had antimalarial activity, with $IC_{50} = 1.9$ µg/ml against the same strain. The isolation of active compounds from the two plants is in process.

CONCLUSIONS

Among bioactive new natural products, saponins and alkaloids are always considered as interesting models. They are mainly distributed in tropical plants. Thus a collaboration with Central and South American countries has been developed not only to find new fundamental structures, but also for future therapeutical applications, especially for major pathologies like parasitic diseases.

REFERENCES

Anton, R., Jiang, Y., Weniger, B., Beck, J.P. and Rivier, L. (1993) Pharmacognosy of *Mimosa tenuiflora* (Willd.) Poiret. *J. Ethnopharmacol.*, **38**, 153–157.

Carballo, A., Germosen Robineau, L., Lagos-Witte, S. and Weniger, B. (1997) *Farmacopea Caribeña*, Germosen Robineau, L. (eds.), Editions Désormeaux, Fort-de-France, Martinique.

Deharo, E., Sauvain, M., Moretti, C., Richard, B., Ruiz, E. and Massiot, G. (1992) Activité antipaludique du n-hentriacontanol isolé de *Cuatresia sp.* (Solanaceae). *Ann. Parasitol. Hum. Comp.*, **67**, 126–127.

Englert, J., Krempp, E., Guillaume, D., Leroy, Y., Weniger, B., Lobstein, A. and Anton, R. (1995a) Triterpenoid saponins from *Mimosa pigra*. *J. Nat. Prod.*, **58**, 1265–1269.

Englert, J. (1995b) *Mimosa pigra* L. (Fabacées). Aspect botanique, étude chimique des hétérosides triterpéniques et essais biologiques. *PhD Thesis*, University Louis Pasteur, Strasbourg.

Fandeur, T., Moretti, C. and Polonsky, J. (1985) *In vitro* and *in vivo* assessment of the antimalarial activity of sergeolide. *Planta Med.*, **51**, 20–23.

Gantier, J.C., Fournet, A., Munos, M.H. and Hocquemiller, R. (1996) The effect of some 2-substitued quinolines isolated from *Galipea longiflora* on *Plasmodium vinckei petteri* infected mice. *Planta Med.*, **62**, 285–286.

Hostettmann, K. and Marston, A. (1995) *Chemistry and Pharmacology of Natural Products. Saponins.* Cambridge University Press.

Jiang, Y. (1991a) Les saponosides du *Mimosa tenuiflora* (Willd.) Poiret (Mimosaceae). Aspects botanique, chimique et biologique. *PhD Thesis*, University Louis Pasteur, Strasbourg.

Jiang, Y., Massiot, G., Lavaud, C., Teulon, J.M., Guéchot, C., Haag-Berrurier, M. and Anton, R. (1991b) Triterpenoid glycosides from the bark of *Mimosa tenuiflora*. *Phytochemistry*, **30**, 2357–2360.

Jiang, Y., Haag-Berrurier, M., Anton, R., Massiot, G., Lavaud, C., Teulon, J.M. and Guéchot, C. (1991c) Structure of a new saponin from the bark of *Mimosa tenuiflora*. *J. Nat. Prod.*, **54**, 1247–1253.

Jiang, Y., Beck, J.P., Italiano, L., Weniger, B., Haag-Berrurier, M. and Anton, R. (1992) Effects of saponins from *Mimosa tenuiflora* on lymphoma cells and lymphocytes. *Phytother. Res.*, **6**, 310–313.

Jiang, Y., Oulad-Ali, A., Guillaume, D., Weniger, B. and Anton, R. (1994) Two triterpenoid saponins from the root of *Sideroxylon cubense*. *Phytochemistry*, **35**, 1013–1015.

Jiang, Y., Weniger, B., Quirion, J.C., Muller, C. and Anton R. (1994) 3α–5α-tetrahydrodeoxycordifoline from *Guettarda ovalifolia*. *Planta Med.*, **60**, 294.

Krugliak, M., Deharo, E., Shalmiev, G., Sauvain, M., Moretti, C. and Ginsburg, H. (1995) Antimalarial effects of C_{18} fatty acids on *Plasmodium falciparum* in culture and on *P. vinckei petteri* and *P. yoelii nigeriensis in vivo*. *Experim. Parasitol.*, **81**, 97–105.

Moretti, C., Deharo, E., Sauvain, M., Jardel, C., Timon David, P. and Gasquet, M. (1994) Antimalarial activity of C_{19} quassinoid cedronine. *J. Ethnopharmacol.*, **43**, 57–61.

Nicolas, G., Oulad-Ali, A., Guillaume, D., Lobstein, A., Weniger, B. and Anton, R. (1995) Triterpenoid saponins from the root of *Sideroxylon foetidissimum*. *Phytochemistry*, **38**, 225–228.

Oulad-Ali, A. (1994). Les saponosides de *Colubrina elliptica* (Sw.) Briz. & Stern. Etude chimique et activités biologiques sur cultures cellulaires animales et humaines. *PhD Thesis*, University Louis Pasteur, Strasbourg.

Oulad-Ali, A., Guillaume, D., Jiang, Y., Weniger, B. and Anton, R. (1993) Mabioside B, a novel saponin from *Colubrina elliptica*. *Nat. Prod. Lett.*, **2**, 203–207.

Oulad-Ali, A., Guillaume, D., Weniger, B., Jiang, Y. and Anton, R. (1994) Mabiosides C-E: triterpenoid saponins from the bark of *Colubrina elliptica*. *Phytochemistry*, **36**, 445–448.

Oulad-Ali, A., Kirchner, V., Lobstein, A., Weniger, B., Anton, R., Guillaume, D. and Charrouf, Z. (1996) Structure elucidation of three triterpene glycosides from the trunk of *Argania spinosa*. *J. Nat. Prod.*, **59**, 193–195.

Peters, W. (1980) *Chemotherapy and Drug Resistance in Malaria.* Vol. 1, 2nd edition. Academic Press, London.

Robineau, L. and Weniger, B. (1991) Naissance d'une pharmacopée caraïbe: fruit d'une interdisciplinarité et d'une continuité entre le recensement, l'évaluation et la diffusion de la médecine traditionnelle. In *Ethnopharmacologie: Sources, Méthodes, Objectifs*, ORSTOM Editions, Paris, pp. 115–122.

Sauvain, M., Moretti, C., Bravo, J.A., Callapa, J., Muñoz, V., Ruiz, E., Richard, B. and Le Men-Olivier, L. (1996) Antimalarial activity of alkaloids from *Pogonopus tubulosus*. *Phytother. Res.*, **10**, 198–201.

Trager, W. and Jensen, J.B. (1976) Human malaria parasites in continuous culture. *Science*, **193**, 674–675.

Weniger, B. (1991) Interest and limitation of a global ethnopharmacological approach. *J. Ethnopharmacol.*, **32**, 37–42.

Weniger, B., Rouzier, M., Daguilh, R. and Anton, R. (1986a) La médecine populaire dans le plateau central d'Haïti. 1ère partie. Etude du système thérapeutique traditionnel dans un cadre socio-culturel rural. *J. Ethnopharmacol.*, **17**, 1–11.

Weniger, B., Rouzier, M., Daguilh, R. and Anton, R. (1986b) La médecine populaire dans le plateau central d'Haïti. 2ème partie. Inventaire ethnopharmacologique. *J. Ethnopharmacol.*, **17**, 12–27.

Weniger, B., Fessler-Marchan, A. and Anton, R. (1993a) Les médecines des Caraïbes insulaires hispanophones. In *Encycl. Méd. Nat., Phytothérapie, Aromathérapie*, Editions Techniques, Paris, A-9-4, 8 p.

Weniger, B., Jiang, Y., Anton, R., Bastida, J., Varea, T. and Quirion, J.C. (1993b) Oxindole alkaloids from *Neolaugeria resinosa. Phytochemistry*, **32**, 1587–1590.

Weniger, B., Anton, R., Varea, T., Quirion, J.C., Bastida, J. and Garcia, R. (1994) Indole alkaloids from *Antirhea portoricensis. J. Nat. Prod.*, **57**, 287–290.

Weniger, B., Rafik, W., Bastida, J., Quirion, J.C. and Anton, R. (1995) Indole alkaloids from *Antirhea lucida. Planta Med.*, **61**, 569–570.

5. BIOASSAY METHODS USEFUL FOR ACTIVITY-GUIDED ISOLATION OF NATURAL PRODUCT CANCER CHEMOPREVENTIVE AGENTS

JOHN M. PEZZUTO[1,2], LYNDA L. SONG[1], SANG KOOK LEE[1], LISA A. SHAMON[1], EUGENIA MATA-GREENWOOD[1], MEISHIANG JANG[1], HYEH-JEAN JEONG[1], EMILY PISHA[1], RAJENDRA G. MEHTA[2] and A. DOUGLAS KINGHORN[1]

[1]*Program for Collaborative Research in the Pharmaceutical Sciences, and Department of Medicinal Chemistry and Pharmacognosy, College of Pharmacy, and* [2]*Department of Surgical Oncology, College of Medicine, University of Illinois at Chicago, Chicago, Illinois 60612*

INTRODUCTION

Based on morbidity, mortality, economic burden, and emotional hardship, cancer may be considered the most onerous health problem afflicting the general population. In the United States, as indicated by the most recent statistics published by the American Cancer Society (1997), the birth to death probability of developing invasive cancer is 38.33% in women (1 in 3) and 48.17% in men (1 in 2). About 1,382,000 new cancer cases are expected to be diagnosed this year, in addition to approximately 900,000 cases of basal and squamous cell skin cancer. Incredibly, approximately 9,000,000 Americans have expired due to the clinical manifestations of cancer since 1990, and the current five-year relative survival rates for several malignant, metastatic cancers remain dismal (e.g., esophageal, liver, lung, pancreatic, and stomach cancers, all 2%; breast cancer, 20%). Even though significant progress has been made in the clinical management of certain cancers (e.g., childhood leukemia, Hodgkin's disease, and testicular cancer), and the consequences of chemotherapy can be managed with improved efficacy through the use of antiemetics and immunostimulants, it is apparent that research programs must continue to focus on the innovation of more effective strategies of managing metastatic diseases.

Obviously, in addition to working on improvements in cancer therapy, primary prevention tactics should be invoked with the greatest possible expediency. It is intuitively clear that all tobacco-related cancers could be completely prevented. Similarly, other types of primary prevention schemes could readily be implemented, such as reducing exposure to harmful levels of sunlight. In the 1980s, a comprehensive plan was developed by the National Cancer Institute in which the stated goal was to reduce the 1985 cancer mortality rate by 50% by the year 2000 (Greenwald *et al.*, 1986, 1987). As we begin to near this target date, it is apparent that the goal of 50% reduction will not be achieved. In fact, it is generally accepted that the death rate due to cancer will actually exceed the death rate due to coronary diseases by the year 2000. Nonetheless, these visionary strategies are extremely valuable for establishing overall priorities in national healthcare. As is now the case with heart disease, prevention must receive greater emphasis as a key management strategy for cancer.

A relatively new area of scientific endeavor concerns dietary or pharmaceutically based approaches for the prevention of cancer. Cancer chemoprevention, a term coined by Dr. Michael B. Sporn as part of his classical work concerning retinoids and cancer prevention (Sporn *et al.* 1976), entails the prevention or delay of carcinogenesis in humans by ingestion of dietary or pharmaceutical agents. In essence, delay of a disease or condition beyond a normal human lifespan is equivalent to a cure. Prevention tactics have been broadly implemented for management of heart disease (Pedersen *et al.* 1994), and it is now generally accepted that cancer can be managed to some extent through chemopreventive intervention.

Various groups of compounds have been classified as cancer chemopreventive agents (Kelloff *et al.* 1994b), largely based on the results of animal studies and epidemiological data (Bjelke 1975; Butterworth 1992; Hirayama 1979; Knekt 1992). Utilizing these agents, a variety of clinical intervention studies are underway with humans at high risk for the development of certain cancer types (Boone *et al.* 1990; Kelloff *et al.* 1994a,b). Although the trend is to evaluate preneoplastic biomarkers in high risk groups, finasteride (Proscar®) is being tested in clinical trials involving 18,000 men over 55 years of age who are ostensibly at risk for prostate cancer (7 years), (Thompson and Coltman 1996), and a large-scale breast cancer prevention trial involving tamoxifen (Ford *et al.*, 1994; Jordan 1994; Powles and Hickesh 1995) has been concluded recently since reduction of approximately 45% was observed. During the course of these large clinical trials, however, some disappointments have been experienced. It is possible that tamoxifen can promote endometrial cancer to a certain extent (Jordan 1995; Powles and Hickish 1995), and β-carotene has failed to reduce the incidence of lung cancer (Blumberg and Block 1994; Omenn *et al.* 1996). Nonetheless, the concept of cancer chemoprevention is firmly entrenched, and it is certain that the full potential of this approach has yet to be realized, especially in the general population.

Our primary research emphasis has focussed on the front-end of the entire process – that of drug discovery. Although a number of organizations concentrate on the discovery and characterization of various therapeutic agents, this is not the case in the area of cancer chemoprevention. We have therefore instituted a multidisciplinary project wherein plant materials procured from throughout the world are used for activity-guided fractionation schemes that yield novel and otherwise unpredictable chemical entities with desirable biologic potential. These agents are fully characterized in terms of structure, and subjected to more advanced mechanistic evaluations. Through coordinated efforts of isolation and chemistry, the lead compounds are characterized in well-established animal models for definition of cancer chemopreventive efficacy. With this knowledge base of structure, mechanism, and chemopreventive activity, the discoveries can logically be considered as candidates for clinical intervention trials. Aside from our work, few if any broad-based systematic searches have been conducted for the discovery of novel cancer chemopreventive agents, and the structural diversity provided by nature presents a unique opportunity for continuous progress.

EXPERIMENTAL DESIGN FOR THE DISCOVERY OF NATURAL PRODUCT CANCER CHEMOPREVENTIVE AGENTS

Since a considerable number of known cancer chemopreventive agents are naturally occurring, it is reasonable to assume additional entities with desirable preventative activities exist in nature. The overall experimental approach we have developed for the discovery of cancer chemopreventive agents has been described in the literature (Kinghorn *et al.* 1997; Pezzuto 1993, 1995, 1996, 1997; Pezzuto *et al.* 1997). In the overall field of natural product drug discovery, a time-tested method for assured success is bioassay-directed fractionation. In this approach, starting materials are selected (e.g., by random collection, information derived from ethnomedical systems of medicine, or literature surveillance), and extracts are prepared that are suitable for biological evaluation. The materials are then tested in bioassay systems, and substances demonstrating a positive response are classified as active leads. After a number of active leads are identified, decisions are made to fractionate the most promising materials. Each fraction is monitored for potential to mediate a positive response in the bioassay test system, and this process continues until a pure active substance is obtained. The resulting substance is then subjected to procedures of structure elucidation. Once an active isolate is obtained, more thorough biological evaluation procedures are often performed and, based on the accumulated data, the material is considered as a candidate for more advanced testing and development.

One question of paramount importance relates to the bioassay test system. In the area of antitumor drug discovery, a large number of *in vitro* test systems and the issues that need to be considered when attempting to interrelate *in vitro* test results with *in vivo* efficacy studies have been described (Suffness and Pezzuto 1991). In general, recent advances in molecular biology have led to a greater understanding of the molecular basis of human disease states. As a correlate, *in vitro* systems that monitor a response that is either closely related to or identical with the molecular event yielding the disease condition can be devised. As summarized previously (Gerhäuser *et al.* 1995; Jang and Pezzuto 1997; Kennelly *et al.* 1997; Mehta *et al.* 1997; Pisha and Pezzuto 1997; Shamon *et al.* 1994; Shamon and Pezzuto 1997; Suh *et al.* 1995), these systems can be employed to assess rapidly and inexpensively the biologic potential of a large number of test materials.

In the field of cancer chemoprevention, definition of the "best" bioassay system remains subjective. Use of *in vitro* assays permits the identification of a suitable number of lead starting materials, and bioassay-directed isolation enables the procurement of active agents from a reasonable number of sources in a finite period of time. However, at least from a superficial viewpoint, there is a dichotomy: it is impossible for short-term *in vitro* systems to fully represent the physiological complexity of mammals, but the ultimate objective of the research is to discover and characterize agents that are useful in humans. As a first step to circumvent these problems, we have assembled a *panel* of bioassay procedures that reflect various stages of carcinogenesis. In our experience, subjecting materials to this broad "net" yields active leads in the range of 2–5% of the number tested. These active materials are then evaluated with a model of greater physiological complexity, i.e., inhibition of carcinogen-induced preneoplastic lesions in mouse mammary organ culture (mmoc). Although this latter model is not suitable for use as a monitor in bioassay-directed fractionation schemes, due to time and expense, the strong positive correlation with chemopreventive activity observed in full-term animal tumorigenesis studies makes the system worthwhile with a limited number of samples. This secondary discrimination step reduces the number of active leads by 2–3-fold (Mehta *et al.* 1997), and these leads are subjected to bioassay-directed fractionation using the *in vitro* assay system in which activity was originally observed as a monitor. It is postulated that the secondary evaluation procedure is able to confirm the relevance of the original observation made with the *in vitro* monitor.

ASSAYS USED FOR BIOASSAY-DIRECTED FRACTIONATION

We have established assays that are suitable for monitoring inhibition of carcinogenesis at the stages of initiation (antioxidant activity, antimutagenic activity, induction of quinone reductase activity in cell culture), promotion (inhibition of phorbol ester-induced ornithine decarboxylase activity in cell culture, inhibition of cyclooxygenase activity), and progression (induction of cell differentiation, anti-estrogenic activity). In every case, the assays were systematically investigated, varying parameters such as incubation time, concentration of substrates, enzymes or receptors, etc. In addition, the possible effects of various solvents (e.g., DMSO) were investigated, and sensitivity and thresholds of activity were normally established by addition of positive control compounds to otherwise inactive plant extracts. Once these types of parameters were fixed, plant extracts with unknown activities were examined. In summary, over 10,000 tests have been performed, and the number of active plants discovered is in the range of 3% of the total. These active plant extracts are generally tested for potential to inhibit carcinogen-induced preneoplastic lesion formation in the mouse mammary organ culture model.

IN VITRO ASSAY RESULTS AND SHORT-TERM ANIMAL STUDIES

Induction of Quinone Reductase (QR) Activity Utilizing Cultured Hepa 1c1c7 Cells

Induction of Phase II drug-metabolizing enzymes, such as glutathione *S*-transferase (GST) and NAD(P)H:quinone reductase (QR), is an important mechanism of cancer chemoprevention (Talalay *et al.* 1995; Zhang *et al.* 1992, 1994). As an approach toward searching for novel cancer chemoprevention agents, we have utilized a rapid, sensitive QR assay and a GST assay to identify potential detoxification enzyme inducers (Gerhäuser *et al.* 1997b). In order to understand the possible mechanism of induction, we have also employed a series of methods to further investigate induction patterns, protein expression, and mRNA expression using transient transfection, western blotting, RT-PCR, and northern blotting techniques (Mehta *et al.* 1995; Song *et al.* 1997a–d). Potent enzyme inducers are evaluated for potential to inhibit carcinogen-induced lesion formation in mouse mammary gland organ culture (Mehta *et al.* 1995). Chemopreventive activities of promising agents are studied for their inhibitory effects in the two-stage mouse skin carcinogenesis model or the rat mammary gland carcinogenesis model (Mehta *et al.* 1995; Song *et al.* 1997b).

To date, we have evaluated 1235 plant extracts and more than 300 compounds in the QR induction assay system using cultured Hepa 1c1c7 cells. Several groups of active principles have been identified, and a series of *in vitro* and *in vivo* studies were performed on selected active agents. A brief summary of these findings is given below.

Plants extracts and active compounds isolated from plants Of the 1235 plant extracts evaluated in the Hepa 1c1c7 QR induction system, 53 were classified as active [CD (concentration required for 2-fold induction of QR specific activity): $<5\,\mu g/ml$; CQ (concentration required for 4-fold induction of QR specific activity): $<20\,\mu g/ml$]. Fifteen of these active extracts were selected for mouse mammary organ culture testing and seven were found to yield a positive response ($>60\%$ inhibition of lesion formation at $10\,\mu g/ml$). Two plants (*Physalis philadelphica* Lam. (Solanaceae) and *Tephrosia purpurea* (L.) Pers. (Leguminosae)) were selected for fractionation using the QR assay as a monitor. One novel compound, 2,3-dihydro-3-methoxywithaphysacarpin [**1**], and two known compounds, withaphysacarpin [**2**] and 24,25-dihydrowithanolide D [**3**] were isolated from *Physalis philadelphica* (Kennelly *et al.* 1997). An isoflavone, 7,4′-dihydroxy-3′,5′-dimethoxyisoflavone [**4**] and a chalcone, (+)-tephropurpurin [**5**], both novel compounds, as well as six constituents of known structure, (+)-purpurin [**6**], pongamol [**7**], lanceolatin B [**8**], (−)-maackiain [**9**], (−)-3-hydroxy-4-methoxy-8,9-methylenedioxypterocarpan [**10**], and (−)-medicarpin [**11**], were obtained as active compounds from *Tephrosia purpurea* (Chang *et al.* 1997a). These compounds demonstrated significant inducing potency in the QR assay system with CD values in a range of 0.15–23 μM. According to their induction pattern, compounds [**1**], [**2**], [**3**], and [**5**] were considered as monofunctional inducers by selectively inducing Phase II enzymes, whereas compounds [**4**, **6–11**] were considered as bifunctional inducers by inducing Phase I and Phase II enzymes. Re-isolation of compounds [**1–3**] from *Physalis philadelphica* is in progress in order to permit the conduct of *in vivo* chemoprevention studies.

Brassinin and analogs Brassinin [**12**], an indole-based dithiocarbamate originally found in cruciferous vegetables such as Chinese cabbage, and several analogs were demonstrated to induce QR and GST activities and mRNA levels in Hepa 1c1c7 mouse hepatoma cells and rat organs (Gerhäuser *et al.* 1997b). This induction was regulated at the transcriptional level by interaction with XRE (xenobiotic responsive element) found in both Phase I and Phase II drug-metabolizing and the ARE (antioxidant responsive element) of QR and GST Ya. On the basis of this induction pattern, brassinin and its analogues were regarded as bifunctional enzyme inducers. In addition, brassinin and its analogue cyclobrassinin [**13**] were shown to mediate significant chemopreventive activities, including inhibition of carcinogen-induced lesion formation in mouse mammary organ culture, reduction of the formation of 7,12-dimethybenz(a)anthracene (DMBA)/phorbol ester-induced papillomas in the two-stage mouse skin carcinogenesis model, and prevention of DMBA-induced mammary gland tumor formation in rats (Mehta *et al.* 1994, 1995).

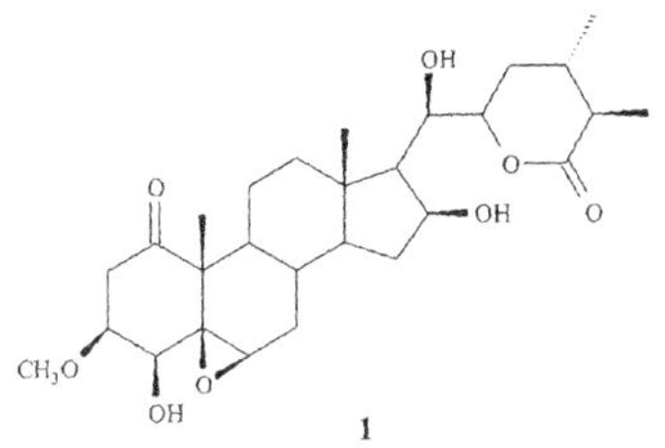

1

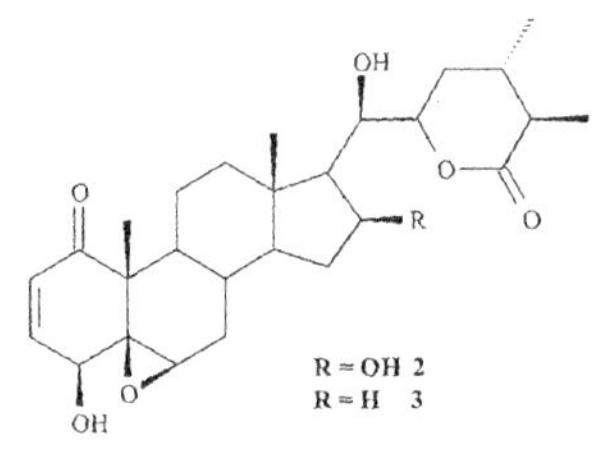

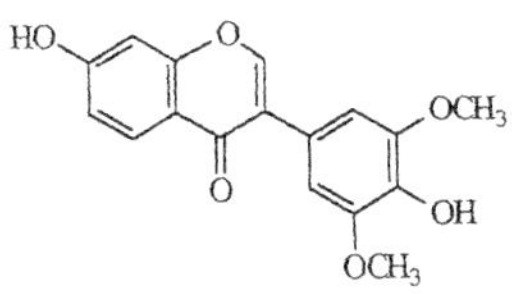

4

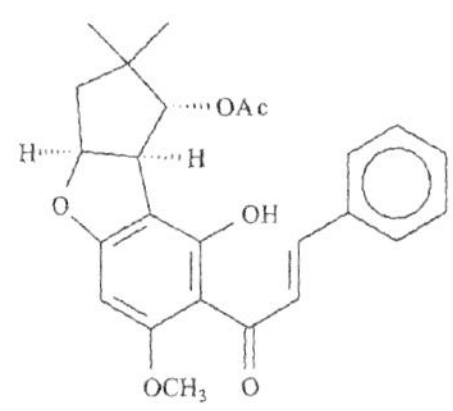

5

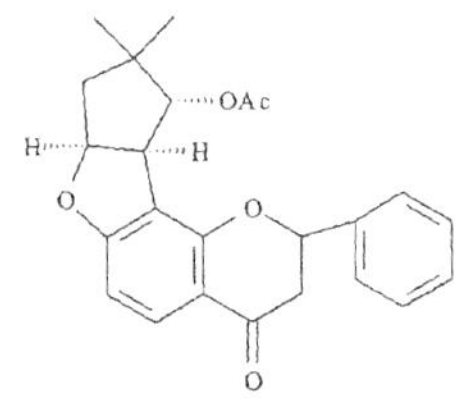

6

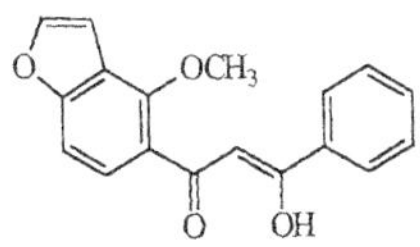

7

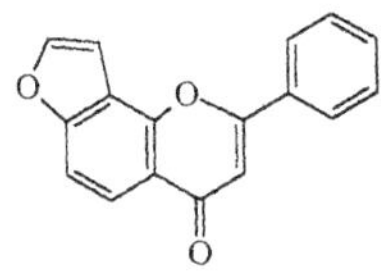

8

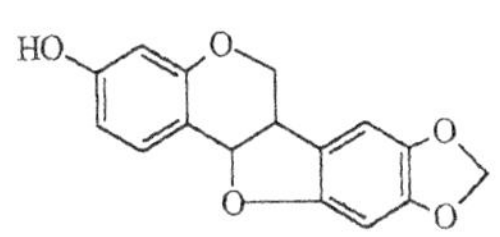

9

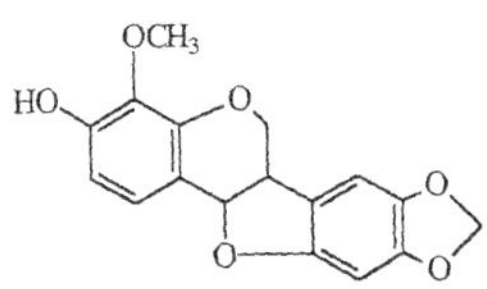

10

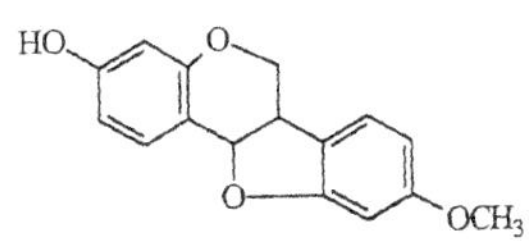

11

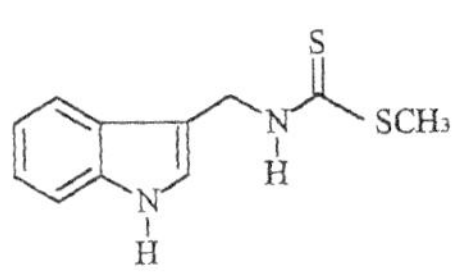

12

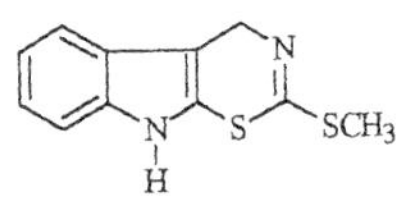

13

Sulforaphane and analogs Sulforaphane [**14**], an aliphatic isothiocyanate isolated from broccoli, was reported as a potent plant-derived monofunctional inducer of Phase II detoxification enzymes and was shown to significantly inhibit DMBA-induced mammary carcinogenesis in rats (Zhang *et al.* 1992, 1994). A novel sulforaphane analog, sulforamate [(±)-4-methylsulfinyl-1-(*S*-methyldithiocarbamyl)-butane] [**15**], was synthesized and studied in our systems (Gerhäuser *et al.* 1997b). Induction potential with the analog was comparable to that observed with sulforaphane (CD of QR: 0.2–1.2 μM), but cytotoxicity was reduced by 3–5-fold (IC_{50}: 30–50 μM). This compound also increased GSH levels about 2-fold in cultured Hepa 1c1c7 mouse and H4IIE rat hepatoma cells. Using western blotting techniques, expression of two major detoxification isoforms of GST (GST Ya: *alpha* type; and GST Yb: *mu* type) were observed in time- and dose-dependent fashions in H4IIE cell culture. This compound was shown to be a monofunctional inducer by interaction with ARE of QR and GST Ya without interaction with XRE. Using northern blotting and RT-PCR techniques, time- and dose-dependent induction of QR mRNA levels were demonstrated in Hepa 1c1c7 cell culture. Moreover, sulforaphane and the analog were identified as potent inhibitors of preneoplastic lesion formation in carcinogen-treated MMOC (75–85% inhibition at 1 μM). Full term chemoprevention studies are currently underway with the sulforaphane analog.

Flavonoids Flavonoids are widely distributed in the plant kingdom. They are found commonly in the human diet and estimated to be consumed in a quantity of approximately 1 g/day. Some flavonoids have been shown to exert a wide variety of biological actions, such as antiallergic, anti-inflammatory, and anticarcinogenic activities. Therefore, we tested about 60 natural or synthetic flavonoids for inducing capacity in the QR assay system (Song *et al.*, in preparation). The compounds flavone [**16**], 3′-bromoflavone [**17**], 4′-bromoflavone [**18**], 3′-chloroflavone [**19**], 4′-chloroflavone [**20**], 3′-fluoroflavone [**21**], 2′-methoxyflavone [**22**], and 2′-methoxyflavanone [**23**] were identified as potent QR inducers, with CD values of 2.5, 0.1, 0.01, 0.27, 0.02, 0.67, 0.63 and 0.63 μM, respectively. Among the active compounds, 4′-bromoflavone was considered as the most interesting (Song *et al.* 1997a). Unlike some other inducers, 4′-bromoflavone not only significantly induced QR activity in Hepa 1c1c7 cell culture (CD: 0.01 μM) and mouse organs (5-fold induction in liver; 3-fold in mammary gland), but also showed no cytotoxicity in Hepa 1c1c7 and H4IIE cells [IC_{50} (concentration required for 50% inhibition of cell growth): > 100 μM; *CI* (*C*hemopreventive *I*ndex; CD/IC_{50}): > 10,000). It was observed to increase GSH levels in H4IIE cell culture (2-fold) and mouse organs (3-fold in liver; 2-fold in mammary gland). Using western blotting techniques, expression of two major detoxification isoforms of GST (GST Ya: *alpha* type; and GST Yb: *mu* type) induced by 4′-bromoflavone was demonstrated in time- and dose-dependent fashions in H4IIE cell culture. To investigate possible regulatory effects on the transcription of QR, we also determined the time- and dose-dependent influence of 4′-bromoflavone on QR mRNA expression in Hepa 1c1c7 cell culture using RT-PCR techniques. In addition, 4′-bromoflavone was identified as a potent inhibitor of chemical-induced mutagenesis (89% inhibition at concentration of 40 μg/ml) with *Salmonella typhimurium* strain TM677. These results suggest that 4′-bromoflavone is a potential chemopreventive agent that is worthy of further study.

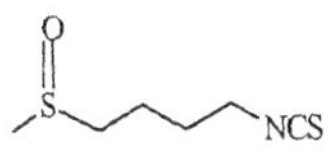

14

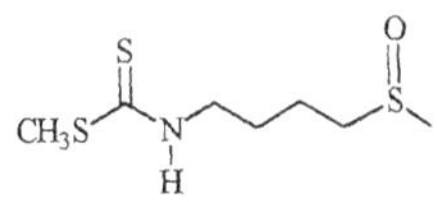

15

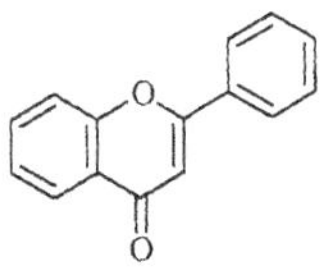

16

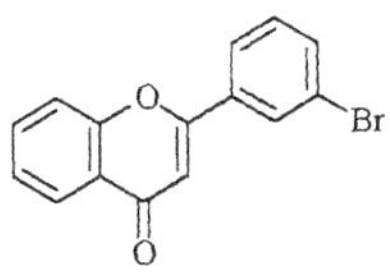

17

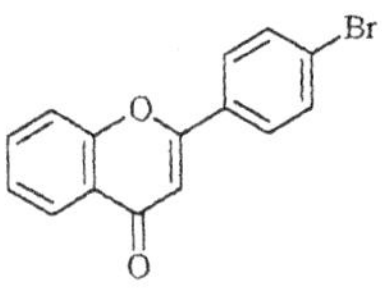

18

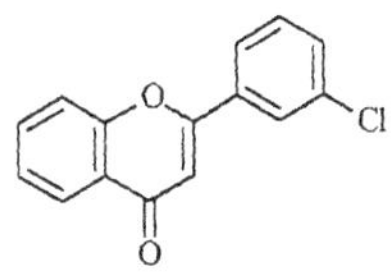

19

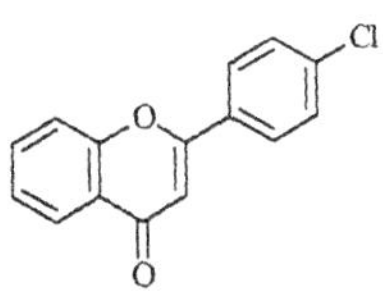

20

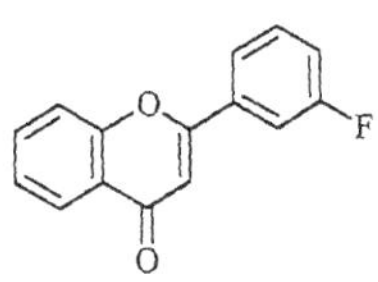

21

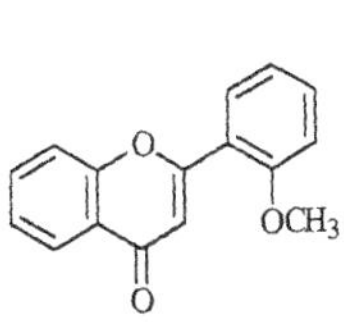

22

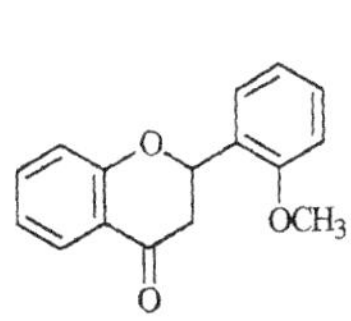

23

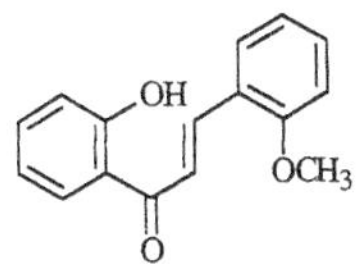

24

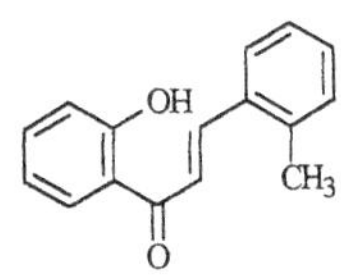

25

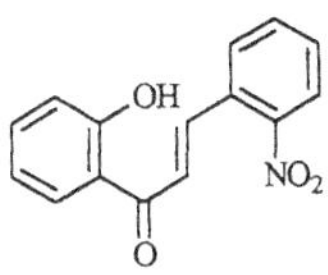

26

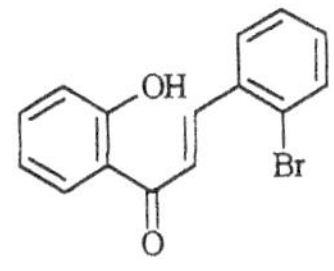

27

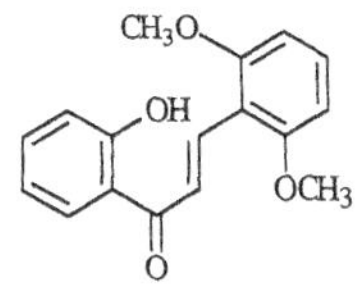

28

We have previously shown that the bromo, chloro or fluoro derivatives of chemopreventive agents do not show any activity in mammary gland organ culture (Mehta *et al.* 1994). Therefore, as expected, 4′-bromoflavone did not show an inhibitory effect in the mammary gland organ culture model. In studies conducted with rats, however, the compound has been shown to induce Phase II enzymes in a number of tissues, and tumor inhibition studies have revealed remarkable chemopreventive activity (Song *et al.*, submitted).

Chalcones Chalcones are ring-open biosynthetic precursors of flavonoids. We have tested naturally occurring chalcone and a series of 30 synthetic derivatives for their potential to induce QR activity in cultured Hepa 1c1c7 cells (Gerhäuser *et al.*, 1996; Song *et al.* 1997b). 2′-Hydroxy-2-methoxychalcone [**24**], 2′-hydroxy-2-methylchalcone [**25**], 2′-hydroxy-2-nitrochalcone [**26**], 2-bromo-2′-hydroxychalcone [**27**], and 2′-hydroxy-2,6-dimethoxychalcone [**28**] were identified as potent inducers of QR with CD values of 0.31, 0.67, 0.30, 0.53, and 0.54 μM, respectively. To investigate possible regulatory effects on the transcription of QR, we have determined the time- and dose-dependent influence of these compounds on QR mRNA expression in Hepa 1c1c7 cell culture using northern blotting and RT-PCR techniques. Preliminary experiments have indicated that these compounds induce QR and GSH in mouse liver and mammary glands.

Antioxidant Activities

Since reactive oxygen radicals play an important role in carcinogenesis, antioxidants present in consumable fruits, vegetables, and beverages have received considerable attention as cancer chemopreventive agents (Muktar *et al.* 1994). Thus, in order to identify antioxidants in plant extracts, activity has been assessed by determining scavenging activity with stable 1,2-diphenyl-2-picrylhydrazyl (DPPH) free radicals (Fujita *et al.* 1988), inhibition of TPA-induced free radical formation with cultured HL-60 human leukemia cells (Sharma *et al.* 1994), and inhibitory activity with a xanthine/xanthine oxidase assay system (Chang *et al.* 1994). In addition, using an electrospray HPLC/MS system, a dereplication method has been developed to directly identify known active antioxidants from plant extracts (Constant and Beecher 1995; Lee *et al.* 1998a). Approximately 700 plant extracts have been evaluated for the potential of antioxidant activity. As a result, 28 plant extracts were found to be active. Based on secondary analyses performed to assess inhibition of DMBA-induced preneoplastic lesion formation with mouse mammary organ culture, *Chorizanthe diffusa* Benth. (Polygonaceae), *Mezoneuron cucullatum* Roxb. (Leguminosae), *Cerbera manghas* L. (Apocynaceae), and *Daphniphyllum calycinum* Blume (Daphniphyllaceae) were selected and subjected to bioassay-guided fractionation. 5,8,3′,4′,5′-Pentahydroxy-3,7-dimethoxyflavone [**29**], 5,8,3′,4′-tetrahydroxy-3,7-dimethoxyflavone [**30**], 5,7,3′,4′-tetrahydroxy-3-methoxyflavone [**31**], and quercetin [**32**] were identified as active principles of *C. diffusa* (Chung *et al.* 1997). Piceatannol [**33**], resveratrol [**34**], apigenin [**35**] and scirpusin A [**36**] were found as the active principles of *M. cucullatum* (Lee *et al.* 1996), olivil [**37**], (−)-carinol [**38**], and

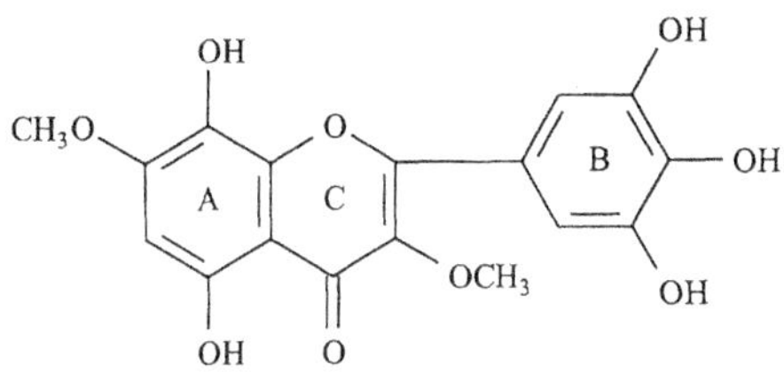

29

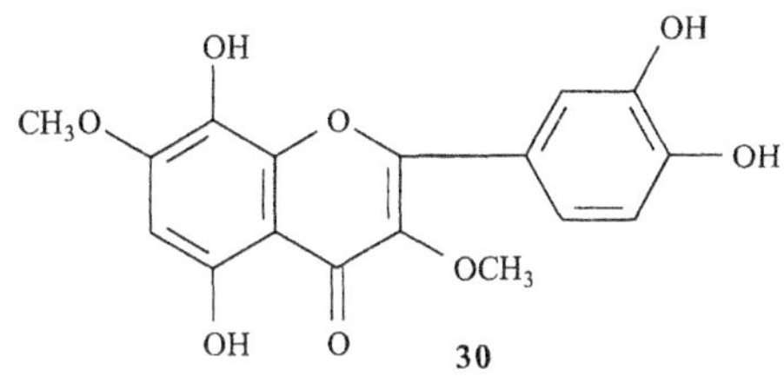

30

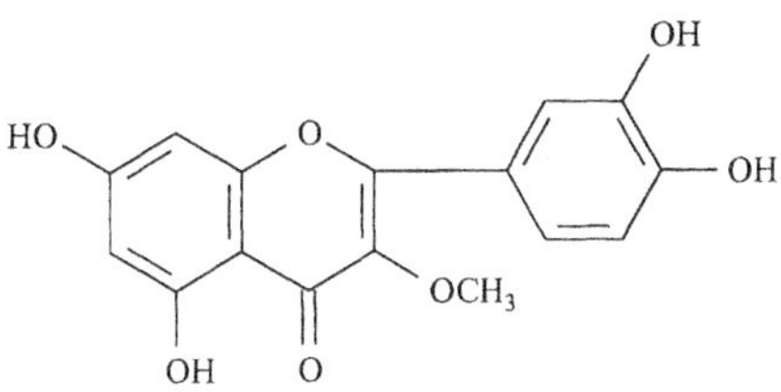

31

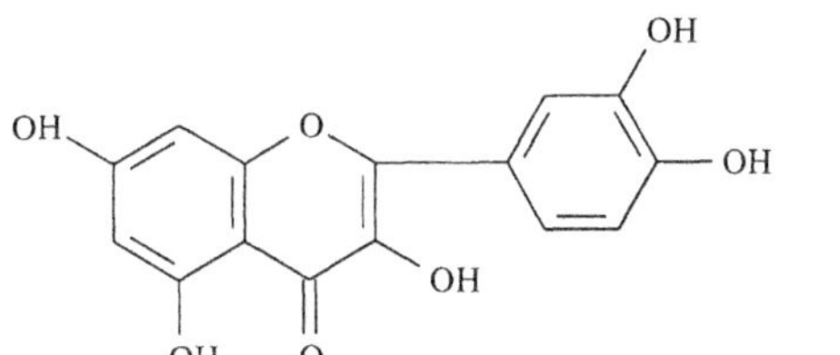

32

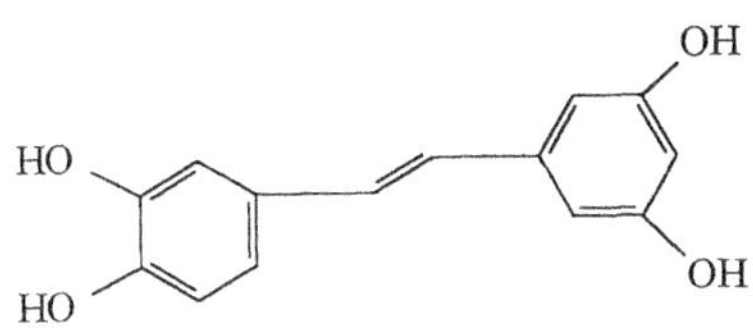

33

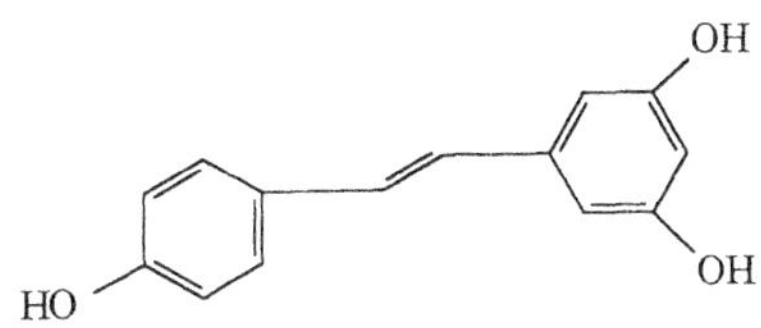

34

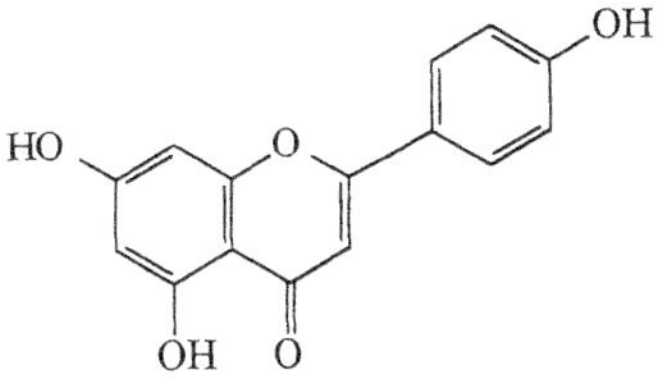

35

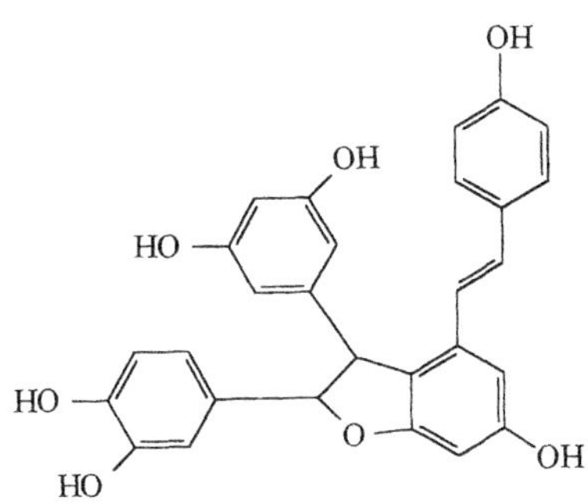

36

(+)-cycloolivil [39] were active principles from *C. manghas* (Pisha *et al.* 1996), and 5,6,7,4′-tetrahydroxyflavonol 3-*O*-rutinoside [40] and kaempferol 3-*O*-neohesperidoside [41] were active principles from *D. calycinum* (Gamez *et al.* 1997). In addition, the hydroxystilbenes piceatannol and resveratrol showed potent inhibitory effects with the mammary organ culture model. Furthermore, utilizing an electro-spray HPLC/MS dereplication system, (−)-epicatechin gallate [42] was identified as an active principle

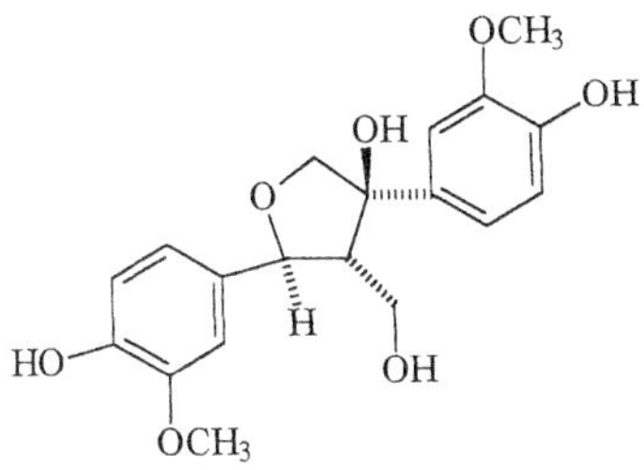

37

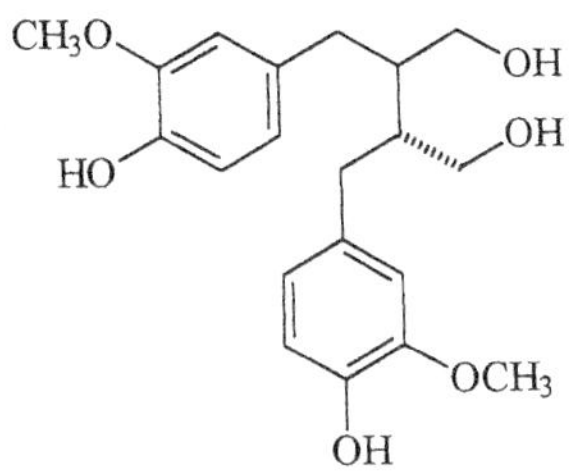

38

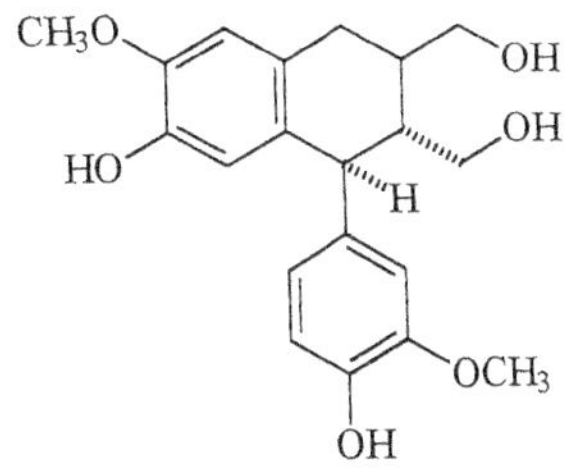

39

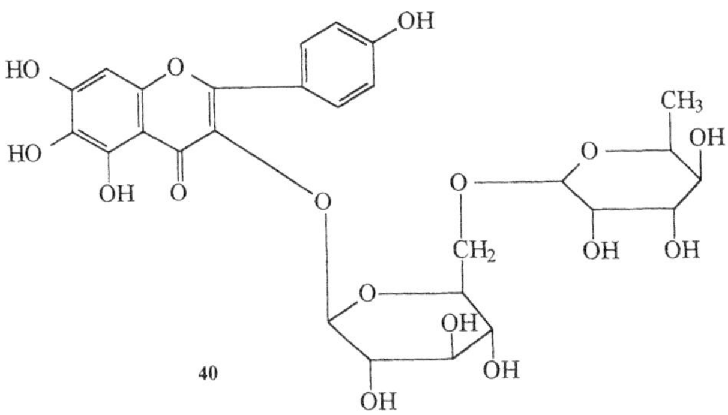

40

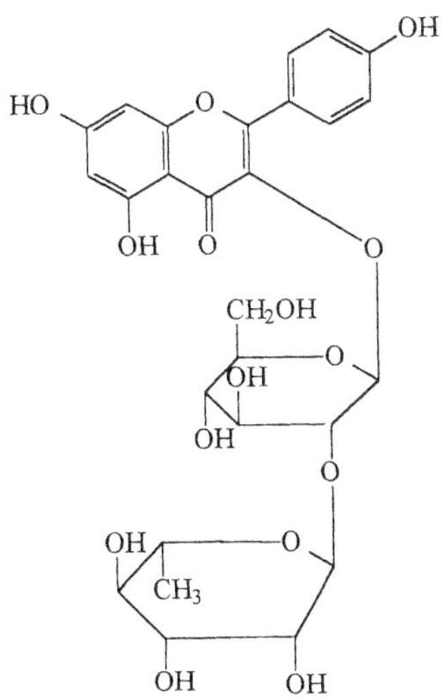

41

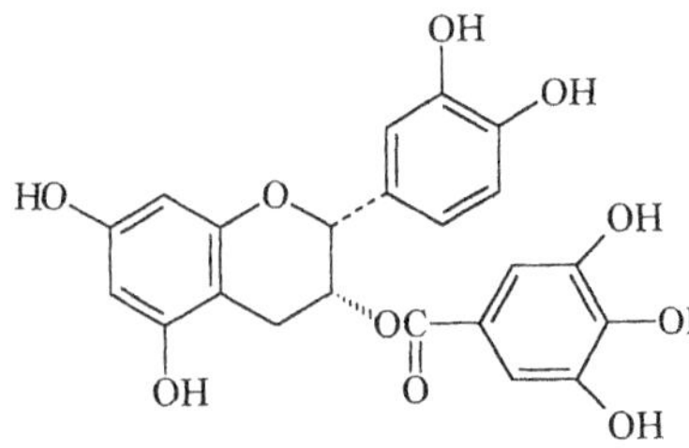
OH
OH
HO
O
OH
OC
OH
O
OH
OH

42

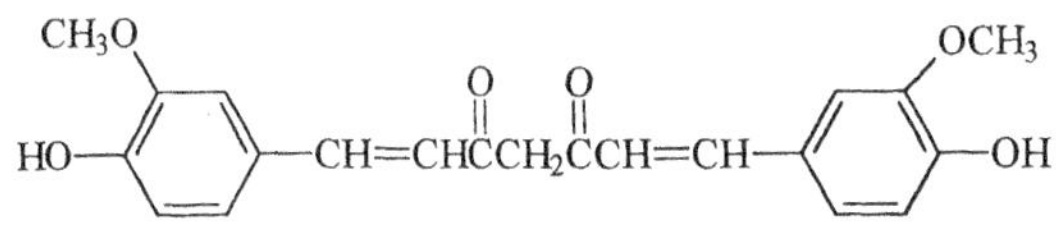
CH3O
OCH3
O
O
HO
CH=CHCCH2CCH=CH
OH

43

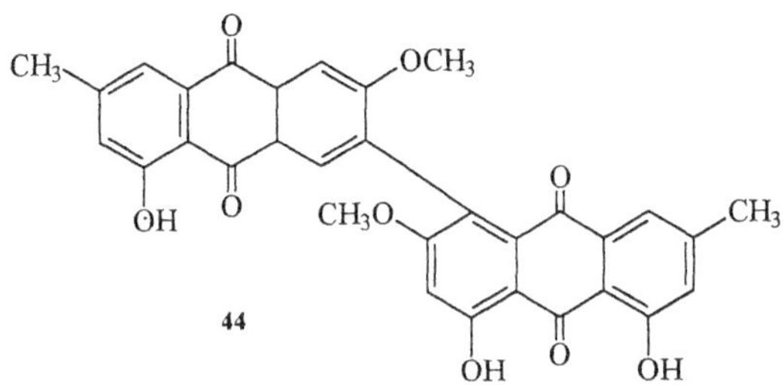
O
CH3
OCH3
O
OH
O
CH3O
CH3
O
OH
OH

44

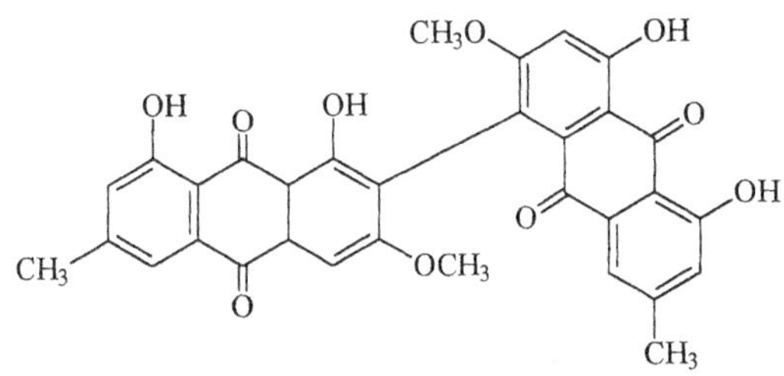
CH3O
OH
OH
O
OH
O
O
OH
CH3
OCH3
O
CH3

45

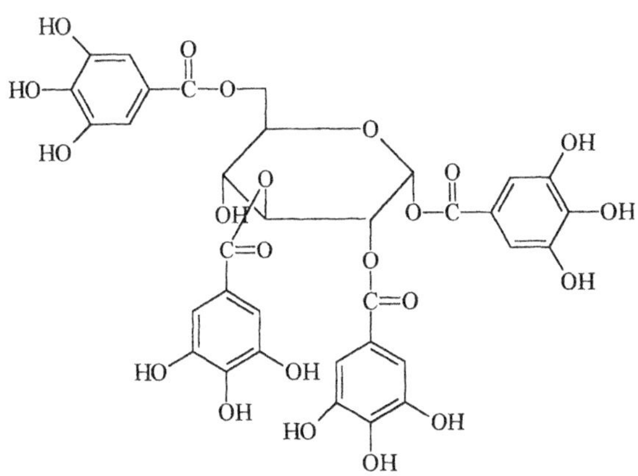
HO
O
HO
C
O
O
HO
OH
O
O
OH
O
C
OH
OH
C=O
O
C=O
OH
HO
OH
OH
HO
OH
OH

46

from *Davidsonia pruriens* Muell. (Davidsoniaceae), curcumin [43] was found in *Curcuma longa* L. (Zingiberaceae), floribundones 1 [44] and 2 [45] were identified in *Cassia quinquangulata* Benth. (Leguminosae), and 1,2,3,6-tetra-*O*-galloyl-*β*-D-glucose [46] was a component of *Paeonia officinalis* L. (Paeoniaceae) (Lee *et al.* 1998a).

Antimutagenesis Studies

As part of our effort to discover natural products that have potential use as cancer chemopreventive agents, plant extracts were screened for their ability to inhibit chemically induced mutation in *Salmonella typhimurium* strain TM677 (Shamon *et al.* 1997). An antimutagenic response may correlate with ability to inhibit the initiation of carcinogenesis, which can result from unrepaired damage to DNA. Prior to this study, this bacterial strain had been used only in a limited capacity in this manner. We have found that activity in this system, either positive or negative, correlates with ability to inhibit DMBA-induced, preneoplastic lesions in a mouse mammary gland organ culture system (Shamon *et al.* 1994).

With the bacterial tester strain, 58 plant extracts have been tested for their potential to inhibit DMBA-induced mutagenesis. As a result, 13 were deemed active (>50% inhibition at 40 μg/mL). Two of these active plants, *Mezoneuron cucullatum* (72% inhibition) and *Casimiroa edulis* (79% inhibition) were subjected to bioassay-directed fractionation using antimutagenesis as a monitor. A total of 16 compounds were isolated from these two plants, including two novel furocoumarins and a novel sesquiterpene. The sesquiterpene, mesocucullin [(4*E*)-7*β*,11-dihydroxymethyl-11-hydroxydodeca-4-en-3-one] [47], was also active in the mouse mammary organ culture assay (88% inhibition at 10 μg/mL). The naturally occurring compounds maesanin [48] and paeoniflorin [49] were also found to be strong antimutagens, with IC_{50} values of 8.8 and 6.6 μM, respectively.

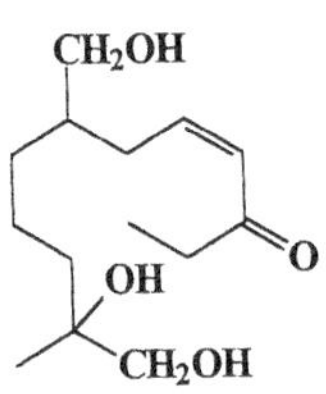

47

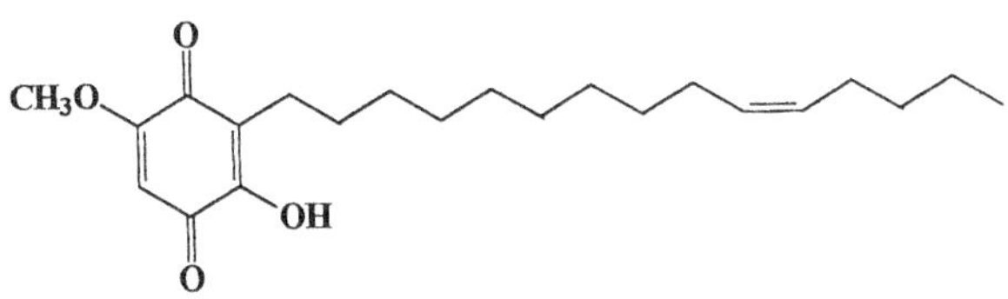

48

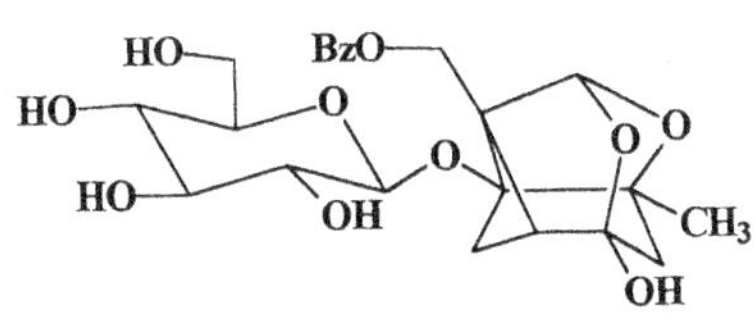

49

Ornithine Decarboxylase Activity

Ornithine decarboxylase (ODC) is the first and rate-limiting enzyme for the biosynthesis of polyamines in mammalian cells (McCann and Pegg 1992). The induction of ODC activity and subsequent polyamine levels by treatment with tumor promoters has been suggested to play a significant role in tumor promotion (O'Brien 1976; Rozhin *et al.* 1984). Indeed, α-difluoromethylornithine (DFMO), a suicide inhibitor of ODC, is a promising cancer chemopreventive agent that is currently in Phase II clinical trials (Kelloff *et al.* 1994). Based on these considerations, we have evaluated the potential of approximately 800 plant extracts to inhibit phorbol ester-induced ODC activity with cultured mouse epidermal 308 cells. As a result, approximately 30 plant extracts were found to demonstrate potent inhibitory activity ($IC_{50} < 4.0\ \mu g/ml$).

With selected plant extracts, further bioassay-guided fractionation led to the identification of the various active principles. With extracts from *Mundulea sericea* Willd. (Leguminosae), the rotenoids, deguelin [**50**] and tephrosin [**51**], were isolated as potent active principles, together with various other classes of compounds (Luyengi *et al.* 1994). The flavonoids zapotin [**52**] and 5,6,2′-trimethoxyflavone [**53**] were identified as active principles from *Casimiroa edulis* Llave et Lex. (Rutaceae), and 4′-demethyldeoxypodophyllotoxin [**54**] was an active principle found in *Amanoa oblongifolia* Muell. (Euphorbiaceae). Deguelin and tephrosin also showed potent inhibitory effects in the mouse mammary organ culture model (Gerhäuser *et al.* 1995), and deguelin inhibited tumor formation in the mouse skin DMBA/TPA two-stage carcinogenesis model and the rat MNU-induced mammary carcinogenesis model (Udeani *et al.* 1997). Based on these results, 16 rotenoids were synthesized and evaluated for potential cancer chemopreventive activity. Structure–activity relationships (SAR) were investigated by studying

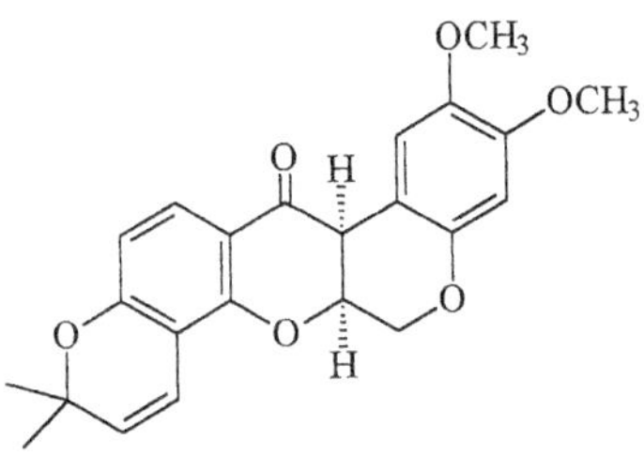

50

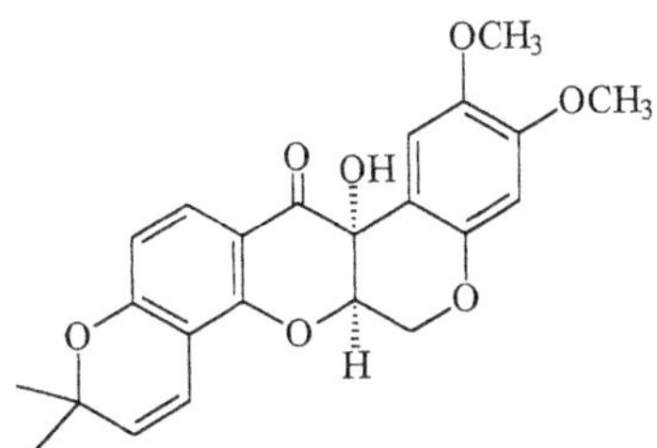

51

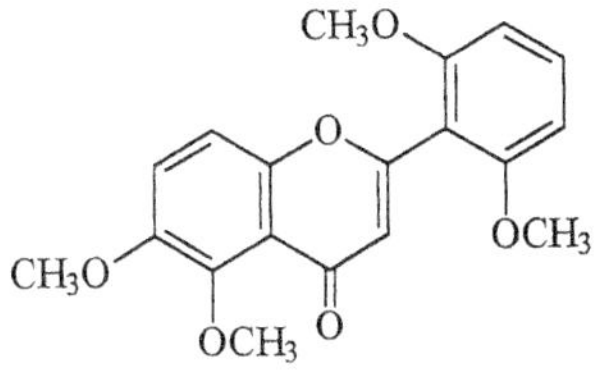

52

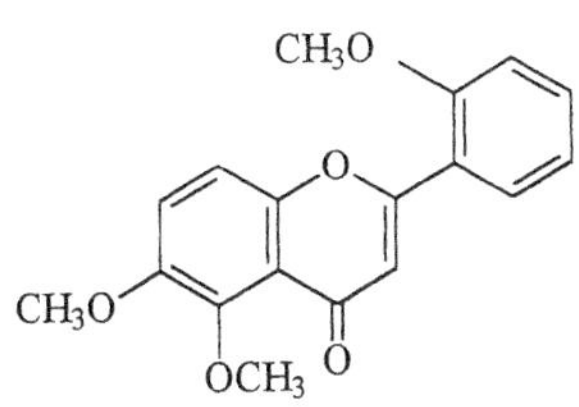

53

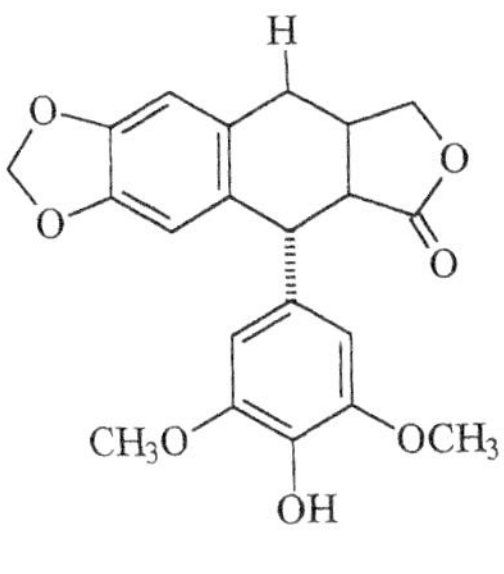

54

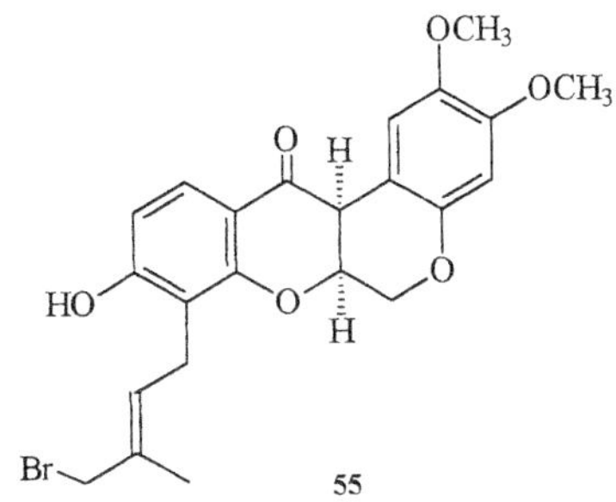

55

TPA-induced ODC activity and cytotoxicity. In general, all of derivatives were less active than the parent molecules deguelin and rotenone. However, 4′-bromorotenoic acid [55] showed good inhibitory activity and reduced cytotoxicity (Lee *et al.* 1997a). In addition, six rotenoids were evaluated for potential to inhibit TPA-induced ODC activity in the skin of female CD-1 mice. In terms of structure-activity relationships, the activity of these test compounds with the *in vivo* mouse skin model correlated with activity observed with the *in vitro* cell culture system. Deguelin and 4′-bromorotenoic acid exhibited potent inhibition of TPA-induced ODC activity in the mouse skin model. Thus, this inhibitory effect may be partially responsible for the inhibition of tumor promotion mediated by rotenoids. As found in mechanistic studies, deguelin, tephrosin and 4′-bromorotenoic acid inhibited TPA-induced ODC mRNA expression. This result indicated that rotenoids can regulate ODC activity at a transcriptional level and thus affect ODC activity. In addition, deguelin inhibited TPA-independent c-Myc-induced ODC activity, and also inhibited NADH dehydrogenase activity and ATP formation in cell culture (Gerhäuser *et al.* 1997a). Therefore, rotenoid-mediated cancer chemopreventive activity might be due to the modulation of energy balance which in turn alters cellular signal transduction mechanisms affected by tumor promoters.

Currently, based on the potent inhibitory activity of 4′-bromorotenoic acid on TPA-induced ODC activity (with mouse epidermal cells in culture and the *in vivo* mouse skin model), the cancer chemopreventive potential of this substance is being assessed with *in vivo* models of carcinogenesis.

Inhibition of Protein Kinase C

Protein kinase C (PKC) has been extensively studied due to its central role in cellular signal transduction (Berridge 1984). There are Ca^{2+}-and phospholipid-dependent forms of the enzyme which utilize *sn*-1,2-diacylglycerol as a lipophilic second messenger. Importantly, as originally shown by Nishizuka, PKC serves as a major phorbol ester receptor (Castagna 1987; Nishizuka 1986), and activation through this interaction is related to tumor promotion (Castagna 1987; Castagna *et al.* 1982; Nishizuka 1986). Therefore, it was surmised that specific inhibitors of PKC could serve as potential cancer chemopreventive agents or anti-tumor promoters. In the search for new natural product cancer chemopreventive agents, approximately 600 plant extracts were

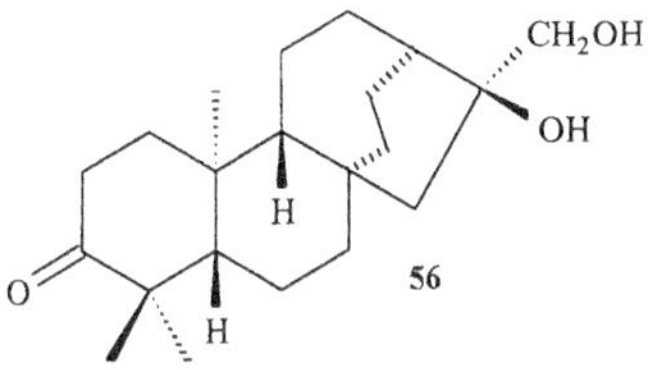

56

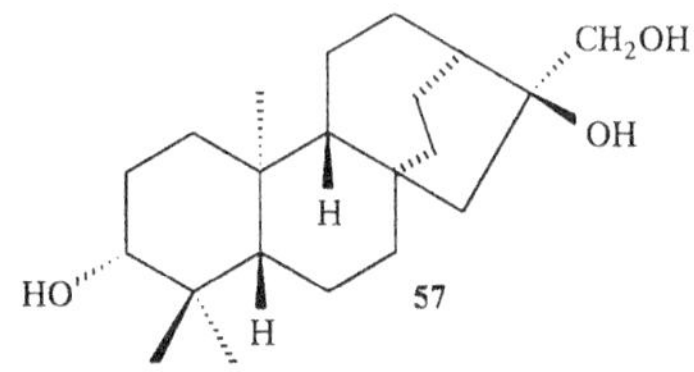

57

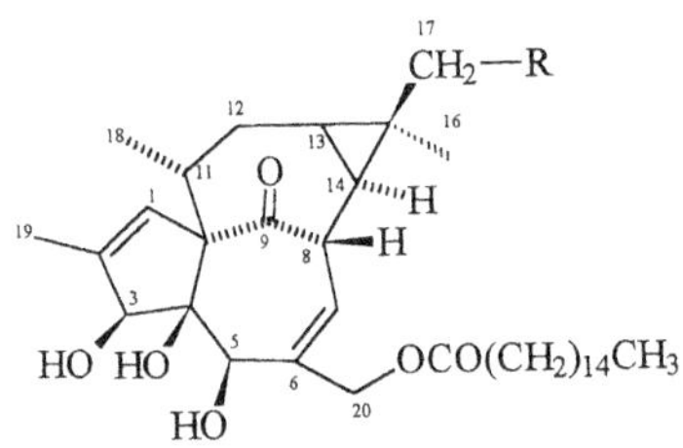

58; **R = OH**

59; **R = H**

evaluated for potential of inhibiting [^{3}H]PDBu-binding to PKC, and follow-up studies monitoring the effects of test substances on kinase activity were conducted using histone IIIs as substrate. Bioassay-directed fractionation of *Euphorbia quinquecostata* Volk. (Euphorbiaceae) led to the isolation of *ent*-16α,17-dihydroxyatisan-3-one [**56**], *ent*-3β,16α,17-trihydroxyatisane [**57**], 17-hydroxyingenol 20-hexadecanoate [**58**] and ingenol 20-hexadecanoate [**59**], which inhibited PDBu-binding to PKC with IC_{50} values of 21.1, 23.4, 0.014, and 3.1 μg/ml, respectively. Similarly, lariciresinol [**60**], sitoindoside II [**61**], genkwanin [**62**], syringaresinol [**63**] and acacetin [**64**] were isolated from *Dirca occidentalis* A. Gray (Thymeleaeceae) extracts, and showed inhibitory effects on PDBu-binding activity with IC_{50} values of 0.4, 0.08, 9.7, 10.0, 25.7 μg/ml, respectively.

Based on the potent inhibitory activity of 17-hydroxyingenol 20-hexadecanoate [**58**], the novel isolate obtained from *E. quinquecostata*, additional studies were performed. 17-Hydroxyingenol 20-hexadecanoate binds to and activates PKC. In addition, similar to phorbol esters, the compound induces ornithine decarboxylase activity in cell culture and mouse skin, stimulates anchorage-independent JB6 cell transformation, induces HL-60 cell differentiation, and produces intracellular translocation of protein kinase C. When compared with the parent alcohol, ingenol, the novel ingenol ester [**58**] showed much greater potency in all of these test systems. Overall, it appears likely that this substance will serve as a tumor promoter (Lee *et al.* 1997b). Mouse skin tumor promotion studies are underway.

Starting with the stem of *Macleaya cordata* (Willd.) R. Br. (Papaveraceae), the benzophenanthridine alkaloid angoline [**65**] was isolated by bioassay-guided fractionation using the PDBu-binding assay as a monitor. This discovery appeared in context since a related a benzophenanthridine alkaloid, chelerythrine [**66**], had been reported to mediate a variety of biological activities, including potent and selective inhibition of PKC (Herbert *et al.* 1990). However, in our studies, angoline was not found to serve as a

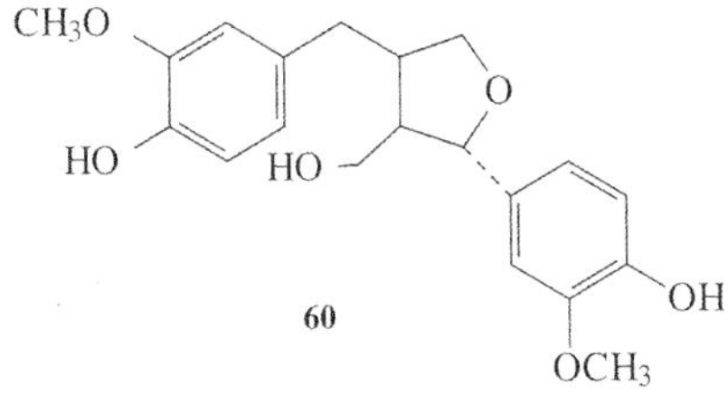

60

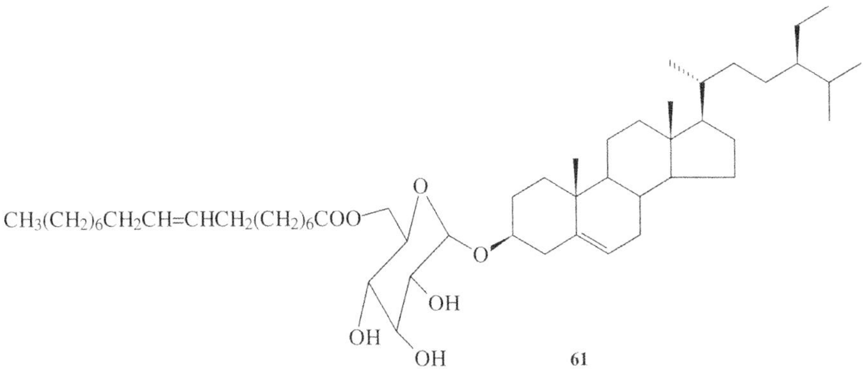

61

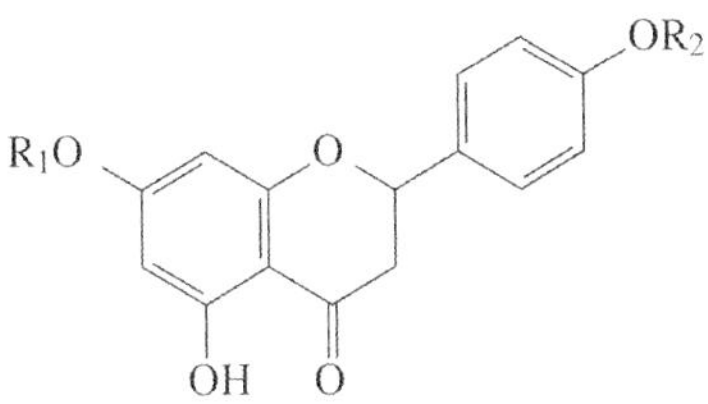

62; R_1=CH_3, R_2=H

64; R_1=H, R_2=CH_3

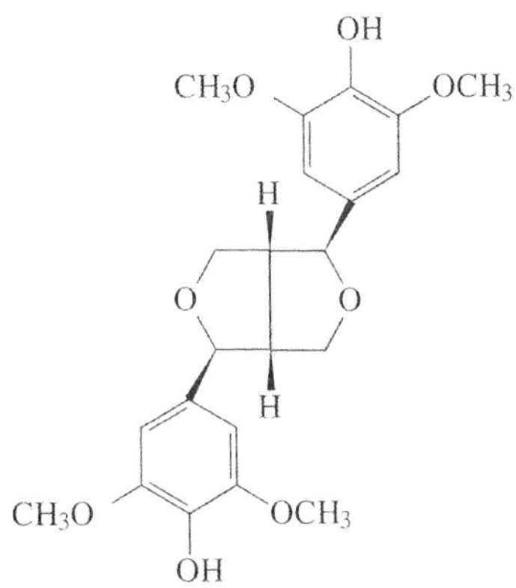

63

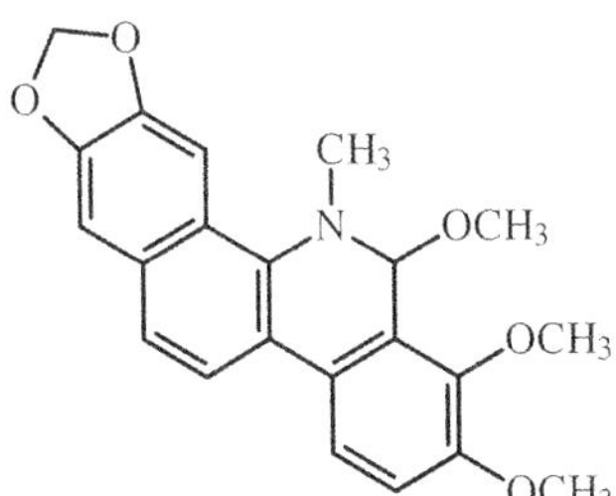

65

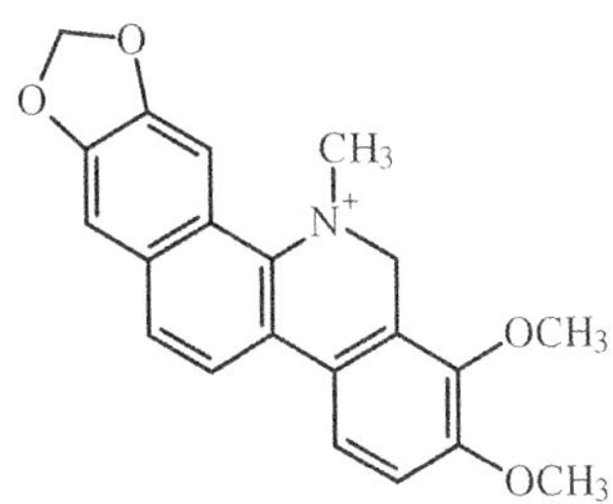

66

potent inhibitor of PKC. Moreover, we were unable to confirm the reported inhibitory activity of chelerythrine. In a series of thorough studies performed with various PKC isozymes derived from a variety of mammalian species, chelerythrine and angoline did not inhibit PKC activity with high potency. Consistent with a recent report (Lombardini, 1996), chelerythrine stimulated PKC activity in the cytosol fraction of rat and mouse brain in concentrations up to 100 μM. In addition, chelerythrine and angoline did not inhibit [^{3}H]PDBu binding to the regulatory domain of PKC at concentrations up to 40 μg/ml, and no significant alteration of PKC translocation was observed with mouse epidermal (ME 308) cells in culture. Chelerythrine did not inhibit TPA-induced ornithine decarboxylase (ODC) activity with mouse 308 cells, but angoline was active in this capacity with an IC_{50} value of 1.0 μg/ml (Lee *et al.* 1998b).

HL-60 Cell Differentiation

Induction of cellular differentiation has been utilized in our laboratory to evaluate agents capable of inhibiting the progression of carcinogenesis (Suh *et al.* 1995a). Recent results are described below.

Brusatol and derivatives The evaluation of 1210 plant extracts has led to the identification of 33 active leads. Bioassay guided fractionation has led to the isolation of brusatol [67] from *Brucea javanica* L. (Merr.) as one of the most potent isolates in this assay system (ED_{50} = 0.04 μg/ml) (Luyengi *et al.* 1996; Suh *et al.* 1995b). However, brusatol shows a poor selectivity ratio (*CI* = 1.6). Thus, as an attempt to find a more selective derivative, a set of 49 analogs (obtained through a collaboration with Professor Paul Grieco, Indiana University) were studied using the HL60 cell culture system. Several features of the quassinoids proved to be of importance for the observed activity, including the presence of an ester side chain at C-15, a 1-hydroxy-2-keto-$\Delta^{3,4}$ system in ring A, and an oxide bridge in ring D. Ten analogs showed higher potency with some having a 10-fold decrease in their ED_{50} values as compared to brusatol. Such analogs possess a hydroxylated-branched side chain at C-15 [68]. Other analogs showed a moderate improvement in the selectivity ratio with a maximum of a 3-fold increase in the *CI* as compared to brusatol. Studies on oncogene expression using western blots revealed these compounds can down-regulate the expression of the *c-myc* oncoprotein, while up-regulating the expression of *c-fos* and *c-jun*. However, the potency and timing of these effects on *c-myc* and *c-jun* varied with different analogs. Brusatol and bruceantin showed a strong induction of *c-jun* between 0.5 and 3 h, while analogs that lacked the

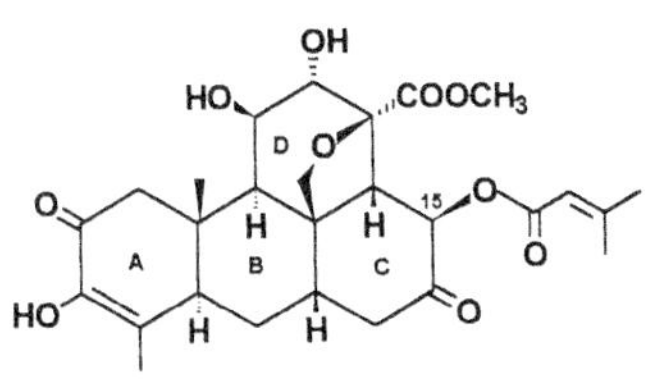

67

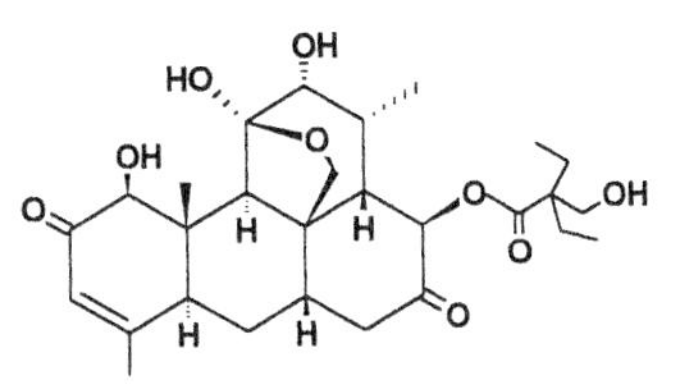

68

side chain showed a late induction at 12–24 h. Brusatol and bruceantin produced a moderate and late decrease in *c-myc* oncoprotein levels, while analogs without the side chain or with a branched-saturated side chain showed a potent and early decrease in the levels of *c-myc* (Mata-Greenwood 1997a,b). These results suggest that the quassinoids might have different affinities for their targeted receptors, which could imply that the synthesis of a specific quassinoid might lead to a less toxic and more potent analog of brusatol. Considering brusatol has shown potent activity in other *in vitro* systems such as inhibition of TPA-induced transformation of JB6 cells and inhibition of DMBA-induced mammary lesions in organ culture, this compound and its analogs are worthy of further study (Mata-Greenwood *et al.* 1997a,b; Suh *et al.* 1995).

Artemesinin and related compounds Artemesinin [69], a natural antiplasmodial agent, was found active in this assay ($ED_{50} = 4\,\mu g/ml$). A set of 42 derivatives of artemisinin was evaluated for both antiplasmodial and cell differentiation induction activities. Both activities have been linked to the ability of these compounds to induce a pro-oxidant state in the cell/parasite. However, artemisinin and its analogs were inactive in the mouse mammary organ culture model (Wongpanich *et al.* 1997).

Flavonoids and chalcones These compounds have shown weak activity as inducers of cellular differentiation (i.e., apigenin, $ED_{50} = 10.1\,\mu g/ml$). However a natural flavone was isolated from *Casimiroa edulis* as a potent inducer of HL-60 cell differentiation (zapotin [52], 5,6,2′,6′-tetramethoxyflavone, $ED_{50} = 0.2\,\mu g/ml$). This finding correlated well with a simultaneous observation that a novel synthetic chalcone, 2,6-dimethoxy-2′-hydroxychalcone [70], showed potent activity as well. These compounds possess methoxy substituents in ring B at sterically hindered positions, and are rare in nature (Mata-Greenwood *et al.* 1996).

Additional active compounds Other active isolates include two benzoquinones [71, 72] with methoxy substituents at *ortho* positions from *Peddia fischeri* ($ED_{50} = 0.2$ and $0.8\,\mu g/ml$), lignans [73] from *Aglaia ponapensis* (a rocaglamide derivative, $ED_{50} = 0.02\,\mu g/ml$) and polysulfides [74] from *Petiveria alliacea* (e.g., dibenzyltrisulfide, $ED_{50} = 0.3\,\mu g/ml$)

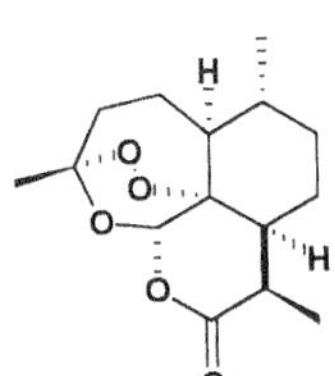

69

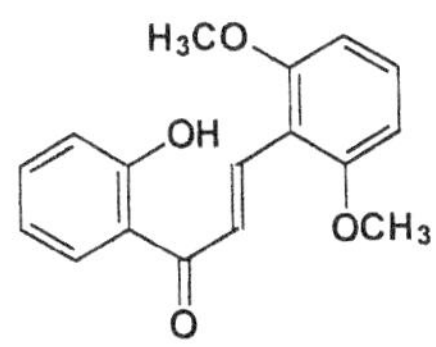

70

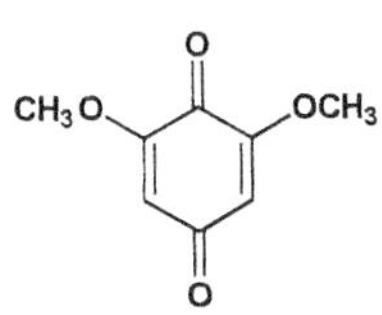

71

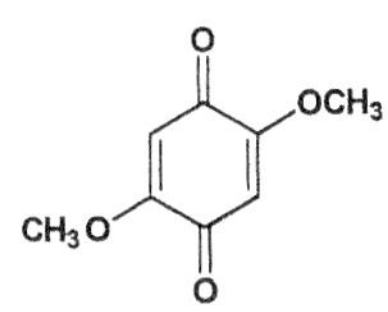

72

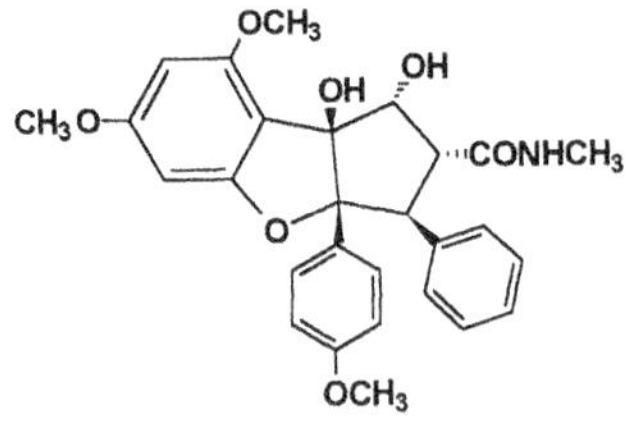

73

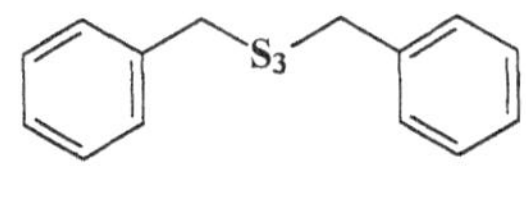

74

(Westenburg *et al.* 1997). Studies with aryl and alkylpolysulfides revealed that monosulfides (i.e., diallylsulfide) were inactive, disulfides were moderately active and trisulfides were potent inducers of cellular differentiation as well as being cytotoxic.

Inhibition of Cyclooxygenase

Cyclooxygenase, a key enzyme in prostaglandin biosynthesis, catalyzes oxygenation of arachidonic acid to prostaglandin G_2 (PGG_2) and reduction of PGG_2 to prostaglandin H_2 (PGH_2) which is the immediate precursor for production of eicosanoids, such as PGD_2, PGE_2, $PGF2\alpha$, PGI_2, and TxA_2, through corresponding synthase activities (Smith, 1989). These arachidonic acid metabolites derived from PGH_2 are important mediators of inflammatory responses, immunological effects, and tumor development (Goodwin and Ceuppens 1985). Non-steroidal anti-inflammatory drugs (NSAIDs) mediate anti-inflammatory and antipyretic effects primarily by blocking PGG_2 formation via inhibition of cyclooxygenase. Moreover, several epidemiological studies have demonstrated that NSAIDs have cancer preventive and tumor regressive effects in the human colon (Logan *et al.*, 1993). Thus, it is of value to study cyclooxygenase inhibitors.

We have thus far evaluated 1340 plant extracts and identified 32 as active ($\geqslant$70% inhibition at 70 μg/ml) in the oxygen consumption assay. We have identified five active principles, resveratrol [34] (IC_{50} = 14.9 μM), *trans*-3,5,3′,5′-tetrahydroxy-4-methoxystilbene [75] (IC_{50} = 20.8 μM), quinquangulin 6-*O*-β-D-glucopyranoside [76] (IC_{50} = 4.3 μM) isolated from *Cassia quinquangulata* Rich. (Leguminosae), as well as afzelechin [77] (IC_{50} = 48.2 μM) isolated from *Daphniphyllum calycinum* Benth. (Daphniphyllaceae) and (+)-catechin [78] (IC_{50} = 50.3 μM) from *Chrysophyllum prunifenum* F. Muell. (Sapotaceae) (Jang and Pezzuto 1997). Among these active isolates, resveratrol [34], demonstrated an ED_{50} value of 3.1 μM in the mouse mammary organ culture model. Subsequently, the compound was found to significantly reduce tumorigenesis in the two-stage mouse skin model (Jang *et al.* 1997). A number of structurally-related compounds have also been evaluated. From a SAR viewpoint, dihydroresveratrol [79] (IC_{50} = 42.2 μM), 2,4,3′,5′-tetrahydroxystilbene [80] (IC_{50} = 8.0 μM), and piceatannol [33] (IC_{50} = >100 μM) are of greatest interest. In addition, we are currently exploring the use of COX-2 for bioassay-directed fractionation.

Inhibition of Aromatase Activity

The aromatase enzyme complex is responsible for the conversion of androgens to estrogens (Santen *et al.* 1990). In the case of estrogen-dependent tumorigenesis,

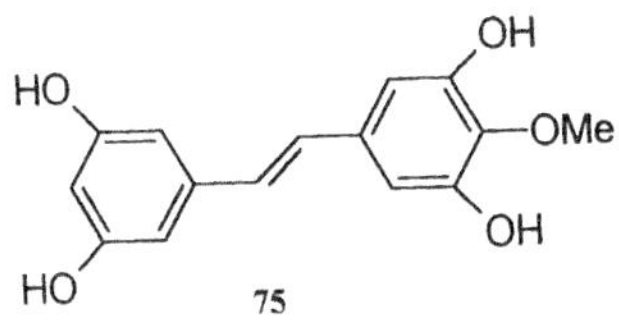

75

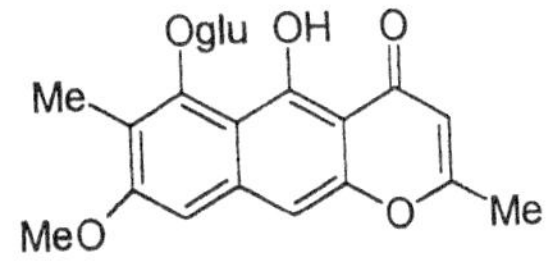

76

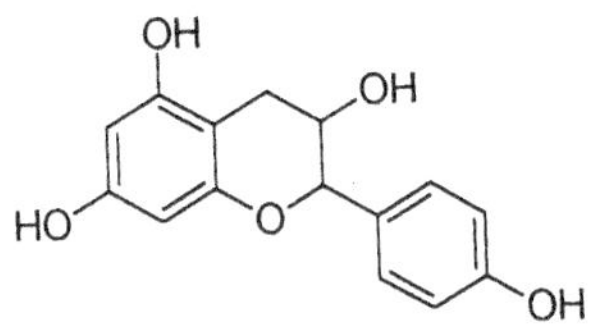

77

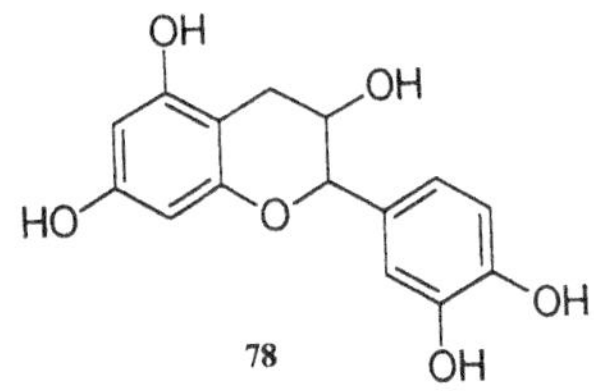

78

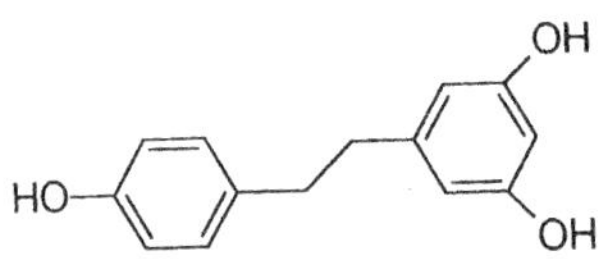

79

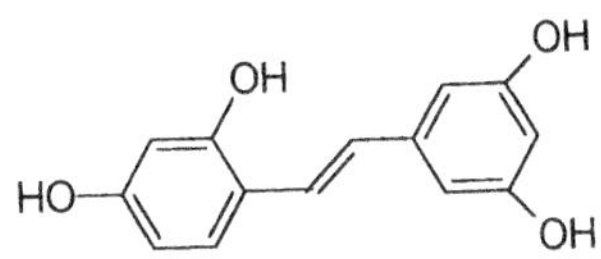

80

aromatase expression may be an important factor in the regulation of tumor growth (Harvey 1996; Yue and Santen 1996). In our continuing search for novel cancer chemoprevention agents, an assay system to identify potential aromatase inhibitors was devised and optimized, and 112 pure compounds and 48 plants extracts were evaluated. In an initial screen, 19 of the 112 compounds produced >80% inhibition at a final concentration of 200 μM. Further evaluations showed miconazole [**81**] and methylene blue [**82**] are very potent aromatase inhibitors with IC_{50} values of 0.04 and 0.55 μM, respectively, whereas the IC_{50} of aminoglutethimide (a positive control) is 1.7 μM.

Of the 48 plant extracts tested in this system, eight were found active. Two of the active plants, *Isodon excisus* var. *coreanus* Kudo (Labiatae) (IC_{50}: 13.7 μg/ml) and *Broussonetia papyrifera* (L.) Vent. (Moraceae) (IC_{50}: 0.72 μg/ml) were selected for bioassay-directed fractionation. From *Isodon excisus* var. *coreanus,* four active compounds have been isolated. Two of them were terpenoids (*ent*-1α-hydroxy-3β,6α-diacetoxy-kaur-16-en-11,15-dione [**83**] and ursolic acid [**84**]) were determined structurally by spectral data interpretation (MS, UV, IR, ^{1}H-NMR, ^{13}C-NMR, APT, HETCOR, ^{1}H-^{1}H COSY, HMBC). Further fractionation and structure elucidation is underway (Jeong *et al.* 1997).

Antiestrogen Binding Site and Ishikawa Cell System

To discover agents that inhibit tamoxifen interaction with the antiestrogen binding site and therefore may potentiate activity at the estrogen receptor, we initiated an investigation for inhibitors of tamoxifen binding to the antiestrogen binding site (AEBS) utilizing [^{3}H]-tamoxifen and a crude receptor preparation from female Sprague-Dawley rat

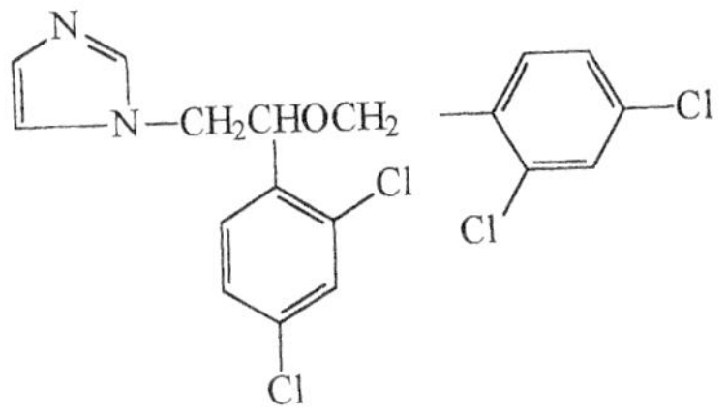

81

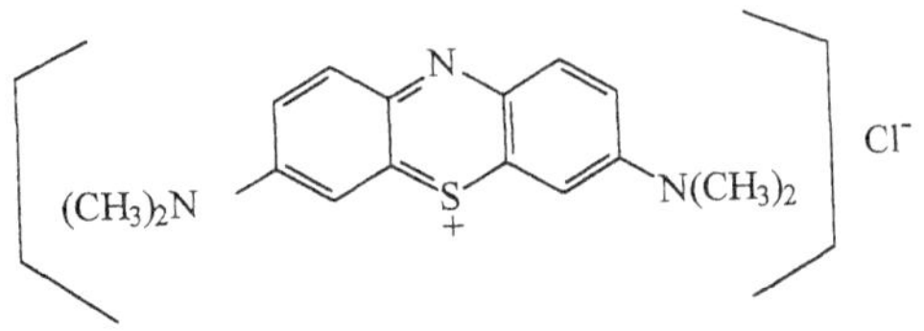

82

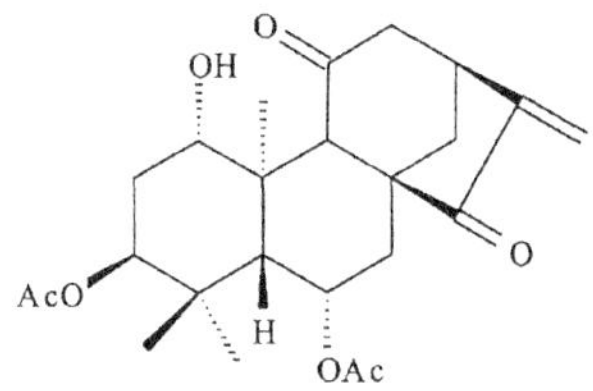

83

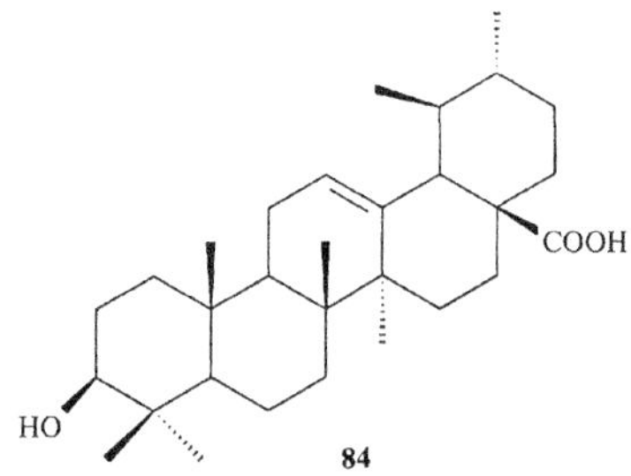

84

liver (Van den Koedijk *et al.* 1992). We have screened approximately 400 plant extracts and have detected five extracts demonstrating ED_{50} values of less than 20 μg/ml: *Pachysandra procumbens*, *Pachysandra terminalis*, *Lithospermum erythrorhizon*, *Pinellia ternata* and *Nelumbo lutea* (Chang *et al.* 1997; Pisha *et al.* 1997a). *Pachysandra procumbens* was further fractionated to isolate a series of potent steroidal alkaloids such as pachyaximine A [**85**], 2,3-didehydropachystermine A [**86**], and spiropachycine [**87**] (Chang *et al.* 1998).

The Ishikawa cell system was employed to determine biological activity employing *in vitro* cell culture techniques (Pisha and Pezzuto 1995, 1997b). Ishikawa cells are of human endometrial carcinoma origin having an estrogen-dependent alkaline phosphatase, induction of which can be estimated kinetically with a microtiter plate reader lending the assay adaptable to large-scale screening on a 96-well plate format. Extracts are deemed estrogenic if the alkaline phosphatase is induced. Antiestrogenic activity is asserted if the extract reduces background induction from the addition of exogenous estradiol. Toxicity of the extracts is also followed by the measurement of protein levels. Of the 925 extracts assessed, 12 demonstrated estrogenic activity, including six that displayed interaction with the estrogen receptor; this correlation indicated the assay is competent at indicating biologically relevant estrogenic activity. Studies with tamoxifen revealed the usefulness of the assay for monitoring antiestrogenic activity. Five extracts demonstrated significant antiestrogen activity having ED_{50} values of <2 μg/ml at subtoxic levels: *Ajuga genevensis*, *Tephrosia purpurea*, *Casimiroa edulis*, *Cerbera manghas* and *Waltheria indica* (Chang *et al.* 1997b; Pisha *et al.* 1997a). *Cerbera manghas* was fractionated to reveal the active antiestrogenic principle, (−)-cycloolivil [**39**], having antiestrogenic and toxic ED_{50} values of 1.9×10^{-6} M and 5.0×10^{-6} M, respectively (Pisha *et al.* 1996). Currently, all active extracts and isolates are undergoing testing in an estrogen/progesterone-dependent mouse mammary organ culture model to determine if the active leads inhibit the formation of ductal hyperplastic lesions which are histologically

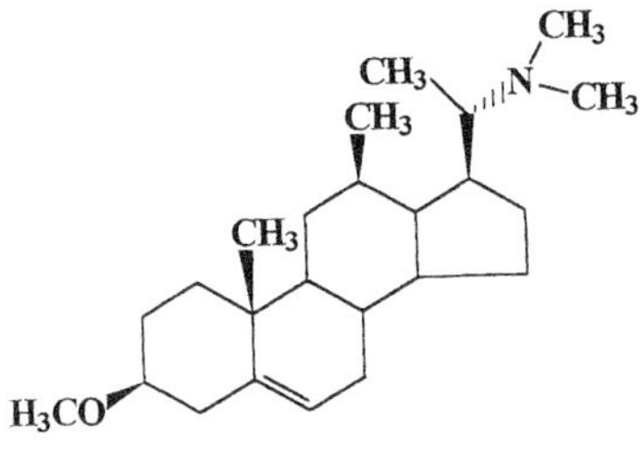

85

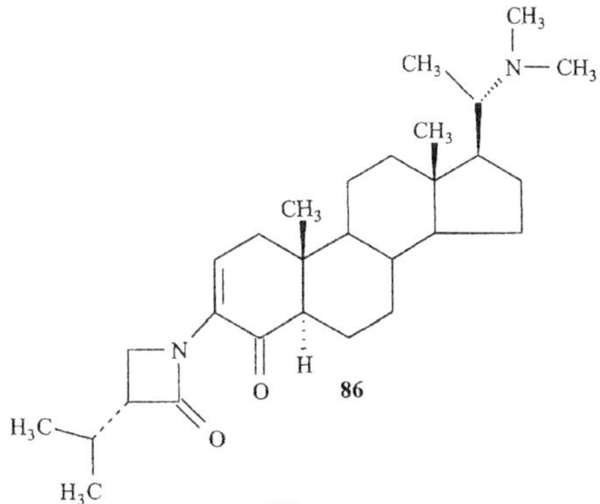

86

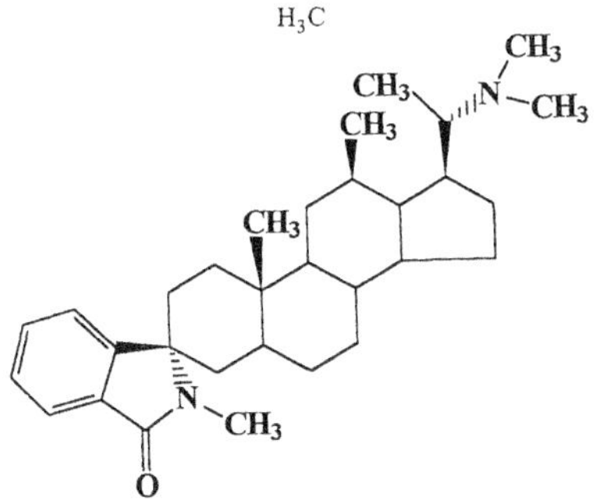

87

similar to human ductal carcinoma *in situ*, the most common form of breast cancer (Hawthorne *et al.* 1996). The active leads are also being evaluated in a short-term mouse mammary organ *in situ* assay to assess inhibition of estrogen-dependent induction of progesterone receptor.

DISCUSSION

The advantages of disease prevention are obvious, especially when dealing with recalcitrant ailments such as cancer. In addition to personal gratification, prevention strategies are consistent with the fiscal objectives of current day managed health care systems. Until recently, pharmaceutically based prevention was largely based on vaccination programs. Presently, however, small-molecular-weight compounds are commonly used for the prevention of certain afflictions such as coronary artery disease and malaria. From a historical perspective, for the prevention of cancer, major emphasis has focussed on primary prevention schemes (e.g., smoking cessation, avoidance of known risk factors). Cancer chemoprevention, the pharmaceutically based strategy for the delay or prevention of cancer, has been limited to dietary measures (e.g., the five-a-day for better health campaign) (Havas *et al.* 1994), or to the treatment of people at high risk for the development of specific types of cancer (Kelloff *et al.* 1994a,b). Prophylactic cancer chemoprevention in the general population is fraught with logistical problems, but due to the magnitude of potential benefits, it is imperative to continue movement in this direction (Pezzuto 1997).

A question of paramount importance, of course, is what chemopreventive formulation should be given to the general population? Although various suggestions can be put forward (Pezzuto 1997), systematic investigations for the discovery of new cancer chemopreventive agents are of utmost importance. As with other drug discovery programs, approaches may be developed on the basis of random screening, combinatorial chemistry, "rational" design, or natural products. In each case, however, a dominant feature of the programs deals with utilizing appropriate *in vitro* models of human disease states. The construct and design of these systems is critical for success. For certain therapeutic agents, the selection of a primary assay systems can be relatively straightforward (e.g., antimicrobial activity as an assay for antibiotics). In the case of cancer chemoprevention, or disease prevention in general, it is more challenging to devise *in vitro* test systems that are indicative of the desirable therapeutic outcome. We have accomplished this through a two-tiered procedure. At the first level, a battery of monitors are utilized that are reflective of various stages of carcinogenesis. In most cases, selectivity or "*in vitro* therapeutic efficacy" is taken into account, through the generation of an index, i.e., the *C*hemopreventive *I*ndex (*CI*).

Once materials are identified as promising leads, they are evaluated in a secondary system of greater physiological complexity. The objective of this secondary discrimination is to further hone down the leads identified in the primary evaluation to yield leads that are more likely to mediate the desirable response with *in vivo* systems. Unfortunately, animal models or *in vivo* test systems cannot be used for bioassay-guided fractionation due to logistical constraints, and it is virtually impossible for *in vitro* test system to fully epitomize the biological intricacies of a mammal. Thus, as an intermediate solution, we have primarily employed a mouse mammary gland organ culture model for secondary evaluation. This model demonstrates a good correlation with full-term mammary carcinogenesis inhibition studies, and meets the general logistical criteria required of a secondary monitor in terms of time, expense, and throughput.

It may be speculated that use of the mouse mammary gland organ culture assay system would direct us toward leads that are exclusively useful for the prevention of breast tumorigenesis. However, since common mechanisms may apply for the genesis of tumors at various organ sites, it is not clear that the mammary organ culture model would be exclusively indicative of breast cancer prevention. In any case, breast cancer is a severe problem wherein efficacious preventive agents are needed, so the model is pertinent. Nonetheless, we are currently expanding the scope of this drug discovery program primarily through utilization of additional secondary discriminators. In particular, we are exploring the use of benzo(a)pyrene-induced transformation of rat tracheal epithelial cells (Arnold *et al.* 1997). Again, since common mechanisms of carcinogenesis will apply, this model may not be exclusively indicative of preventing respiratory cancers, but it is necessary to perform experiments in order to establish the full potential of the test system. Similar to the mammary organ culture model, there are readily available animal models to investigate the potential of lead molecules to inhibit lung cancer, so follow-up studies with active leads are apparent.

The multi-tiered investigative approach described herein has proven successful for the discovery of promising new cancer chemopreventive agents. Thus far, brassinin

(Mehta *et al.* 1995), deguelin (Gerhäuser *et al.* 1995; Udeani *et al.* 1997), resveratrol (Jang *et al.* 1997), and 4′-bromoflavone have been shown to mediate cancer chemopreventive activity in full-term carcinogenesis studies, and they are being evaluated in more advanced test systems. As noted above, additional agents have demonstrated promising activity with *in vitro* or short-term animal studies, and full-term carcinogenesis inhibition work is slated. In two cases (a sulforaphane analog, and 4′-bromorotenoic acid), sufficient material is currently available and tumor inhibition studies are underway. Various other promising leads have been identified and are currently being processed. We are confident that one or more of the agents discovered during the course of this project will be found worthy of a clinical intervention trial and, hopefully, in the longer term, proven useful for the prevention of cancer in humans.

ACKNOWLEDGEMENTS

Progress summarized in this overview results from the efforts of an eminent team of investigators including Drs. N.R. Farnsworth, H.H.S. Fong, D.D. Soejarto, R.C. Moon, R.M. Moriarty, and S. Hedayat, all affiliated with the University of Illinois at Chicago. In addition, we gratefully acknowledge the assistance of plant suppliers from throughout the world, Steven Totura, and the various graduate students and postdoctoral associates listed in the references of this manuscript, and collaborators who have generously supplied cell cultures, test compounds, and reagents. This work is supported by the National Cancer Institute under the auspices of program project grant P01 CA48112.

REFERENCES

American Cancer Society. *Cancer Facts & Figures-1997.*

Arnold, J.T., Wilkinson, B.P., Sharma, S. and Steele, V.E. (1997) Chemoprevention of carcinogen-induced transformation in primary rat tracheal epithelial cells. *Meth. Cell Sci.*, **19**, 3–8.

Berridge, M.J. (1984) Inositol triphosphate and diacylglycerol as second messengers. *Biochem. J.*, **220**, 345–360.

Bjelke, E. (1975) Dietary vitamin A and human lung cancer. *Int. J. Cancer*, **15**, 561–565.

Blumberg, J. and Block, G. (1994) The alpha-tocopherol, beta-carotene cancer prevention study in Finland. *Nutr. Rev.*, **52**, 242–250.

Boone, C.W., Kelloff, G.J. and Malone, W.E. (1990) Identification of candidate cancer chemopreventive agents and their evaluation in animal models and human intervention trials: A review. *Cancer Res.*, **50**, 2–9.

Butterworth, Jr., C.E. (1992) Effect of folate on cervical cancer. *Ann. N.Y. Acad. Sci.*, **669**, 293–299.

Castagna, M. (1987) Phorbol esters as signal transducers and tumor promoters. *Biol. Cell.*, **59**, 3–14.

Castagna, M., Takai, Y., Kaibuchi, K., Sano, K., Kikkawa, U. and Nishizuka, Y. (1982) Direct activation of calcium-activated, phospholipid-dependent protein kinase by tumor-promoting phorbol esters. *J. Biol. Chem.*, **257**, 7847–7851.

Chang, L.C., Gerhäuser, C., Song, L.L., Farnsworth, N.R., Pezzuto, J.M. and Kinghorn, A.D. (1997a) Activity-guided isolation of constituents of *Tephrosia purpurea* with potential to induce the phase II enzyme quinone reductase. *J. Nat. Prod.*, **60**, 869–873.

Chang, L.C., Bhat, K.P.L., Pisha, E., Kennelly, E., Fong, H.H.S., Pezzuto, J.M. and Kinghorn, A.D. (1998) Activity-guided isolation of steroidal alkaloid antiestrogen-binding site inhibitors from *Pachysandra procumbens*. *J. Nat. Prod.*, in press.

Chang, L.-C., Pisha, E., Kennelly, E.J., Fong, H.H.S., Pezzuto, J.M. and Kinghorn, A.D. Antiestrogenic steroidal alkaloids from *Pachysandra procumbens*. 38th Meeting of the American Society of Pharmacognosy, Iowa City, Iowa, July 26–30, 1997b.

Chang, W.-S., Chang, Y.-H., Lu, F.-J. and Chiang, H.-C. (1994) Inhibitory effects of phenolics on xanthine oxidase. *Anticancer Res.*, **14**, 501–506.

Chung, H.S., Chang, L.C., Lee, S.K., Shamon, L.A., van Breemen, R.B., Mehta, R.G., Farnsworth, N.R., Pezzuto, J.M. and Kinghorn, A.D. (1998) Flavonoid consitutents of *Chorizanthe diffusa* with potential cancer chemopreventive activity. In preparation.

Constant, H.L. and Beecher, C.W.W. (1995) A method for the dereplication of natural product extracts using electrospray HPLC/MS. *Nat. Prod. Lett.*, **6**, 193–196.

Ford, L.G., Brawley, O.W., Perlman, J.A. and Nayfield, S.G., Johnson, K.A. and Kramer, B.S. (1994) The potential for hormonal prevention trials. *Cancer*, **74**, 2726–2733.

Fujita, Y., Uehara, I., Morimoto, Y., Nakashima, M., Hatano, T. and Okuda, T. (1988) Studies on inhibition of mechanism of antioxidation by tannins and flavonoids. II. Inhibition of mechanism of caffeetannins isolated from leaves of *Artemisia* species on lipoxigenase dependent lipid peroxidation. *Yakugaku Zasshi*, **108**, 129–135.

Gamez, E.J.C., Luyengi, L., Lee, S.K., Zhu, L.F., Zhou, B.-N., Fong, H.H.S., Pezzuto, J.M. and Kinghorn, A.D. Potential cancer chemopreventive agents from *Daphniphyllum calycinum*. 38th Annual Meeting of the American Society of Pharmacognosy, University of Iowa, Iowa City, Iowa, July 26–30, 1997.

Gerhäuser, C., Kosmeder, J.W., Song, L., Hawthorne, M., Mehta, R., Moon, R.C., Moriarty, R.M. and Pezzuto, J.M. Chalcones induce drug-metabolizing enzymes *in vitro* and *in vivo*. 37th Annual Meeting of the American Society of Pharmacognosy, Santa Cruz, CA, July 27–31, 1996.

Gerhäuser, C., Lee, S.K., Kosmeder, J.W., Moriarty, R.M., Hamel, E., Mehta, R.G., Moon, R.C. and Pezzuto, J.M. (1997a) Regulation of ornithine decarboxylase induction by deguelin, a natural product cancer chemopreventive agent. *Cancer Res.*, **57**, 3429–3435.

Gerhäuser, C., Mar, W., Lee, S.K., Suh, N., Luo, Y., Kosmeder, J., Luyengi, L., Fong, H.H.S., Kinghorn, A.D., Moriarty, R.M., Mehta, R., Constantinou, A., Moon, R.C. and Pezzuto, J.M. (1995) Rotenoids mediate potential cancer chemopreventive activity through transcriptional regulation of ornithine decarboxylase. *Nature Med.*, **1**, 260–266.

Gerhäuser, C., You, M., Liu, J., Moriarty, R.M., Hawthorne, M., Metha, R.G., Moon, R.C. and Pezzuto, J.M. (1997b) Cancer chemopreventive potential of sulforamate, a novel analog of sulforaphane that induces phase 2 drug-metabolizing enzymes. *Cancer Res.*, **57**, 272–278.

Goodwin, J.S. and Ceuppens, J.L. (1985) Prostaglandins, cellular immunity and cancer. In J.S. Goodwin, (ed.), *Prostaglandins and Immunity*, Martinus Nijhoff Publishing, Boston, pp. 1–34.

Greenwald, P., Cullen, J.W. and McKenna, J.W. (1987) Cancer prevention and control from research through applications. *J. Natl. Cancer Inst.*, **79**, 389–400.

Greenwald, P, Sondik, E. and Lynch, B.S. (1986) Diet and chemoprevention in NCI's research strategy to achieve national cancer control objectives. *Annu. Rev. Public Health*, **7**, 267–291.

Havas, S., Heimendinger, J., Reynolds, K., Baranowski, T., Nicklas, T.A., Bishop, D., Buller, D., Sorensen, G., Beresford, S.A.A., Cowan, A. and Damron, D. (1994) Five a day for better health: A new research initiative. *J. Am. Diet. Assoc.*, **94**, 32–36.

Harvey, H.A. (1996) Aromatase inhibitors in clinical practice current status and a look to the future. *Semin. Oncol.*, **23** (Suppl 9), 33–38.

Hawthorne, M.E., Singh, M., Thompson, H.J., Steele, V., Kelloff, J.G. and Mehta, R.G. Induction of DMBA-induced mammary ductal lesions in organ culture: a model for human DCIS.

87th Annual Meeting of the American Association for Cancer Research, Washington, D.C., April 20–24, 1996.

Herbert, J.M., Augereau, J.M., Gleye, J. and Maffrand, J.P. (1990) Chelerythrine is a potent and specific inhibitor of protein kinase C. *Biochem. Biophys. Res. Commun.*, **172**, 993–999.

Hirayama, T. (1979) Diet and cancer. *Nutr. Cancer*, **1**, 67–81.

Jang, M., Cai, L., Udeani, G.O., Slowing, K., Thomas, C.F., Beecher, C.W.W., Fong, H.H.S., Farnsworth, N.R., Kinghorn, A.D., Mehta, R.G., Moon, R.C. and Pezzuto, J.M. (1997) Cancer chemopreventive activity of resveratrol, a natural product derived from grapes. *Science*, **275**, 218–220.

Jang, M.-S. and Pezzuto, J.M. (1997) Assessment of cyclooxygenase inhibitors using *in vitro* assay systems. *Meth. Cell Sci.*, **19**, 25–31.

Jeong, H.-J., Song, L., Kinghorn, A.D. and Pezzuto, J.M. Aromatase inhibitors from *Isodon excisus* var. *coreanus*. 38th Meeting of the American Society of Pharmacognosy, Iowa City, Iowa, July 26–30, 1997.

Jordan, V.C. (1994) *Long-term Tamoxifen Treatment for Breast Cancer*, University of Wisconsin Press, Madison.

Jordan, V.C. (1995) Tamoxifen and carcinogenicity: A predictable concern. *J. Natl. Cancer Inst.*, **87**, 623–626.

Kelloff, G.J., Boone, C.W., Crowell, J.A., Steele, V.E., Lubet, R.A. and Sigman, C.S. (1994a) Chemopreventive drug development: Perspective and progress. *Cancer Epidem. Biomarkers Prevent.*, **3**, 83–98.

Kelloff, G.J., Crowell, J.A., Boone, C.W., Steele, V.E., Lubet, R.A., Greenwald, P., Alberts, D.S., Covey, J.M., Doody, L.A., Knapp, G.G., Nayfield, S., Parkinson, D.R., Prasad, V.K., Prorok, P.C., Sausville, E.A. and Sigman, C.C. (1994b) Clinical development plans for cancer chemopreventive agents. *J. Cell. Biochem.*, Suppl. **20**, 55–299.

Kennelly, E.J., Gerhäuser, C., Song, L.L., Pezzuto, J.M. and Kinghorn, A.D. (1997) Induction of quinone reductase by withanolides isolated from *Physalis philadelphica* (tomatillos). *J. Agric. Food Chem.*, **35**, 3771–3777.

Kinghorn, A.D., Fong, H.H.S., Farnsworth, N.R., Mehta, R.G., Moon, R.C., Moriarty, R.M. and Pezzuto, J.M. (1997) Cancer chemopreventive agents discovered by activity-guided fractionation: A Review: In K. Hostettmann, (ed.), *Current Organic Chemistry*, Bentham Science Publishers, Hilversum, in press.

Knekt, P. (1992) Vitamin E and cancer: Epidemiology. *Ann. N.Y. Acad. Sci.*, **669**, 269–279.

Lee, S.K., Constant, H.L., Graham, J.G., Mbwambo, Z.H., Farnsworth, N.R., Kinghorn, A.D., Beecher, C.W.W. and Pezzuto, J.M. Evaluation of the antioxidant potential of plant extracts and identification of active principles using HPLC/MS dereplication analysis. 37th Annual Meeting of the American Society of Pharmacognosy, Santa Cruz, CA, July 27–31, 1996.

Lee, S.K., Kosmeder II, J.W., Gerhäuser, C., Mar, W., Luyengi, L., Moriarty, R.M., Fong, H.H.S., Kinghorn, A.D. and Pezzuto, J.M. (1997a) Structure-activity studies of rotenoids that inhibit phorbol ester-induced ornithine decarboxylase activity in cell culture. In preparation.

Lee, S.K., Mbwambo, Z., Suh, N., Kinghorn, A.D. and Pezzuto, J.M. (1997b) Potential tumor promoter activity mediated by a novel ingenol ester isolated from *Euphorbia quinquecostata*. In preparation.

Lee, S.K., Mbwambo, Z.H., Chung, H.S., Luyengi, L., Gamez, E.J.C., Mehta, R.G., Kinghorn, A.D. and Pezzuto, J.M. (1998a) Evaluation of the antioxidant potential of natural products. *Combinatorial Chemistry and High Throughput Screening*, **1**, 35–46.

Lee, S.K., Qing, W.G., Mar, W., Luyengi, L.A., Kinghorn, A.D. and Pezzuto, J.M. (1998b) Chelerythrine and angoline, benzophenanthridine alkaloids that do not inhibit protein kinase C. *J. Biol. Chem.*, **273**, 19829–19833.

Logan, R.F., Little, J., Hawlin, P.G. and Hardcastle, J.D. (1993) Effect of aspirin and non-steroidal anti-inflammatory drugs on colorectal adenomas: case-control study of subjects participating in the Nottingham fecal occult blood screening program. *Brit. Med. J.*, **307**, 285–289.

Lombardini, J.B. (1996) Stimulation by chelerythrine of the phosphorylation of the amino acid serine in a ~20 kda protein present in the mitochondrial fraction of the rat retina. *Biochem. Pharmacol.*, **52**, 253–257.

Luyengi, L., Lee, I.-S., Mar, W., Fong, H.H.S., Pezzuto, J.M. and Kinghorn, A.D. (1994) Rotenoids and chalcones from *Mundulea sericea* that inhibit phorbol ester-induced ornithine decarboxylase activity. *Phytochemistry*, **36**, 1523–1526.

Luyengi, L., Suh, N., Fong, H.H.S., Pezzuto, J.M. and Kinghorn, A.D. (1996) A lignan and four terpenoids from *Brucea javanica* that induce differentiation with cultured HL-60 promyelocytic leukemia cells. *Phytochemistry*, **43**, 409–412.

Mata-Greenwood, E., Grieco, P.A. and Pezzuto, J.M. Cell Differentiation and regulation of oncogene expression by selected quassinoids. 38th Meeting of the American Society of Pharmacognosy, Iowa City, Iowa, July 26–30, 1997a.

Mata-Greenwood, E., Mehta, R.G., Ito, A., Yu, B., Kosmeder, II, J.W., Moriarty, R.M., Kinghorn, A.D. and Pezzuto, J.M. Flavone and chalcone inducers of HL-60 cell differentiation. 37th Annual Meeting of the American Society of Pharmacognosy, Santa Cruz, CA, July 27–31, 1996.

Mata-Greenwood, E., Suh, N., Mehta, R.G., Grieco, P.A. and Pezzuto, J.M. (1997b) Effects of antineoplastic quassinoids on HL-60 cell proliferation, viability, and differentiation: down regulation of *c-myc* and up-regulation of *c-fos* and *c-jun* are early molecular events. In preparation.

McCann, P. and Pegg, A.E. (1992) Ornithine decarboxylase as an enzyme target for therapy. *Pharmacol. Ther.*, **54**, 195–215.

Mehta, R.G., Jang, M., Pisha, E., Mata-Greenwood, E., Song, L. and Pezzuto, J.M. (1997) Utilization of carcinogen-treated murine mammary glands in organ culture as a secondary descriminator for natural product cancer chemopreventive drug discovery. In preparation.

Mehta, R.G., Liu, J., Constantinou, A., Hawthorne, M., Pezzuto, J.M., Moon, R.C. and Moriarty, R.M. (1994) Structure-activity relationships of brassinin in preventing the development of carcinogen-induced mammary lesions in organ culture. *Anticancer Res.* **14**, 1209–1214.

Mehta, R.G., Liu, J., Constantinou, A., Thomas, C.F., Hawthorne, M., You, M., Gerhäuser, C., Pezzuto, J.M., Moon, R.C. and Moriarty, R.M. (1995) Cancer chemopreventive activity of brassinin, a phytoalexin from cabbage. *Carcinogenesis*, **16**, 399–404.

Muktar, H., Katiyar, S.K. and Agarwal, R. (1994) Green tea and skin-anticarcinogenic effects. *J. Dermatol.*, **102**, 3–7.

Nishizuka, Y. (1986) Studies and perspectives of protein kinase C. *Science*, **233**, 305–312.

O'Brien, T.G. (1976) The induction of ornithine decarboxylase as an early, possibly obligatory event in mouse skin carcinogenesis. *Cancer Res.*, **36**, 2644–2653.

Omenn, G.S., Goodman, G.E., Thornquist, M.D., Balmes, J., Cullen, M.R., Glass, A., Keogh, J.P., Meyskens, F.L., Jr., Valanis, B., Williams, J.H., Jr., Barnhart, S., Cherniack, M.G., Brodkin, C.A. and Hammar, S. (1996) Risk factors for lung cancer and for intervention effects in CARET, the beta-carotene and retinol efficacy trial. *J. Natl. Cancer Inst.*, **88**, 1550–1559.

Pedersen, T.R., Scandinavian Simvastatin Survival Study Group (1994) Randomised trial of cholesterol lowering in 4444 patients with coronary heart disease: The Scandinavian simvastatin survival study (4S). *Lancet*, **344**, 1383–1389.

Pezzuto, J.M. (1993) Cancer chemopreventive agents: From plant materials to clinical intervention trials. In A.D. Kinghorn and M.F. Balandrin, (eds.), *Human Medicinal Agents from Plants* (ACS Symposium Series No. 534), American Chemical Society Books, Washington, pp. 205–217.

Pezzuto, J.M. (1995) Natural product cancer chemopreventive agents. In J.T. Arnason, R. Mata, and J.T. Romeo (eds.), *Recent Advances in Phytochemistry*, **29**, *Phytochemistry of Medicinal Plants*, Plenum Press, New York, pp. 19–45.

Pezzuto, J.M. (1996) Bioassay-directed isolation of natural product drugs. *ICAST 96 Proceedings. The 12th International Conference on Advanced Science and Technology*, **12**, 23–28, Chicago.

Pezzuto, J.M. (1997) Plant-derived anticancer agents, *Biochem. Pharmacol.*, **53**, 121–133.

Pezzuto, J.M., Angerhofer, C.K. and Mehdi, H. (1997) *In vitro* models of human disease states, In Atta-ur-Rahman (ed.), *Studies in Natural Products Chemistry*, **19**, Elsevier Science Publishers, Amsterdam, in press.

Pisha, E., Chang, L.-C., Kinghorn, A.D., Hawthorne, M., Mehta, R.G. and Pezzuto, J.M. Identification of novel anti-estrogenic chemopreventive agents from natural products. 88th Annual Meeting of the American Association for Cancer Research, San Diego, CA, April 12–16, 1997a.

Pisha, E. and Pezzuto, J.M. Cell-based assay development for the detection of estrogenic and anti-estrogenic compounds. 36th Annual Meeting of the American Association of Pharmacognosy, University of Mississippi, Oxford, MS, July 23–27, 1995.

Pisha, E. and Pezzuto, J.M. (1997b) Cell-based assay for the determination of estrogenic and anti-estrogenic activities. *Meth. Cell Sci.*, **19**, 37–43.

Pisha, E., Luyengi, L., Lee, S.K., Farnsworth, N.R., Mehta, R.G., Moon, R.C., Kinghorn, A.D. and Pezzuto, J.M. Potential cancer chemopreventive lignans from *Cerbera manghas*. 20th IUPAC Symposium on the Chemistry of Natural Products, Chicago, IL, Sept. 15–20, 1996.

Powles, T.J. and Hickish, T. (1995) Tamoxifen therapy and carcinogenic risk. *J. Natl. Cancer Inst.*, **87**, 1343–1345.

Rozhin, J., Wilson, P.S., Bull, A.W. and Nigro, N.D. (1984) Ornithine decarboxylase activity in the rat and human colon. *Cancer Res.*, **44**, 3226–3230.

Santen, R.J., Manni, A. and Harvey, H.A. (1990) Endocrine treatment of breast cancer in women. *Endocr. Rev.*, **11**, 221–265.

Shamon, L.A., Chen, C., Mehta, R.G., Steele, V., Moon, R.C. and Pezzuto, J.M. (1994) A correlative approach for the discovery of antimutagens that demonstrate cancer chemopreventive activity, *Anticancer Res.*, **14**, 1775–1778.

Shamon, L.A. and Pezzuto, J.M. (1997) Assessment of antimutagenic activity with *Salmonella typhimurium* strain TM677. *Meth. Cell Sci.*, **19**, 57–62.

Sharma, S., Stutzman, J.D., Kelloff, G.J. and Steele, V.E. (1994) Screening of potential chemopreventive agents using biochemical markers of carcinogenesis. *Cancer Res.*, **54**, 5848–5855.

Smith, W.L. (1989) The eicosanoids and their biochemical mechanisms of action. *Biochem. J.*, **25**, 315–324.

Song, L.L., Kosmeder, J.W., Gerhäuser, C., Lantvit, D., Moon, R.C., Moriarty, R.M. and Pezzuto, J.M. (1997a) Potential cancer chemopreventive activity mediated by 4′-bromoflavone. Submitted.

Song, L., Kosmeder, J.W., Gerhäuser, C., Hawthorne, M., Mehta, R.G., Moon, R.C., Moriarty, R.M. and Pezzuto, J.M. Potent induction of phase II enzymes by flavones and chalcones. 88th Annual Meeting of the American Association for Cancer Research, San Diego, CA, April 12–16, 1997b.

Song, L.L., Lantvit, D., Lubet, R.A., Steele, V.E., Kelloff, G.J., Pezzuto, J.M. and Moon, R.C. Chemoprevention for mammary carcinogenesis by inducers of phase II enzymes. 88th Annual Meeting of the American Association of Cancer Research, San Diego, CA, April 12–16, 1997c.

Song, L.L., Lantvit, D., Lubet, R.A., Steele, V.E., Kelloff, G.J., Pezzuto, J.M. and Moon, R.C. (1997d) Chemoprevention for mammary carcinogenesis by inducers of phase II enzymes. In preparation.

Sporn, M.B., Dunlop, N.M., Newton, D.L. and Smith, J.M. (1976) Prevention of chemical carcinogenesis by vitamin A and its synthetic analogs (retinoids). *Fed. Proc.*, **35**, 1332–1338.

Suffness, S.M. and Pezzuto, J.M. (1991) Assays for cytotoxicity and antitumor activity. Chapter 4. In: K. Hostettmann (ed.), *Methods of Plant Biochemistry*, **9**, Academic Press, London, pp. 71–133.

Suh, N., Luyengi, L., Fong, H.H.S., Kinghorn, A.D. and Pezzuto, J.M. (1995a) Discovery of natural products as chemopreventive agents utilizing HL-60 cell differentiation as a mdel. *Anticancer Res.*, **15**, 233–240.

Suh, N., Luyengi, L., Fong, H.H.S., Kinghorn, A.D., Mehta, R.G., Moon, R.C. and Pezzuto, J.M. Potent Chemopreventive activity by brusatol isolated from *Brucea javanica*. 36th Annual Meeting of the American Association of Pharmacognosy, University of Mississippi, Oxford, MS, July 23–27, 1995b.

Talalay, P., Fahey, J.W., Holtzclaw, W.D., Prestera, T. and Zhang, Y. (1995) Chemoprotection against cancer by phase II enzyme induction. *Toxicol. Lett.*, **82–83**, 173–179.

Thompson, I.M. and Coltman, C.A. (1996) Screening for prostate cancer: Opportunities for prevention. *Sem. Urol. Oncol.*, **14** (Suppl. 2), 4–10.

Udeani, G.O., Gerhäuser, C., Thomas, C.F., Moon, R.C., Kosmeder, J.W., Kinghorn, A.D., Moriarty, R.M. and Pezzuto, J.M. (1997) Cancer chemopreventive activity mediated by deguelin, a naturally occurring rotenoid. *Cancer Res.*, **57**, 3424–3428.

Van den Koedijk, C.D.M.A., Vis Van Heemst, C., Elsendoorn, G.M., Thijssen, J.H.H. and Blankenstein, M.A. (1992) Comparative affinity of steroidal and non-steroidal antioestrogens, cholesterol derivative and compounds with a dialkyamino side chain for the rat liver antioestrogen binding site. *Biochem. Pharmacol.*, **43**, 2511–2518.

Westenburg, H.E., Mata-Greenwood, E., Song, L., Castillo, J.J., Soejarto, D.D., Pezzuto, J.M. and Kinghorn, A.D. Naturally occurring and synthetic aryl polysulfides with activity in HL-60 cell differentiation and quinone reductase induction assays. 38th Meeting of the American Society of Pharmacognosy, Iowa City, Iowa, July 26–30, 1997.

Wongpanich, V., Mata-Greenwood, E., Suh, N., Chai, H., Fong, H.H.S., Pezzuto, J.M. and Angerhofer, C.K. Correlation of the antimalarial activity of artemisinins and their potential to induce cell differentiation. 38th Meeting of the American Society of Pharmacognosy, Iowa City, Iowa, July 26–30, 1997.

Yue, W. and Santen, R.J. (1996) Aromatase inhibitors rationale for use following antiestrogen therapy. *Semin. Oncol.*, **23** (Suppl. 9), 21–27.

Zhang, Y., Kensler, T.W., Cho, C.G., Posner, G.H. and Talalay, P. (1994) Anticarcinogenic activities of sulforaphane and structurally related synthetic norbornyl isothiocyanates. *Proc. Natl. Acad. Sci. USA*, **91**, 3147–3150.

Zhang, Y., Talalay, P., Cho, C.G. and Posner, G.H. (1992) A major inducer of anticarcinogenic protective enzymes from broccoli: isolation and elucidation of structure. *Proc. Natl. Acad. Sci. USA*, **89**, 2399–2403.

6. BIODIVERSITY CONSERVATION AND DRUG DEVELOPMENT IN SURINAME

DAVID G.I. KINGSTON[1], A.A. LESLIE GUNATILAKA[1], BING-NAN ZHOU[1], MAGED ABDEL-KADER[1], JOHN BERGER[1], HENDRIK VAN DER WERFF[2], RANDALL EVANS[2], LISA FAMOLARE[3], RUSSELL MITTERMEIER[3], STAN MALONE[4] and JAN H. WISSE[5]

[1]*Department of Chemistry, Virginia Polytechnic Institute and State University, Blacksburg, Virginia 24061-0212 USA*
[2]*Missouri Botanical Garden, P.O. Box 299, St. Louis, MO 63166-0299, USA*
[3]*Conservation International, 2501 M Street, Suite 200, Washington, DC 20037, USA*
[4]*Conservation International Suriname, Burenstr. 7, Paramaribo, Suriname*
[5]*Bedrijf Geneesmiddelen Voorziening Suriname, Commissaris Roblesweg 156, Paramaribo, Suriname*

INTRODUCTION

The search for bioactive products from the universe of natural products has been remarkably successful, with over half of the world's 25 best-selling pharmaceuticals as of 1991 being either natural products or derived from natural products (O'Neill and Lewis 1993). The rosy periwinkle of Madagascar for treating Hodgkins's disease and childhood leukemia, quinine for malaria, and curare as a muscle relaxant for certain types of surgery are three well-known examples of pharmaceuticals derived from plants. Although recent competition for resources from the emerging disciplines of combinatorial chemistry (Plunkett and Ellman 1997; Borman 1996) and combinatorial biochemistry has limited the effort that various pharmaceutical companies have been able to put into drug discovery by the natural products pathway, this pathway remains an important and productive one (Cragg *et al.* 1997); with more than 250,000 species of flowering plants identified, we have just begun to tap a rich lode for biotechnology.

In spite of the success of the natural products approach to drug discovery, a number of concerns must be addressed in implementing a successful drug discovery program. Foremost among these is the question of the supply of material for screening purposes; this concern is particularly serious in the case of plant material, because much of the world's biodiversity is associated with tropical rain forests, and these forests are rapidly being lost to deforestation. It is estimated that an area of tropical forest the size of Florida is lost to timber, mining, grazing, and other destructive forces every year; in total, an area larger than the size of France has been deforested in the Amazon region alone, over 12% of the worlds most diverse ecosystem.

This exponential rate of destruction leads to the marginalization and decline of indigenous cultures and the irreversible loss of genetic resources, two valuable national resources. A part of the reason for this deforestation is that the forests are undervalued; a recent study has placed a value of $33 trillion per year on the world's ecosystem, in comparison with a global gross national product of $18 trillion per year (Costanza *et al.* 1997). However, most of this value is outside of the market and thus largely unrecognised by individuals and governments. Ideally, a successful drug discovery program would provide a means of realizing at least a small part of the value of the ecosystem within the market system, and this leads to the next consideration.

The second concern is that plant collections must be carried out according to the best current practices. The ethical and legal aspects of the natural product drug discovery process have recently undergone a profound and long overdue change. In years past it was considered acceptable practice to collect biological specimens (whether terrestrial or marine, microbial, plant or animal) without any compensation to the source country other than possible fees for collecting permits and the costs of collection. This is no longer the case today. For reasons explained in part above, scientists and collectors the world over now realize that the biodiversity in any country is the valid property of that country, and this understanding has been codified in international law by the Rio Treaty on Biodiversity (Reid *et al.* 1993). In the United States of America the scientific community and major government institutions such as the National Science Foundation and the National Institutes of Health have recognized this changed situation, even though the U.S. Congress has not yet formally ratified the Rio Treaty.

The International Collaborative Biodiversity Group (ICBG) Program was created out of a desire to provide a mechanism to put the principles and requirements of the Rio Treaty into effective practice.

THE ICBG PROGRAM

The concept of the ICBG Program was developed over the period 1989–1991 as a collaborative effort between the U.S. National Institutes of Health, the National Institute of Mental Health, U.S. National Science Foundation, and the U.S. Agency for International Development. Funding for the program became available in 1992, and the administration of the Program was assigned to the Fogarty International Center at the National Institutes of Health. A Request for Applications (RFA) was released on June 12, 1992, with an application deadline of November 17, 1992 (Anonymous 1992). After scientific review of the approximately thirty proposals that were received in response to the RFA, an initial group of three proposals was funded in September 1993, with an additional two proposals funded in 1994.

The original RFA specified four objectives for applicants to the Program.

1. The discovery, isolation, and preclinical evaluation of agents from natural sources to treat or prevent cancer, infectious diseases including AIDS, cardiovascular diseases, mental disorders, and other diseases or medical conditions of primary concern to developing countries.
2. To undertake inventories of biological diversity and develop collection practices compatible with conserving biodiversity, and to produce documentation of all collected material in the form of museum catalogues, published works, and/or databases.

3. To support research training targeted towards the needs of the developing country or other countries represented within the Group and related to the scope of the work of the RFA, and to augment field experience and training of U.S. scientists in areas of knowledge unique to the developing country.
4. To assist in improving the scientific infrastructure within the developing country(ies) where the biodiversity resources are found.

In addition to these four scientific and development goals, applicants were required to provide a mechanism for obtaining patent protection on all patentable discoveries, with appropriate benefit sharing with the host country according to the provisions of the Rio Treaty. It was also required that at least one of the associate programs be directed by host country participants.

THE SURINAME ICBG PROGRAM

Following discussions among the participants described below, a Group was assembled to work in Suriname. The Group consisted of David Kingston at Virginia Polytechnic Institute and State University as the Group Leader (Principal Investigator) and Associate Programs from Missouri Botanical Garden (MBG; Henk van der Werff and then Jim Miller), Conservation International (CI, Mark Plotkin and then Russell Mittermeier), Bedrijf Geneesmiddelen Voorziening Suriname (BGVS; Rob Verhage and then Jan Wisse) and Bristol-Myers Squibb Pharmaceutical Research Institute (BMS, Terrence Doyle and then Dinesh Vyas). This application was successful in the competition for funding, and an award was made to begin work in September 1993.

The Republic of Suriname (formerly Dutch Guiana) is located in the northeastern corner of South America, between Guyana and French Guiana. The second of the Guianas in size and population, it has a human population of approximately 400,000 in an area of 166,000 square kilometers, making it one of the least densely populated tropical countries in the world. Furthermore, 95% of the population is concentrated in the capital city of Paramaribo, and small towns of the coastal region. Only about 5% of the population lives in the interior, most of it in small scattered villages along the three major rivers.

Suriname is also home to a diverse variety of biodiversity including 674 species of birds, 200 species of mammals, 130 species of reptiles, 99 species of amphibians, and roughly 5,000 species of plants. Furthermore, while Suriname is a relatively small country, it is internationally quite significant because of its extensive tropical forest cover. The country has one of the highest percentages of tropical forest cover in the world with nearly 90% of the total area covered by forests. Indeed, 14,855,000 hectares of tropical forest gives Suriname nine times the forest cover of Costa Rica, and 75% of the total forest cover in all of Central America. It also has a unique culture, with two distinct peoples occupying the interior. The Amerindians are descendants of the original native inhabitants of the region, and form one people group with distinct settlements and culture. The other people group is that of the so-called Bushnegroes, who are descendants of African slaves who escaped into the interior and were recognized as a separate people by the Dutch. These Bushnegroes have thus established an "African" culture in South America. This existence of two distinct tribal cultures offered attractive possibilities for ethnobotanical studies.

Suriname is in a unique global position. It currently maintains a relatively vast fortune of tropical forest, a resource that is increasingly threatened throughout the world. The pressures that normally threaten tropical forests such as burgeoning migratory populations and intense forest resource demand are scarce in Suriname. However, this situation is rapidly changing. In 1993, Suriname took the desperate step of inviting Asian investors to explore possibilities for establishing multi-million hectare logging concessions in the country's interior. By mid-1994, at least five proposals were on the table and in negotiation: one from a Malaysian gambling tycoon; two from Indonesian logging consortia; and two from state-owned Chinese enterprises. Total investments proposed are over $500 million, and somewhere between 25 to 40 percent of the country's land area could be logged.

SURINAME ICBG OBJECTIVES AND STRUCTURE

The objectives of the Suriname ICBG closely paralleled those specified in the original RFA. The first objective was biodiversity assessment, and this task was undertaken by Missouri Botanical Garden and its highly competent staff. The second objective of the documentation and dissemination of ethnobotanical knowledge was undertaken by Conservation International, who had already established a "Shaman's Apprentice" program to assist in the preservation and transmission of ethnomedical knowledge to the next generation. The training of Surinamese nationals was a third objective which was shared by all members of the Group, with each member providing training appropriate to its expertise.

A new objective was that of comparing traditional botanical and ethnobotanical approaches to collecting plants for drug discovery purposes. Since MBG could best collect on a standard botanical basis, while CI had a specific interest in collecting ethnobotanical samples, a comparison of these two approaches appeared to be a reasonable objective, so that the most effective approach for drug discovery could be used in the future.

The objective of new drug discovery and development was shared by all the partners. The program was set up so that the plants collected either by MBG or by CI would be extracted by BGVS, and the extracts would then be forwarded to VPISU and BMS. Based on the bioassay results at these two Associate Programs, plants of interest would be recollected and large extracts prepared, and these extracts would be fractionated to yield pure active compounds. Those compounds with sufficient activity would then be developed further by BMS, with the ultimate goal of development of a new drug.

The final objective was that of economic development in Suriname. Our objectives here were twofold. The short-term goal was to contribute to the economy through the development of microenterprises at the local level – through small projects such as the development of ecotourism and native woodcarving enterprises. Funding for these projects came from the Forest Peoples' Fund. This fund is designed to support local communities in the interior of Suriname in projects involving community development, biodiversity conservation, and health-care, and to act as a mechanism by which any up-front compensation and future royalties provided by new drug development will be returned to the tribal people of Suriname. The fund was initiated with a $50,000 donation from Bristol-Myers Squibb Pharmaceuticals as an initial up-front compensation to the forest people of Suriname for use of their natural resources in cooperation with the ICBG project.

The long-term objective was, of course, to make a major contribution to the income of local peoples and organizations through royalties paid on a drug developed from the Surinamese flora. Clearly the chance of this objective becoming reality is much less than 100%, but the potential benefits to Suriname and to its forest peoples would be significant if it were to occur.

LEGAL ASPECTS

According to the Rio Biodiversity Treaty, "The authority to determine access to genetic resources rests with the national government". This is simple enough, and it meant that a collection permit from the government of Suriname was required before we could begin work. However, since we were to be collecting on lands in the interior which the tribal peoples considered to be their property, it was also necessary to obtain approval from the appropriate tribal authorities before beginning any collections. This was not a simple matter, since it involved extensive face-to-face negotiations between the staff of CI-Suriname, representing the group as a whole, and the tribal leaders. These negotiations were successfully concluded, and a mutually acceptable agreement was duly drawn up between the Granman (paramount chief) of the Asindopo region and representatives of CI-Suriname.

The collections permit from the Government of Suriname turned out to be more of a problem then we had anticipated, but a one-year permit was eventually issued after an initial delay of almost a year. When this permit expired there was a further delay of almost a year before it was renewed, and so collecting activities have been more restricted than we had initially anticipated. Such delays are inherent in working in a new country with a fluid political structure, but we anticipate that future permits will be issued in a timely fashion as the participants and the Suriname Government develop improved relationships.

The agreement drawn up between the Group members calls for future royalties to be paid on a scale that depends on whether the drug was initially developed from a botanical lead or an ethnobotanical lead, with the latter commanding a somewhat higher royalty rate. The rate is reduced if the actual drug substance is a chemical derivative (other than a pro-drug form) of the natural product, and is reduced still further if the actual drug substance is a synthetic substance that is designed using the natural product as a starting pharmacophore. Even in this case, however, when the connection between the actual drug substance and the natural product is fairly tenuous, some significant royalties would accrue to Suriname.

A further feature of the agreement is that it contains provisions for a share of the royalties to accrue to the Forest Peoples' Fund; other shares go to various government agencies, CI-suriname, and BGVS. In the event of a licensed drug, there is thus an assurance that a significant portion of the income would end up being used for the benefit of the forest peoples of Suriname. The division of these benefits among Surinamese parties is outlined in a separate benefit-sharing agreement; some of the innovative contractual provisions discussed by Rubin and Fish (1994) are included in these documents.

PLANT COLLECTION

The first part of the on-ground work in Suriname involved plant collections. Although our initial idea had been to lay out well-defined study plots and to collect all the species in a given plot, this concept was found to be impracticable. A major reason for this was that collections could only be carried out at certain times, due to both logistical and climatic considerations, and the use of the study plot approach would have unnecessarily limited the plant species that could have been collected during these limited collecting periods. As an example of the considerations involved, collections could be made most effectively during the relatively dry season in Suriname, since wet plant materials spoil very rapidly in the tropical heat; collections made in the wet season had to be taken back to the village and dried under cover. The concept of collecting in research plots was thus abandoned, and both MBG and CI made their collections in similar regions around the major villages of central Suriname.

All plant collections made were properly vouchered, with voucher samples deposited at the Natural Herbarium of Suriname and at Missouri Botanical Garden. Dried plants were shipped to Paramaribo for grinding and extraction by BGVS.

At this point approximately 500 species have been collected by MBG on a botanical basis and 323 by CI on an ethnobotanical basis. These have yielded 961 "botanical" samples and 505 "ethnobotanical" samples, since each plant will usually yield at least two different plant samples (leaves and stems, for example).

The biodiversity assessment part of our overall objectives has been postponed so as to concentrate the available resources on the plant collection aspects of the project, since this was deemed to be of a higher immediate priority for the initial years of the project. It is planned to complete a preliminary survey of Suriname's biodiversity over the next two years, however.

PLANT EXTRACTION

Plant extraction is carried out by BGVS in Paramaribo, the capital of Suriname. Extracts are prepared by standard procedures using ethyl acetate followed by methanol, so that two extracts are prepared for each sample. Extracts are evaporated on a rotary evaporator and transferred to storage vials from which samples are transferred to shipping vials for shipment to VPISU and to BMS. Shipment is made by courier service to minimize shipping delays, and samples are coded prior to shipment to maintain confidentiality of the extracts. A total of 2688 extracts has been prepared and shipped to date.

In addition to making plant extracts, BGVS personnel have received training at VPISU in bioassay techniques, and are beginning a separate program of isolation chemistry with the aim of isolating compounds with antimicrobial activity.

SCREENING

Screening for bioactive constituents is carried out both at VPISU and at BMS. Screening at BMS is done with a battery of automated mechanism-based screens; these screens are proprietary and are also undergoing constant change, so no detailed descriptions can be given here, but they cover the major disease areas of interest to BMS, including the anticancer, antiviral, antiinfective, cardiovascular, and CNS areas.

The screens at VPISU are currently all yeast-based. The most reliable screen has been one using a yeast strain designated Sc-7. This strain, originally developed at Bristol-Banyu in Japan, responds preferentially to cytotoxic compound and thus serves as an initial indicator of cytotoxicity or potential cytotoxicity.

One consequence of the fact that the BMS screens change rapidly was the conclusion that any comparison of botanical vs ethnobotanical collecting strategies would necessarily be very tentative, since there could be no assurance that all samples from the two strategies had been compared against the same set of assays. In addition, the fact that the ethnobotanical collections were made on a disease by disease basis, while the screening at BMS was for mechanism-based actives, made a direct and valid test of the value of the ethnobotanical approach essentially impossible.

In spite of these limitations, it may still be possible to examine whether ethnobotanical collections might have higher concentrations of active constituents or greater numbers of active constituents, on the average, than their botanical counterparts. These data should show up as a higher frequency of hits *on the average* from the ethnobotanical collections in the BMS screens.

Regrettably, it is too early in the project to draw even this limited conclusion from the available data, since the number of confirmed and dereplicated hits in the BMS screens is too small to give a statistically significant answer. All that can be said at the present time is that there is no obvious advantage in favor of the ethnobotanical collections when evaluated in this limited way. It should be noted however, that even if this tentative conclusion is corroborated by future results, it still does not discredit ethnobotanical collections as a route for finding extracts that would be active in disease-based (e.g. whole animal) bioassays, and we are planning some direct experiments to test this hypothesis in the future.

ISOLATION OF BIOACTIVE COMPOUNDS

Work at Virginia Polytechnic Institute and State University has led to the isolation of several new and known bioactive compounds, which will be described briefly here; full details have been published elsewhere (Abdel-Kader *et al.* 1997a; 1998; Zhou *et al.* 1997).

Renealmia alpinia (Rott) Maas (Zingiberaceae)

This plant is used in traditional medicine in Suriname, and was detected as an active using the Sc-7 yeast screen. Bioassay-guided fractionation (Figure 1) led to the isolation of the three bioactive labdane diterpenoids **1–3**, of which the most active compound **1** is a new compound. The stereochemistry of **1** at the C-position was determined by reduction and derivatization of **1** to the Mosher triesters **5** and determination of their NMR spectra. The dialdehyde **6** was also formed by ozonolysis of **2**, and its CD spectrum confirmed the absolute stereochemistry shown (Figure 2). The enhanced cytotoxicity of **1** as compared with **2** and **3** suggested that the hemiacetal linkage might be important; consistent with this hypothesis, the reduced analog **4** was less active than **1**.

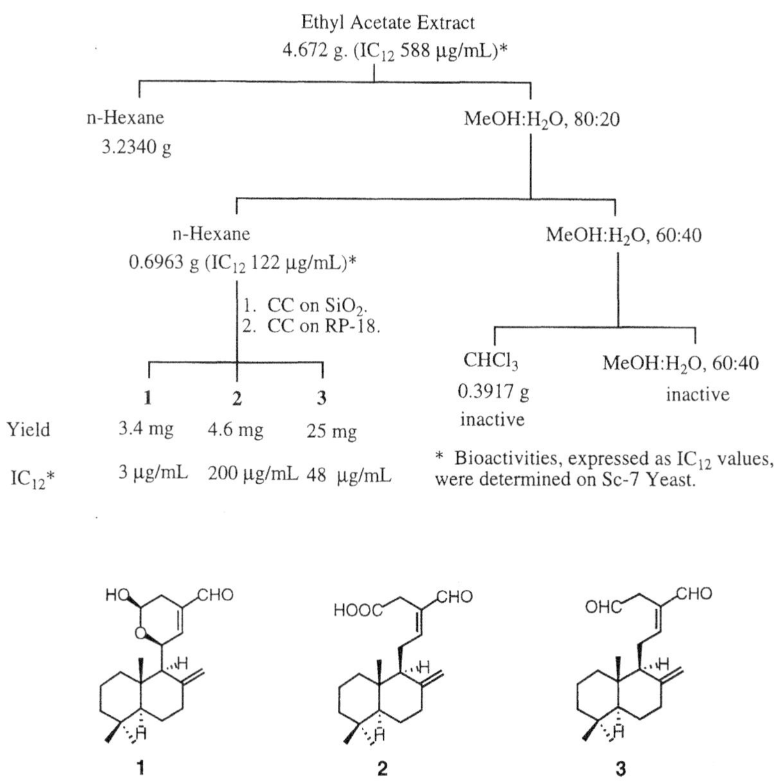

Figure 1 Isolation of bioactive constituents from *Renealmia alpinia*

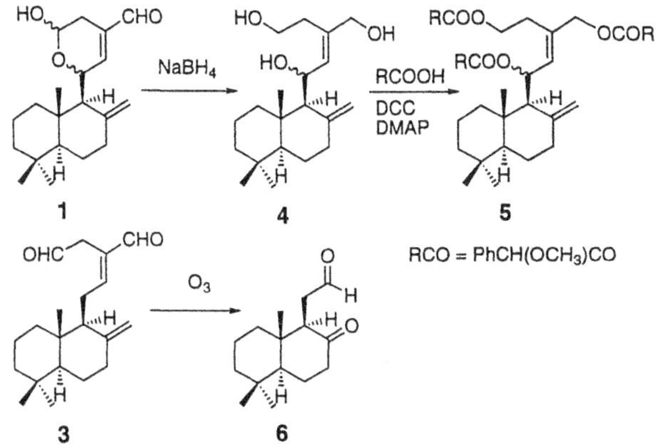

Figure 2 Stereochemical assignment of the new labdane diterpenoid

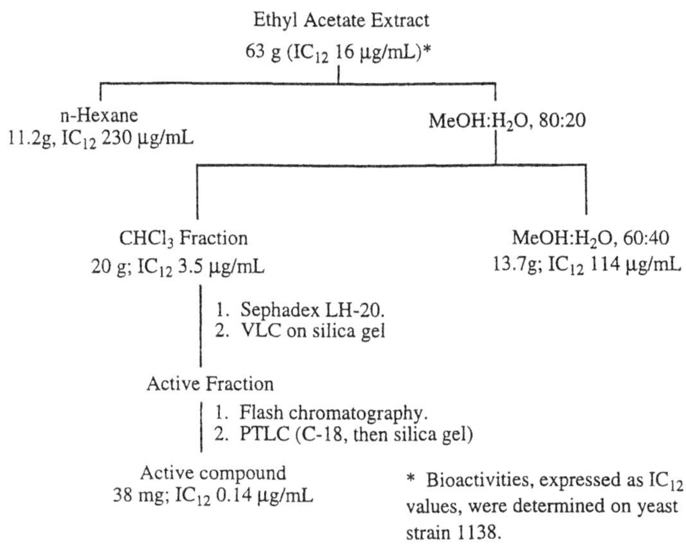

Figure 3 Isolation of bioactive constituents from *Eclipta alba*

Eclipta alba (L.) Hassk. (Asteraceae)

Bioassay-directed fractionation of this plant (Figure 3) led to the isolation of the known alkaloid verazine (**7**) as the bioactive constituent. Since there is no obvious reason other than the imino group why verazine should show a significant activity in the Sc-7 assay, we prepared a number of simple analogs containing this group (Figure 4). To date none of them has shown as much activity as verazine, suggesting that the molecule may be acting to interfere physically with the lipid bilayer in the yeast cells.

Himatanthus fallax (Muell.Arg.) Plumel and *Allamanda cathartica* L. (Apocynaceae)

Fractionation of *Allamanda cathartica* yielded the toxic iridoids plumericin (**8**) and isoplumericin (**9**) as the active constituents, while a similar fractionation of *Himatanthus*

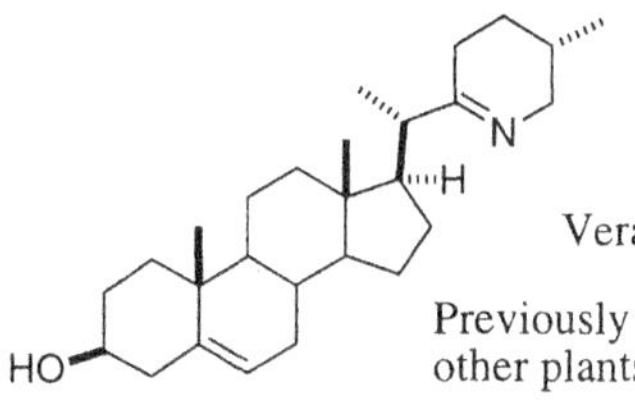

Verazine

Previously isolated from Veratrum album and other plants: Tetrahedron **1967**, *23*, 167.

7

Synthesis of analogs:

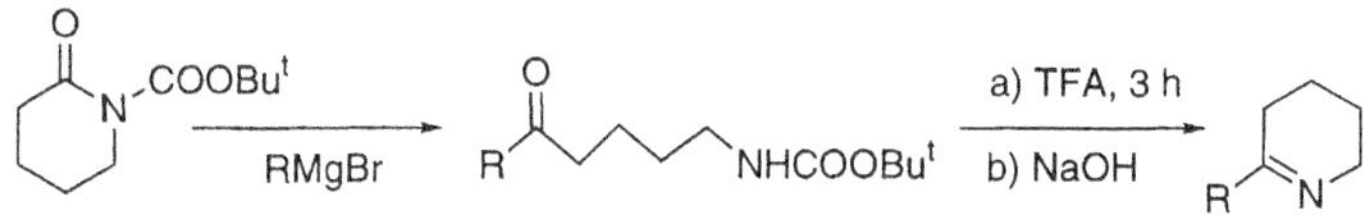

Initial indications are that simple analogs (R = Ph, Pr, p-ClC_6H_4, etc.) are not active in the yeast assay

Figure 4 Structure and analogs of the bioactive constituent of *Eclipta alba*

Ethyl Acetate Extract of *Himatanthus fallax*
16.5 g (IC_{12} 2,130 μg/mL)*

- n-Hexane
  10.6g, inactive
- MeOH:H_2O, 80:20
  - $CHCl_3$ Fraction
    3.1 g; IC_{12} 1470 mg/mL
    - Sephadex LH-20.
      Active Fraction
      1.2 g; IC_{12} ,1000 μg/mL
      - 1. VLC.
        2. PTLC (C-18, then silica gel)
        - Active compounds
          8 2.6 mg
          9 1.1 mg;
        - Inactive compound
          10
  - MeOH:H_2O, 60:40
    2.7g; inactive

* Bioactivities, expressed as IC_{12} values, were determined on Sc-7 Yeast.

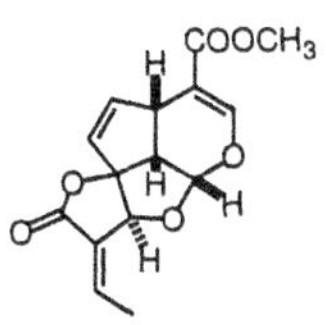

8
Plumericin
IC12 74μg/mL

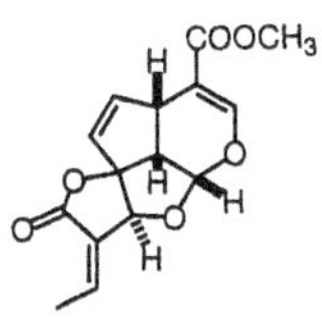

9
Isoplumericin
IC12 42μg/mL

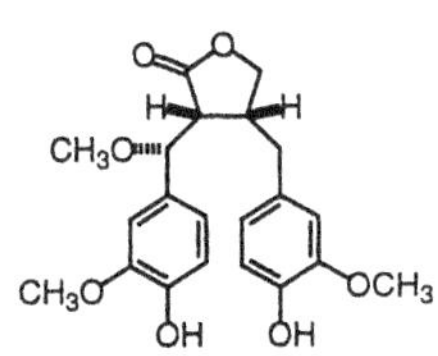

10

H. fallax and *A. cathartica*

H. fallax

Figure 5 Isolation of bioactive constituents from *Himatanthus fallax*, and compounds isolated from *Himatanthus fallax* and *Allamanda cathartica*

fallax gave the same two iridoids together with the novel but inactive lignan **10** (Figure 5). This is the first reported isolation of lignan derivatives from a *Himatanthus* species.

Miconia lepidota DC (Melastomataceae)

The plant *Miconia lepidota* yielded the alkyl benzoquinones **11** and **12** as bioactive constituents (Figure 6). Since the activity of these quinones was greatest for the one with the longest alkyl chain, we prepared some additional analogs by synthesis using a literature procedure (Konig *et al.* 1993) (Figure 7). The analog **13**, with a significantly longer

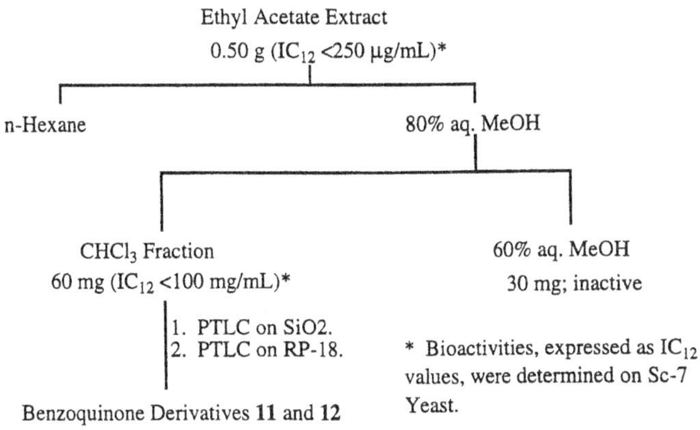

Figure 6 Isolation of bioactive constituents from *Miconia lepidota*

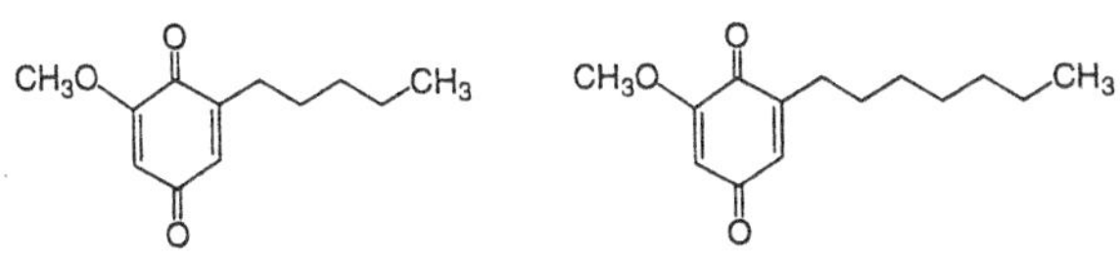

11 **12**

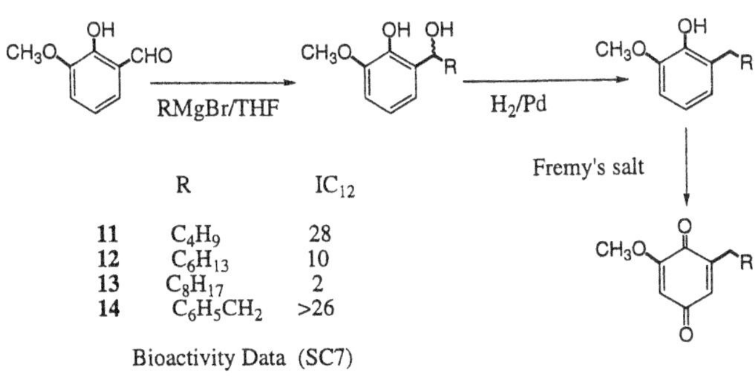

Figure 7 Structure and synthesis of substituted benzoquinones from *Miconia lepidota*

alkyl chain, was the most active compound of those tested, while the benzyl analog **14** was much less active.

Other Results

Work at Bristol-Myers Squibb Pharmaceutical Research Institute has been carried out on a range of bioactive extracts, but only a few have passed a rigorous dereplication process and entered fractionations. The extract that is furthest along is one with cytotoxic activity, and it has yielded three pure compounds with activity in the nanomolar range. The outlook is thus encouraging for finding a good lead compound or compounds from this plant.

ACCOMPLISHMENTS OF THE ICBG PROGRAM

It is appropriate at this time to summarize briefly the accomplishments of the Suriname ICBG program, so that it can be evaluated as a whole and not as isolated and unrelated parts. The discussion will cover each of the major areas of the ICBG program in turn.

Administrative and General

The Suriname ICBG program has shown that it is possible for very different organizations with different missions to work together for common goals. At the beginning of the project the members of CI and BMS, for example, had very different perspectives about several issues, but the fact of working together on this project educated each group to the other group's perspectives, and ended up enriching all the participants. This is not to say that there have been no differences of opinion, but for the most part these differences have been resolved as the group members learned to respect and trust each other.

Biodiversity Conservation

The group has made a number of significant beginnings in this area. It has created a national botanical database at the National Herbarium of Suriname; it has provided biological information to begin an assessment of Suriname's biological resources; and it has provided critically needed supplies and repairs to the National Herbarium of Suriname.

Perhaps most significant, however, is the role the project has played in providing credible alternatives to timber harvesting in Suriname. In 1993 several major Pacific Rim forestry enterprises requested logging concessions in Suriname which, if approved, would have opened up approximately 25% of the interior to logging. Although the final outcome of these requests has not yet been seen, it appears probable that the size of the concessions has been significantly reduced due in part to representations made by CI on the value of retaining sustainable utilization of the rain forest for a variety of products including pharmacological ones.

Training and Economic Development

The establishment of the Forest Peoples' Fund with a generous donation from Bristol-Myers Squibb was extremely important to the economic development aspects of the project, since it enabled numerous small-scale projects to be supported by the fund. These include such things as equipment purchases, training workshops, small-scale tropical agriculture projects, meeting travel for tribal leaders to discuss common concerns, and renovations of an ecotourism lodge.

In the training area, several Surinamese nationals have been trained in skills ranging from botanical and ethnobotanical techniques to GIS operation and bioassay methods.

Drug Discovery

In this area over 800 plant species have been collected, over 2,500 extracts prepared, and some 75,000 bioassays run. Approximately 50 active extracts have been resupplied, and nine bioactive compounds have been isolated. Most encouragingly, two of these are highly cytotoxic leads under investigation at BMS, with a reasonable probability of being of sufficient interest for further development. Four analogs of the compounds isolated at VPISU have been prepared, and one of these has shown improved activity as compared with its lead compound.

SUMMARY

In summary, the ICBG program provides a unique mechanism for combining the skills and resources of several different institutions to address the common goals of biodiversity conservation, economic development, and drug discovery. The program has been successful in raising awareness of biodiversity issues among the natural products and pharmaceutical communities in the USA, and it provides a powerful example of bioprospecting that is in harmony with the Convention on Biological Diversity. Although the program has only been in operation for a little over three years, preliminary indications are that it has been successful in its initial aims, and even better results can be anticipated as the program matures.

ACKNOWLEDGMENTS

The work described was supported by NIH grant number 5 UO1 TW/CA00313, and this support is gratefully acknowledged. It is also a pleasure to acknowledge the support and cooperation of the staff of the Fogarty International Center, and in particular the ICBG coordinator Dr. Joshua Rosenthal, as well as the U.S. Government Scientific Advisory Committee and its Chairperson, Dr. James Rodman. The collaboration of the staff of Bristol-Myers Squibb Pharmaceutical Research Institute, and especially of Dr. D. Vyas, Dr. K. Wright, and Dr. Y.-Z. Shu, is gratefully acknowledged.

REFERENCES

Abdel-Kader, M.S., Wisse, J., Evans, R., van der Werff, H. and Kingston, D.G.I. (1997a) Bioactive iridoids and a new lignan from *Allamanda cathartica* and *Himatanthus fallax* from the Suriname rain forest. *J. Nat. Prod.*, **60**, 1294–1297.

Abdel-Kader, M.S., Malone, S., Werkhoven, M.C.M., van Troon, F., David, M., Wisse, J.H. and Kingston, D.G.I. (1998) DNA damaging steroidal alkaloids from *Eclipta alba* from the Suriname rain forest. *J. Nat. Prod.*, in press.

Anonymous (1992) RFA TW-92-01, Fogarty International Center, National Institutes of Health, Bethesda, MD.

Borman, S. (1996) Combinatorial chemists focus on small molecules, molecular recognition, and automation. *Chem. Eng. News*, **74**(7), 29–54.

Costanza, R., d'Arge, R., de Groot, R., Farber, S., Grasso, M., Hannon, B., Limburg, K., Naeem, S., O'Neill, R.V., Paruelo, J., Raskin, R.G., Sutton, P. and van den Belt, M. (1997) The value of the World's ecosystem services and natural capital. *Nature*, **387**, 253–260.

Cragg, G.M., Newman, D. and Snader, K. (1997) Natural products in drug discovery and development. *J. Nat. Prod.*, **60**, 52–60.

König, W.A., Faasch, H., Heitsch, H., Colberg, C. and Hausen, B.M. (1993) Synthesis of side-chain-modified analogues of the allergen primin. *Z. Naturforsch., B.*, **48**, 387–393.

O'Neill, M.J. and Lewis, J.A. (1993) The renaissance of plant research in the pharmaceutical industry. In A.D. Kinghorn and M.F. Balandrin, (eds.), *Human Medicinal Agents from Plants*, American Chemical Society, Washington, DC; ACS Symposium Series 534, pp. 48–55.

Plunkett, M.J. and Ellman, J.A. (1997) Combinatorial chemistry and new drugs. *Scientific American*, April, 68–73.

Reid, W.V., Laird, S.A., Meyer, C.A., Gamez, R., Sittenfeld, A., Janzen, D.H., Gollin, M.A. and Juma, C. (1993) A new lease on life. In *Biodiversity Prospecting*, World Resources Institute: Washington, DC., pp. 1–52.

Rubin, S.M. and Fish, S.C. (1994) Biodiversity prospecting: using innovative contractual provisions to foster ethnobotanical knowledge, technology and conservation. *Colo. J. Intl. Envtl. L. & Policy* **5:23**, 23–58.

Zhou, B.-N., Baj, N.J., Glass, T.E., Malone, S., Werkhoven, M.C.M., van Troon, F., David, Wisse, J.H. and Kingston, D.G.I. (1997) Bioactive labdane diterpenoids from the Suriname rain forest. *J. Nat. Prod.*, **60**, 1287–1293.

7. ANTIMICROBIAL ACTIVITIES OF PHYTOCHEMICALS FROM BRITISH COLUMBIAN MEDICINAL PLANTS

G.H.N. TOWERS[1], A.R. McCUTCHEON[1], H. MATSUURA[1], J. PAGE[1], G. SAXENA[2], S. FARMER[3], E. GIBBONS[4], T.E. ROBERTS[4], L.A. BABIUK[4], L.M. THORSON[5], R.W. STOKES[5], P. SOKOL[6] and H. KLINGEMANN[7]

[1]*Department of Botany, The University of British Columbia, Vancouver, B.C., Canada, V6T 1Z4*
[2]*TerraGen Diversity Inc., Suite 300-2386 East Mall, UBC, Vancouver, B.C., V6T 1Z3*
[3]*Department of Microbiology, UBC, Vancouver, B.C., Canada, V6T 1Z3*
[4]*Veterinary Infectious Disease Organization, Saskatoon, Saskatchewan, Canada*
[5]*Infectious and Immunological Diseases, B.C. Children's Hospital, Vancouver, B.C.*
[6]*Department of Microbiology and Infectious Diseases, University of Calgary, Alberta*
[7]*Terry Fox Labs, B.C. Cancer Research Center, Vancouver, B.C., Canada*

INTRODUCTION

A checklist of the vascular plants of British Columbia (B.C.) indicates that there are about 3137 taxa including 662 introduced species (Taylor and MacBride, 1977). Many of the endemic species have never been subjected to phytochemical or pharmacological studies although they have and continue to be used as medicines by First Nations or aboriginal peoples. Through the auspices of the Canadian Bacterial Diseases Network, we have initiated a systematic study of the native flora of B.C. for antibacterial, antifungal and antiviral activities concentrating on those species whose traditional applications suggest that they have been used to treat infections.

There is an increasingly urgent need for novel antibiotic compounds as drug resistance is rapidly becoming a major obstacle in the treatment of bacterial infections. It was because of the very encouraging results of our initial antibiotic screen, that the project was subsequently expanded to include other types of infectious organisms for which new drugs are also urgently needed.

In the first phase of this research, methanolic extracts of 100 plant species were screened for antibacterial, antifungal, anti-mycobacterial and antiviral activity. For each screening, an ethnopharmacological analysis was also conducted, based on the traditional medicinal uses of these plants obtained from the ethnobotanical literature.

SCREENING FOR ANTIBACTERIALS

The methanolic plant extracts were screened against 11 bacterial strains. Eighty-nine per cent were found to have significant antibiotic activity (McCutcheon, 1992). The extracts with the broadest spectra of activity were from *Alnus rubra* (bark and catkins), *Fragaria chiloensis* (leaves), *Moneses uniflora* (aerial parts), and *Rhus glabra* (branches).

The ethnopharmacological analysis showed that 94% of the plants designated as potentially antibiotic and 77% of the plants designated as possible antibiotics, based on their traditional usage, exhibited antibiotic activity (McCutcheon *et al.*, 1996). These results suggested that the selection criterion used was effective in targeting a high percentage of plants with antibiotic activity. The work was extended to a larger group including non-medicinal plants, to further test the hypothesis.

Methanolic extracts of 185 new samples of western North American plants were screened for antibiotic activity against seven bacterial strains. Overall, 85% of the extracts exhibited activity. Out of the 158 active extracts, 50 extracts showed strong broad spectrum activity (active against a minimum of 5 bacterial species) which was equal to or better than the performance of the positive control, gentamicin. Extracts from nine plant species were active against all seven bacterial species: *Abies grandis*, *Elliottia pyroliflorus*, *Geum triflorum*, *Horkelia fusca*, *Paxistima myrinites*, *Paeonia brownii*, *Phyllodoce empetriformis*, *Picea sitchensis* and *Pseudotsuga menzesii*.

There was a significant difference in the degree of antibiotic activity between the traditional plant medicines (83% active) and the non-medicinal plants (22% active). Ninety-one percent (91%) of the plants classified as potential antibiotics based on their traditional usage were found to have antibiotic activity. The taxa with the highest percentage of active plants were the Filinicae (ferns) and Gymnospermae (conifers) of which 100% were active, followed by the Angiospermae (flowering plants) of which 89% were active. The analysis of antibiotic activity by plant habitat also provided some intriguing results as the plants collected from saltwater and those collected from freshwater wetlands were found to exhibit the lowest percentage of antibiotic activity, 17% and 68% respectively, while 100% of the plants collected in sub-alpine regions exhibited activity.

The results of these screenings not only provided many new leads in the search for novel antibiotic compounds but also gave valuable insights on ways to improve the design of future screenings. The ethnopharmacological analyses results support one of the underlying assumptions of ethnopharmacology, namely the inherent value of focusing investigations on traditional medicines. Furthermore, the data also support the hypothesis that this information may be effectively used to select species with a high probability of activity. From a cultural perspective, the results of the ethnopharmacological analysis provide scientific evidence of the potential efficacy of North American traditional herbal medicines.

SCREENING FOR ANTIFUNGALS

Human pathogenic fungi are becoming an increasing public-health hazard (Sternberg, 1994), as increasing numbers of immunosuppressed and immunocompromised people are stricken with life threatening fungal infections. A lack of antifungal drugs, increasing fungal resistance, the growing list of fungal pathogens and lagging research have been cited as factors contributing to the emergence of fungi as serious pathogens

(Sternberg, 1994). In their recent review of human mycoses, Georgopapadakou and Walsh (1994) concluded that new approaches and new drugs are urgently needed. Exploring traditional treatments for fungal infections provides a route to the discovery of such drugs.

The methanolic plant extracts were screened against nine fungal species. Eighty-one were found to have antifungal activity and 30 extracts showed activity against four or more of the fungi assayed (McCutcheon, 1994). All of the ferns and gymnosperms assayed exhibited antifungal activity. The extracts with the greatest fungal inhibition were from *Alnus rubra* (catkins), *Artemisia ludoviciana* (aerial parts), *A. tridentata* (aerial parts), *Geum macrophyllum* (roots), *Mahonia aquifolium* (roots), and *Moneses uniflora* (aerial parts). In addition to these, extracts of the following also exhibited antifungal activity against all nine fungi: *Balsamorhiza sagittata* (roots), *Empetrum nigrum* (branches), *Fragaria chiloensis* (leaves), *Glehnia littoralis* (roots), *Ipomopsis aggregata* (aerial parts and roots), and *Rhus glabra* (branches).

An ethnopharmacological analysis was made to determine how such information may best be used to guide future research projects. Given the fairly ambiguous nature of the ethnobotanical descriptions, the results far exceeded expectations. One hundred percent of the plants classified as potential antifungals based on their traditional usage exhibited activity compared to 33% of the plants with other types of medicinal usages. These results certainly suggest that it would be most profitable for future antifungal screenings to focus on plants used specifically to treat fungal infections.

With the dual objectives of testing this hypothesis and identifying more leads, a second screening of 185 new plant extracts against 7 fungi was performed. One hundred and nine extracts (59%) exhibited some antifungal activity. The taxa with the highest percentage of active plants were the Gymnospermae (conifers) of which 100% were active.

The most active broad spectrum antifungal extracts were from *Boykinia occidentalis* (whole plant), *Chimaphila umbellata* (aerial parts), *Drosera rotundifolia* (whole plant), *Epilobium angustifolium* (aerial parts and roots), *Geum triflorum* (aerial parts), *Horkelia fusca* (aerial parts), *Oenothera villosa* (aerial parts and roots), *Paeonia brownii* (root bark), *Potentilla norwegica* (whole plant), *Trillium ovatum* (whole plant), and *Woodsia scopulina* (whole plant).

Analysis of the ethnopharmacological data showed that those species which were used in traditional medicine exhibited a much greater degree of activity (35%) than did the non-medicinal plants (5%). Seventy-five percent of the plants classified as potential antifungals and 48% of those classified as possible antifungals were active, while only 28% of the plants with other medicinal uses were active. As the majority of these plant medicines are applied topically to fungal infections of the skin or mucosa, it is possible that these traditional remedies are efficacious.

SCREENING FOR ANTIMYCOBACTERIALS

Tuberculosis, an ancient disease which most people believed had been conquered by modern antibiotics, has staged a resurgence in recent years due to the emergence of multiple drug resistant strains. In western countries, the success of this resurgence has been greatly aided by the large pool of super-susceptible immunocompromised patients facilitating the dissemination of the multiple drug resistant strains. In addition to *Mycobacterium tuberculosis*, *M. avium* and *M. intracellulare* (*M. avium* complex or MAC) have emerged as important human pathogens because of the increased incidence associated with AIDS and their natural resistance to the common anti-mycobacterial drugs. In order to stave off this potentially epidemic situation, new anti-mycobacterial drugs are desperately required.

The 100 methanolic plant extracts were screened for activity against *Mycobacterium tuberculosis* and *Mycobacterium avium*. Nineteen extracts exhibited some activity against *M. tuberculosis* and 16 extracts showed some activity against *M. avium* (McCutcheon *et al.*, 1997). There was a correlation (0.945) between anti-mycobacterial activity and strong activity against the fast growing, non-pathogenic *Mycobacterium phlei* which was used in the antibiotic screening. Extracts made from *Heracleum maximum* (roots), *Moneses uniflora* (aerial parts) and *Oplopanax horridus* (inner bark) completely inhibited the growth of both organisms at a concentration equivalent to 20 mg dried plant material/disc. Extracts of *Alnus rubra* (bark and catkins), *Empetrum nigrum* (branches), *Glehnia littoralis* (roots), and *Lomatium dissectum* (roots) completely inhibited the growth of both *M. tuberculosis* and *M. avium* at a concentraton equivalent to 100 mg dried plant material/disc.

It is noteworthy that three of the extracts with the greatest activity against the mycobacteria are members of the Apiaceae, namely *G. littoralis*, *L. dissectum* and *H. maximum*. A pair of unstable tetronic acids were identified as the antimicrobial constituents of *L. dissectum* (Cardellina and Vanwagenen, 1985; Vanwagenen and Cardellina, 1986) but it is not yet known if these compounds are also responsible for this plant extract's anti-mycobacterial activity. The Apiaceae is well-known for its phototoxic furano-coumarins and these compounds may be responsible for some degree of the anti-mycobacterial activity observed in these family members.

Can the traditional usage of a plant to treat tuberculosis also be used as an effective selection criterion for anti-mycobacterial screenings? A comparison of the ethnobotanical literature and the screening results shows that 13 of the 19 active extracts (68%) were prepared from plant species which were specifically reported to have been used for the treatment of tuberculosis. Of the 100 extracts screened, a total of 37 samples were prepared from plant species which were reported to have been used to treat tuberculosis or consumption (McCutcheon *et al.*, 1997). Analyzed in this context, it may be concluded that roughly one-third of traditional tuberculosis remedies assayed did exhibit *in vitro* anti-mycobacterial activity.

These results suggest that there may be a positive correlation between traditional usage in the treatment of tuberculosis and anti-mycobacterial activity. It seems reasonable to suggest that some of these traditional tuberculosis remedies, representing the cumulative knowledge from hundreds of years of experience, may have indeed been efficacious.

ANTIVIRAL SCREENINGS

There have not been many successes in the development of new antivirals. Antiviral research has been greatly accelerated in the face of the current AIDS epidemic. While the search for anti-human immunodeficiency virus (HIV) agents has been vigorous in recent years (De Clercq, 1988), progress in the development of useful new antivirals has been painstakingly slow (Galasso, 1988). Frequencies of viral resistance to the few available drugs are increasing (De Clercq, 1993) and the problem of viral latency, the greatest obstacle to treatment of some viral infections, remains unsolved.

Ethnopharmacological screenings provide an alternative avenue to discovery from the current mainstream approach of attempting to design narrow spectrum drugs for specific molecular targets. The ethnopharmacological approach has equal potential for identifying new antiviral compounds, yet relatively few antiviral screenings of plant ethnomedicines have been conducted to date. It seems prudent, if not imperative, that researchers continue to investigate these sources before the knowledge or the plants themselves are lost.

Therefore, the 100 methanolic plant extracts from the project were screened for antiviral activity against seven viruses. Twelve extracts were found to have antiviral activity at the non-cytotoxic concentrations tested (McCutcheon *et al.*, 1995). As has been found in previous antiviral screenings (Hudson, 1990), none of the extracts exhibited broad spectrum activity. Each active extract was effective against only one of the seven viruses screened.

Three of the most active extracts in this study were members of the same plant family. The extracts made from *Rosa nutkana* and *Amelanchier alnifolia*, both members of the Rosaceae, completely inhibited the cytopathic effects of an enteric coronavirus. The extract of another member of the Rosaceae, *Potentilla arguta*, completely inhibited respiratory syncytial virus. Coronavirus and respiratory syncytial virus are similar in that they are both single-stranded RNA viruses which infect mucosal surfaces.

A branch tip extract of *Sambucus racemosa* also completely inhibited the cytopathic effects of respiratory syncytial virus while an inner bark extract of *Oplopanax horridus* exhibited partial inhibition. The extract of *Ipomopsis aggregata* completely inhibited cytopathology induced by parainfluenza virus type 3, another single-stranded RNA virus which causes respiratory disease. None of the extracts was effective against the fourth single-stranded RNA virus used in the screening, vesicular stomatitis virus.

Rotavirus is a double-stranded RNA virus that causes gastroenteritis, one of the major infectious diseases in the world today, as judged by mortality statistics (Vesikari *et al.*, 1988). The only extract which exhibited activity against this serious pathogen was *Lomatium dissectum* (roots) which completely inhibited the cytopathic effects.

In addition to these, extracts prepared from the following plants exhibited antiviral activity against herpesvirus type 1: *Cardamine angulata*, *Conocephalum conicum*, *Lysichiton americanum*, *Polypodium glycyrrhiza*, and *Verbascum thapsus*.

Given the pressing need for new antiviral agents and the inherent limitations of *in vitro* antiviral testing for such agents, the results of this screening are promising. It is possible that the elucidation of the active constituents in these plants may provide useful leads in the development of antiviral therapeutics. It is interesting to note that 10 of these 12 active plant species were traditionally used to treat what are now known as viral ailments. Eight of the active plant species have been used to treat the specific diseases or symptoms caused by the virus that they exhibited activity against.

PROTEASE AND ELASTASE SCREENINGS

Preliminary protease and elastase assays have been conducted on 18 of the phase two extracts. Three extracts (*Geum triflorum*, *Horkelia fusca* and *Picea sitchensis*) showed quite promising results in the protease assay, reducing activity by more than 98.5% compared to the control. The extract of *Geum triflorum* and an extract of *Epilobium angustifolium* completely eliminated activity in the elastase assay. The active constituent in *E. angustifolium* has been partially isolated. Further screenings and chemical isolation work is in progress.

ACTINEOPLASTIC SCREENINGS

Preliminary Con A stimulation and mixed lymphocyte stimulation screenings have been conducted on nine extracts. An extract of *Artemisia tripartita* showed promising results in both assays and further work is ongoing. Many plants traditionally used to treat cancers remain to be tested.

ETHNOPHARMACOLOGICAL AND CHEMICAL ANALYSES OF SPECIFIC PLANT SPECIES OF B.C.

Alnus rubra Bong. (Betulaceae) Red Alder

A bark decoction is drunk in Spring to purify the blood. It is used throughout the year as a stomach tonic and treatment for tuberculosis and other respiratory ailments. Externally, it is also used as a wash for skin infections and wounds (Pojar and MacKinnon, 1994). The bark is also used in poultices for sore eyes, as a wash for scabby skin or allergic rashes (Turner and Hebda, 1990). Catkins are used in poultices for toothache (Turner and Hebda, 1990). Other uses include treatment of children's mouth sores and as a purgative (Turner, 1973), for asthma and eczema (Turner and Bell, 1973) and for internal injuries, broken bones, and bruises (Turner *et al.*, 1983).

Although a variety of compounds have reported from species of *Alnus* (Lee *et al.* 1992) these have not been examined for their antimicrobial activity. Our bioassay-guided fractionation resulted in the isolation of a diarylheptanoid (**1**) and its glycoside (**2**) as antimicrobial components. Positive activity was displayed against *Staphylococcus aureus* (MIC 31.2 μg/ml) and *Pseudomonas aeruginosa* (MIC 125 μg/ml) (Saxena *et al.*, 1995).

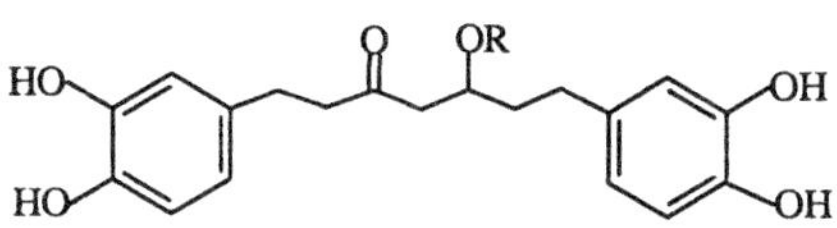

1 R = H; (5*S*)-1,7-bis(3,4-dihydroxyphenyl)-5-hydroxyheptane-3-one
2 R = -β-D-xylopyranose; Diarylheptanoid xyloside (oregonin)

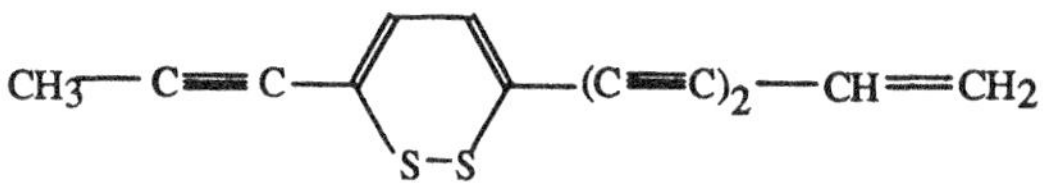

3 Thiarubrine A

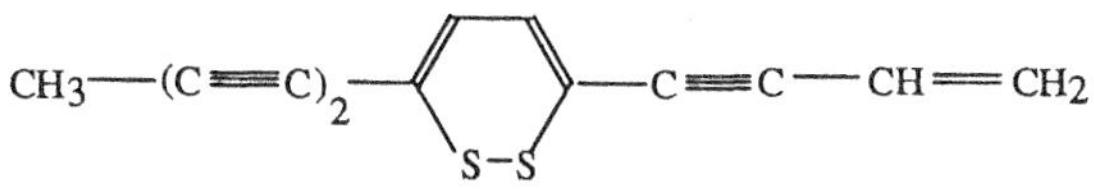

4 Thiarubrine B

Ambrosia chamissonis (Less.) Greene var. *chamissonis* (Asteraceae) Ragweed

Leaves are used in the treatment of rheumatism and internal pains (Turner, unpublished Haida mss.) and for the treatment of skin eruptions, infected toes, scalp disease, sores and to prevent blood poisoning (Moerman, 1986). Our investigations have shown that the roots and, to a lesser extent, the leaves contain thiarubrines including thiarubrine A (**3**) and thiarubrine B (**4**) (Balza *et al.*, 1989), an unusual group of red dithiacyclohexadiene polyyne pigments with pronounced antifungal and antibacterial properties (Towers *et al.*, 1985). Thiarubrines exhibit sensitivity to light, and are found to co-occur in *A. chamissonis* with their thiophene photoproducts. Thiarubrine-derived thiophenes are photosensitizers that exhibit UV-mediated antimicrobial activity (Constabel and Towers, 1989).

Balsamorhiza sagittata (Pursh) Nutt. (Asteraceae) – Arrow Leaf Balsamorhiza, Spring Sunflower

Powdered leaves are applied directly to severe burns to ease the pain and promote healing. A root decoction is drunk for rheumatism. Powdered root applied directly on burns and sores. A root poultice is applied to bruised or painful areas, and on sores (Turner *et al.*, 1980). Root decoction drunk for tuberculosis, whooping cough and to increase urinary flow (Hart, 1974). Other reported uses include fevers, sore throat, consumption, venereal disease, syphilitic sores and wounds (Moerman, 1986). The seeds are eaten as a remedy for dysentery (Turner *et al.*, 1990).

Flavonoids (Bohm *et al.*, 1989), guaianolides, heliangolides, diterpenes and cycloartenol derivatives (Bohlmann *et al.*, 1985) have been reported from this species, although research from an ethnobotanical view point has not been carried out so far. Our search for antimicrobial compounds resulted in the isolation of 7,10-epithio-7,9-tridecadiene-3,5,11-triyne-1,2-diol (**5**) for the first time from *Balsamorhiza sagittata* (Matsuura *et al.*, 1996a) although this compound was previously isolated from another Asteraceae, *Ambrosia chamissonis* (Balza *et al.*, 1989). Compound **5** was found active against Gram-positive bacteria viz., *Staphylococcus aureus* methicillin resistant strain and *Bacillus subtilis* at an MIC of 100 μg/ml and 50–100 μg/ml, respectively. The MIC decreased to 25 μg/ml when this compound was exposed to UV-A light.

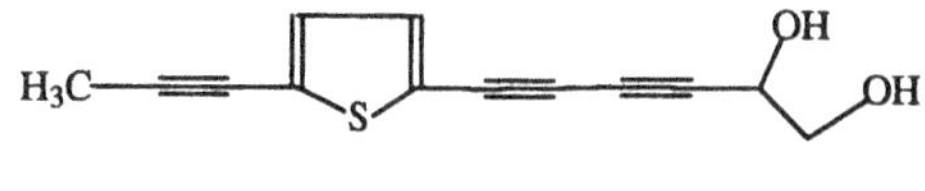

5 7,10-epithio-7,9-tridecadiene-3,5,11-triyne-1,2-diol

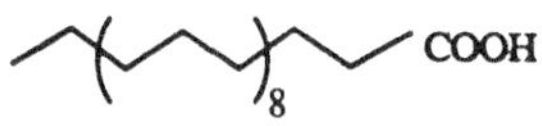

6 Octacosanoic acid

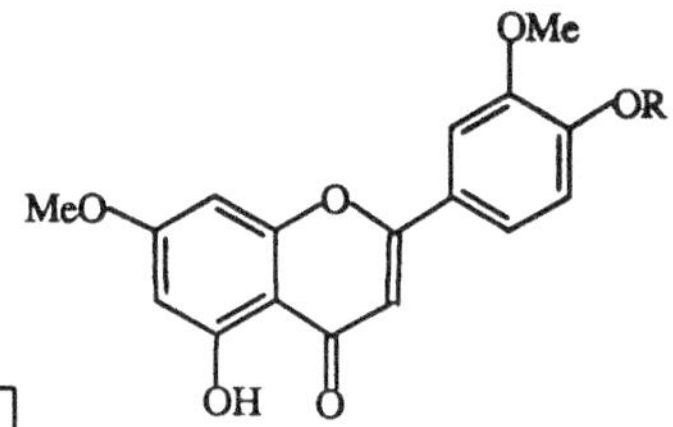

7 R = H; Velutin
8 R = Me; 4'-methylvelutin

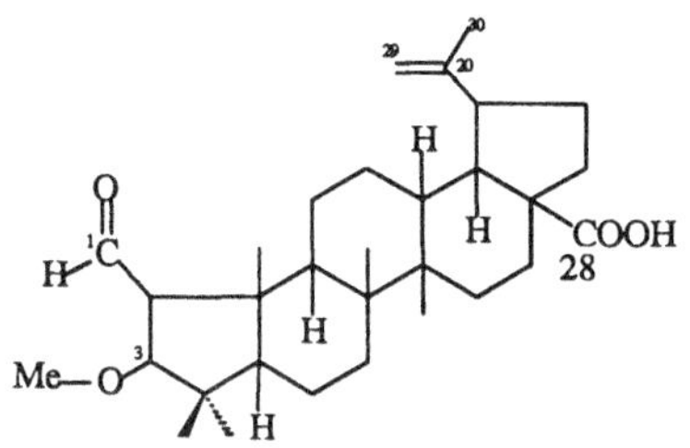

9 2-formyl-3-methoxy-A(1)-norlup-20(29)-en-28-oic acid

Ceanothus velutinus Dougl. ex Hook (Rhamnaceae) Snow-brush, Squaw Tea

Infusions are used to treat sores and eczema; a dried leaf poultice is applied to sores and a tea taken to help mend broken bones (Turner *et al.*, 1980). Branch decoctions are used as a wash for rheumatism, aching muscles or drunk as a treatment for gonorrhea, cancer and diarrhea (Turner *et al.*, 1990). The Kootenay made the leaves into a tea which was said to be good for tuberculosis (Pojar and MacKinnon, 1994). Uses specific to its antifungal activity include treatment of dandruff, prevent diaper rash, as a skin conditioner and baby powder (Turner *et al.*, 1980).

Nonacosane, 1-hexacosanol, velutin, cinnamic acid, ceanothenic acid, betulinic acid and β-sitosterol have been identified from this species (Craig *et al.*, 1971). The major antibacterial compounds which we isolated are octacosanoic acid (**6**), 4′,5-dihydroxy-3′,7-dimethoxyflavone (velutin) (**7**), 5-hydroxy-3′,4′,7′-trimethoxyflavone (4′-*O*-methyl velutin) (**8**), 2-formyl-3-methoxy-A(1)-norlup-20(29)-en-28-oic acid (**9**) (unpublished), a new compound with strong activity against *Bacillus subtilis* at an MIC of 25 µg/ml. With this zizyberanolic acid, acetyl zizyberanolic acid (Das *et al.*, 1970) was also isolated and shown to be active against *B. subtilis* and *S. aureus*.

Chaenactis douglasii (Hook.) H. & A. (Asteraceae) – False Yarrow, Hoary Chaenactis

A preventative medicine "to avoid contracting consumption" was made by boiling the roots; this tea was drunk by the surviving members after a death in the family

(Turner *et al.*, 1980). A decoction of the entire plant was drunk as a remedy for any kind of swellings, although some said it should not be applied to external boils or tumors. A milder decoction was drunk as a tonic for the stomach and for lassitude. A stronger decoction was used as a wash for chapped or cracked hands, pimples, or any skin eruptions (Turner *et al.*, 1990).

We have found high amounts of the antimicrobial polyynes thiarubrines A (**3**) and B (**4**) in the roots of *Chaenactis douglasii* and *Eriophyllum lanatum* (Norton *et al.*, 1985). These compounds are particularly active against the yeast *Candida albicans*, an agent responsible for candidiasis.

Empetrum nigrum L. (Empetraceae) Crowberry

Branch decoctions are taken for tuberculosis and other ailments (Turner, unpublished Haida mss.). Used as a diuretic, especially for children with fevers (Leighton, 1985), for diarrhea, stomach problems, kidney trouble and to heal sore eyes (Kari, 1987).

We isolated four antibacterial compounds, viz., 3,3′-dihydroxy-5-methoxybibenzyl (**10**: Batatasin III), 3,3′-dihydroxy-4′-5-dimethoxybibenzyl (**11**), 3′-*O*-methylbatatasin III (**12**) and 7-hydroxy-2,4-dimethoxy-9,10-dihydrophenanthrene (**13**) (Matsuura *et al.*, 1995a). Although these compounds were reported from several plant species e.g. Sachdev and Kulshreshtha (1986), they have not been reported from *E. nigrum* before.

Glehnia littoralis F. Schmidt ssp. *leiocarpa* (Mathias) Hult. (Apiaceae) Beach Carrot

The roots of this perennial herb, growing on sandy beaches of Pacific countries, is described in the Chinese and Japanese Pharmacopoeias as a diaphoretic, antipyretic and

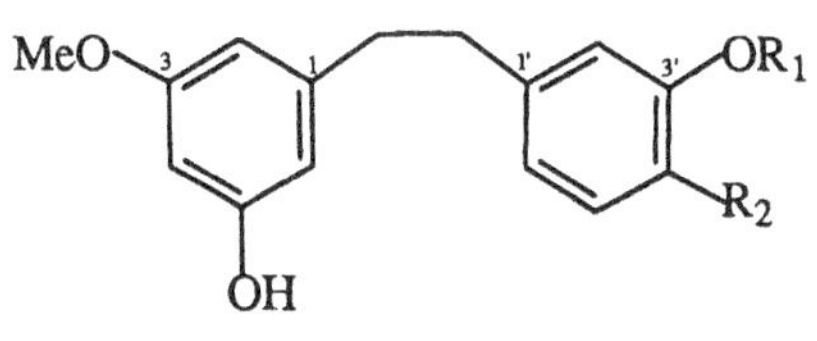

10 $R_1 = R_2 = H$; 3,3'-dihydroxy-5-methoxybibenzyl (Batatasin III)
11 $R_1 = H$, $R_2 = OMe$; 3,3'-dihydroxy-4',5-dimethoxybibenzyl
12 $R_1 = Me$, $R_2 = H$; 3'-*O*-methylbatasin III

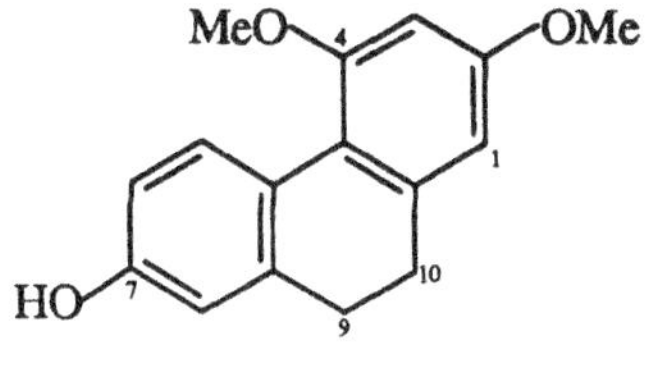

13 7-hydroxy-2,4-dimethoxy-9,10-dihydrophenanthrene

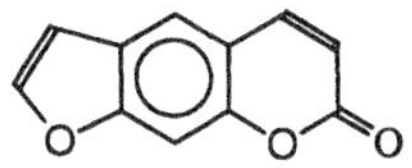

14 Psoralen

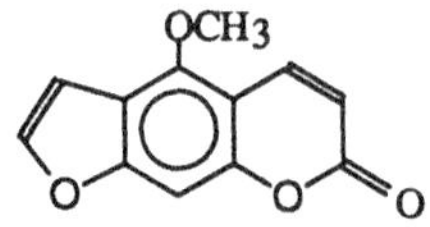

15 Bergapten

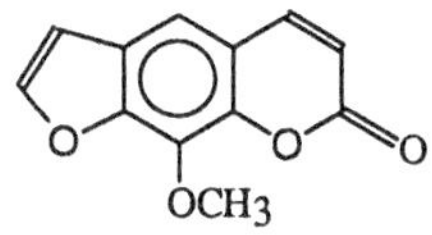

16 Xanthotoxin

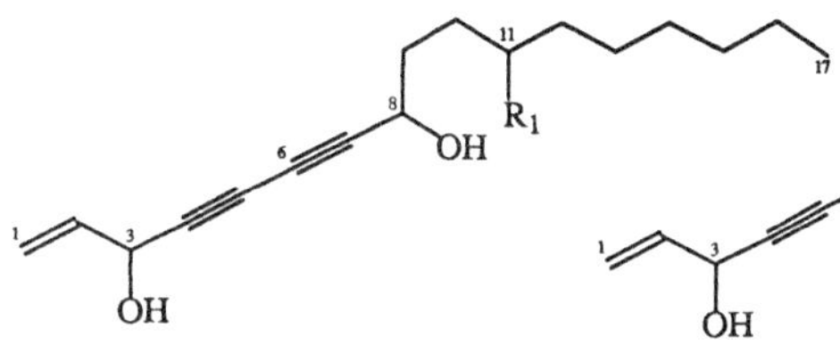

17 R_1 = H; Falcarindiol

18 R_1 = OH; (9*Z*)1,9-heptadecadiene-4,6-diyne-3,8,11-triol

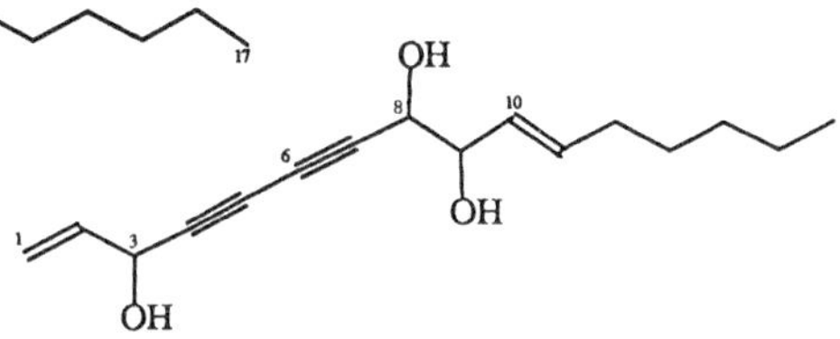

19 (10*E*) 1,10-heptadecadiene-4,6-diyne-3,8,9-triol

analgesic medicine for bladder infections. The Haida people of B.C. used the roots of this herb to treat bladder infections and other ailments (Turner, personal communication). We isolated six active compounds by activity guided fractionation using a TLC-overlay bioassay technique (Saxena *et al.*, 1995b). These were psoralen (**14**), bergapten (**15**), xanthotoxin (**16**), falcarindiol (**17**), (9*Z*)-1,9-heptadecadiene-4,6-diyne-3,8,11-triol (**18**) and (10*E*)-1,10-heptadecadiene-4,6-diyne-3,8,9-triol (**19**). Polyines **18** and **19** are novel compounds with antimicrobial MICs around 200–400 μg/ml against *S. aureus* methicillin sensitive, *S. aureus* methicillin resistant, *P. aeruginosa*, *E. coli*, *B. subtilis* and *C. albicans*. Psoralen (**14**), bergapten (**15**) and xanthotoxin (**16**) are photosensitizers and did not show growth inhibitory effects on incubation in dark. Falcarindiol (**17**) which has been previously reported as an antibiotic by Muir *et al.* (1982), showed a growth inhibitory effect against *E. coli* DC2 (MIC 2.5 μg/ml), *S. aureus* methicillin sensitive (MIC 12.5 μg/ml), *S. aureus* methicillin resistant (MIC 25.0 μg/ml) and *B. subtilis* (MIC 12.5 μg/ml) (Matsuura *et al.*, 1996b).

Ipomopsis aggregata (Pursh) Grant var. *aggregata* (Polemoniaceae) (*Cantua aggregata* Pursh, *Gilia aggregata* (Pursh) Spreng.) Scarlet Gilia, Sky-rocket

This is the most wide-ranging species of the genus distributed along the whole west coast, from British Columbia to Mexico. A root infusion of *Ipomopsis aggregata* was drunk for high fever by the Okanagan-Colville people. An infusion of the aerial parts

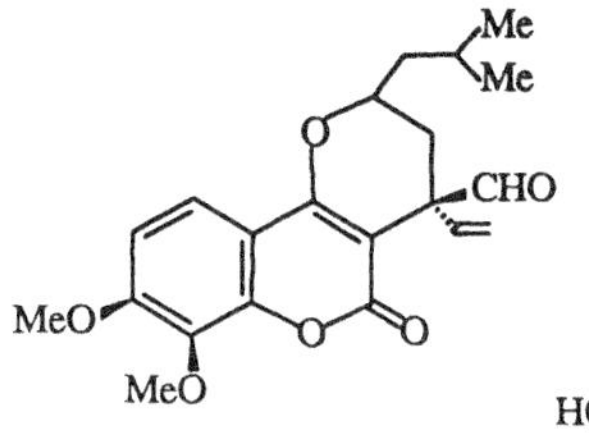

20 Giliacoumarin

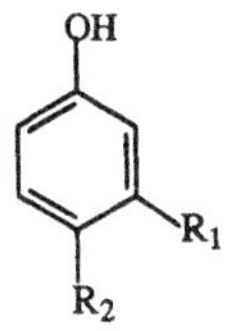

21 R_1 = OH, R_2 = H; Resorcinol
22 R1 = H, R2 = -O-glucose; Hydroquinone glucoside

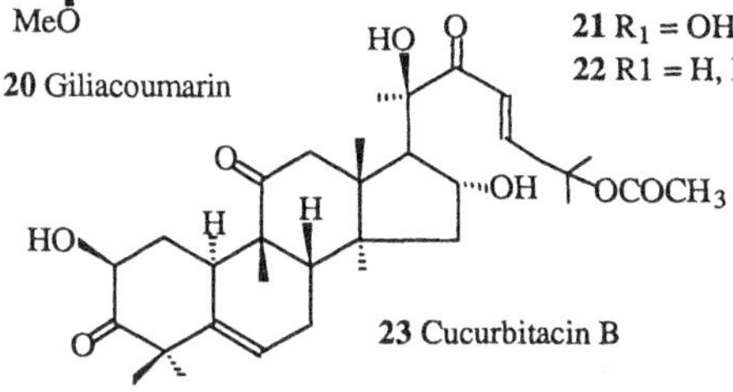

23 Cucurbitacin B

of *I. aggregata* was used to "clean out your system" and as a laxative (Turner *et al.*, 1980). Moerman (1986) cited reports of the Navaho, Paiute, and Shoshone also using *I. aggregata* for colds, stomach disease, gonorrhoea, syphilis, rheumatism, and as a cathartic, emetic, physic and blood tonic.

Methanolic extracts showed good antibacterial and antifungal activity (McCutcheon *et al.*, 1992; 1994). The root extract also strongly inhibited bovine parainfluenza virus type 3 (McCutcheon, 1995). Six tumor inhibiting compounds (three cucurbitacins, ipomopsin, hydroquinone and schottenol glycoside) had previously been identified from *I. aggregata* (Arisawa *et al.*, 1984; 1985).

We isolated four antimicrobial compounds (**20–23**) from *I. aggregata*, the most active of which was a novel coumarin which we named giliacoumarin (Saxena *et al.*, 1996a). Its MIC against *S. aureus* was 25 μg/ml and 200 μg/ml against *E. coli*, *P. aeruginosa*, *B. subtilis* and *C. albicans*. Resorcinol (**21**), hydroquinone glucoside (**22**) and cucurbitacin B (**23**) displayed MICs ranging from 200–800 μg/ml. It has not been determined whether any of these compounds are also responsible for the antiviral activity exhibited by the crude extract.

Moneses uniflora (L.) A. Gray (Ericaceae) Single Delight

Used to draw pus from boils and abscesses, as a cancer treatment (Turner and Bell, 1973) and for sore throat (Compton, 1993; Kari, 1987). A decoction of vegetative parts drunk for cancer (Turner, unpublished Haida mss.). It is also used to alleviate cough and colds, reduce pain and cure cancer and paralysis (Turner *et al.*, 1980; Arnason *et al.*, 1981).

Extracts of *Moneses* exhibited good antibacterial, antifungal and antiviral activities. A new naphthoquinone, chlorochimaphilin (**24**), along with chimaphilin (**25**) and hydroxy-chimaphilin (**26**) were isolated in addition to two naphthalene derivatives viz., 2,7-dimethyl-1,3-dihydroxynaphthyl 4-*O*-α-L-rhamnopyranoside (**27**) and 2,7-dimethoxy-1,4,8-trihydroxy naphthalene (**28**) (Saxena *et al.*, 1996b). Chimaphilin (**25**), a well-known naphthoquinone, has long been known to have antifungal, antibacterial, antihemorrhagic

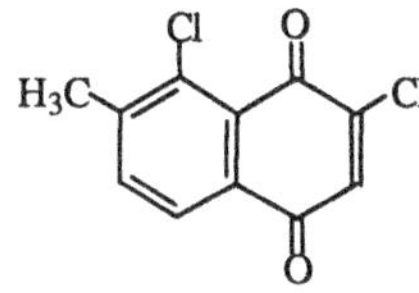

24 8-chlorochimaphilin

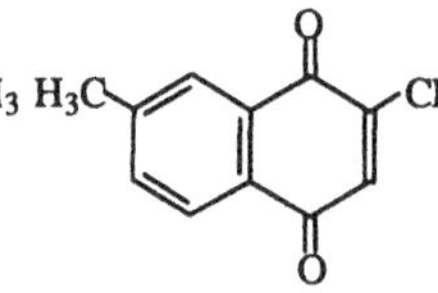

25 Chimaphilin

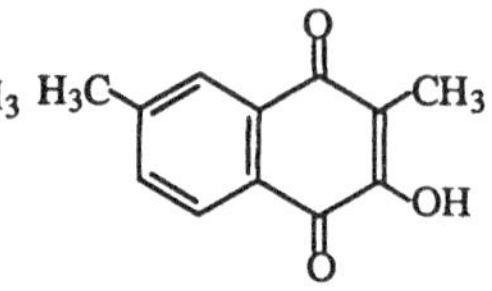

26 3-hydroxychimaphilin

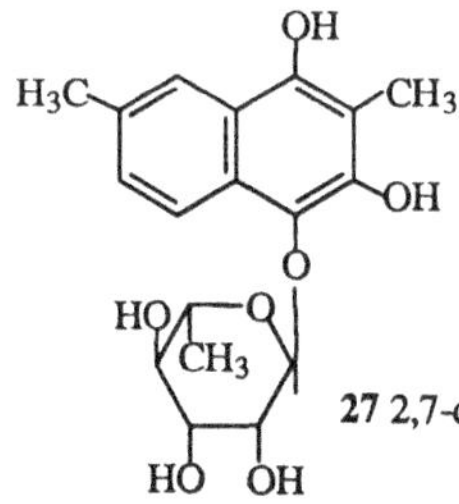

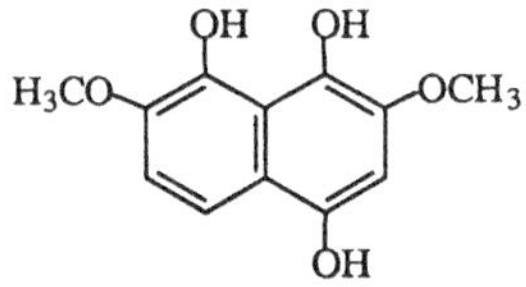

28 2,7-dimethoxy-1,4,8-trihydroxy naphthalene

27 2,7-dimethyl-1,3-dihydroxynaphthyl 4-O- α-L-rhamnopyranoside

from *Moneses uniflora*

and vitamin K-like activities (Hausen and Schiedermair, 1988) and to display anti-inflammatory and analgesic properties (Kosuge *et al.*, 1985). Compounds 24–26 were active against Gram-positive and Gram-negative bacteria as well as against variety of fungi while compounds 27 and 28 were inactive. The MIC of chloro-chimaphilin was 12.5 μg/ml against *S. aureus*, *B. subtilis* and *Fusarium tricuictum*, respectively and 50 μg/ml against *C. albicans*. Against *Microsporum gypseum*, chloro-chimaphilin showed activity at 3.1 μg/ml. It is known that naphthoquinone derivatives, in contrast to naphthalene derivatives, can act as inhibitors of bacterial growth by participating competitively in electron transport, with synthetic naphthoquinone derivatives (Amborgi *et al.*, 1970). Crude extracts as well as compounds 24, 25, and 26 were inactive against *Herpes simplex virus-1* and *Respiratory syncytial virus*. Chimaphilin was originally isolated from the related genus, *Chimaphila*. According to Sheth *et al.* (1967) the Wasco people of the Warm Springs Reservation of Oregon as well as the Sahaptin-speaking aboriginal people were using decoctions of the whole plant of *Chimaphila umbellata* for "Consumption" or "tuberculosis". We tested the activity of compounds 24, 25, and 26 against *Mycobacterium tuberculosis* 107 and *Mycobacterium avium* 724 and found that there was no inhibition at a concentration of 100 μg/disk (Saxena *et al.*, 1996b).

Oplopanax horridus Miq. (Araliaceae) (*Echinopanax horridus* Decne et Planch., *Fatsia horrida* Benth. et Hook.) Devil's Club

O. horridus was a very important spiritual and physical medicine among the aboriginal peoples of western North America. The extensive traditional uses of Devil's Club were reviewed by Turner (1982). The inner bark was used fresh, dried, or ashed in a wide variety of preparations. For external applications, it was applied directly, as a salve or as a poultice for burns, swellings, sores, boils, swollen glands and sore areas. Poultices were also applied to wounds to prevent blood poisoning and to broken bones, toothaches and abrasions to relieve the pain and swelling. Devil's Club was used in steambaths for arthritis, rheumatism, body pain, lameness, stomach trouble and as a skin tonic (Turner, 1982).

For internal use, the inner bark was either chewed and the juice swallowed or a decoction or infusion was drunk. One of the primary medicinal uses was in the treatment of respiratory ailments such as colds, coughs, chest pains, blood spitting and tuberculosis. Devil's Club was also used internally to treat a variety of diseases including: influenza, measles, diabetes, cancer, arthritis and rheumatism. It was often taken as a tonic or blood purifier for any general sickness and especially those with fever, weight loss and general debility. In higher doses, it was used as a purgative, emetic and cathartic (Turner, 1982).

An inner bark infusion was also drunk for stomach troubles such as stomach pain, ulcers, gallstones, indigestion and constipation. It was commonly used to relieve internal pains caused by ailments as varied as broken bones to black eyes. Women drank a decoction to help expel the afterbirth, to start menstruation post partum and to treat uterine cancers (Turner, 1982; Turner and Bell, 1973; Moerman, 1986; Turner and Hebda, 1990). Kari also found that Kenai Tanainas of Alaska were using the inner bark tea of this plant for the treatment of tuberculosis and for other ailments too (Kari, 1977). An analysis of the 33 ethnobotanical references which we have uncovered on the uses of Devil's Club shows that the treatment of tuberculosis is the tradional use most frequently cited (McCutcheon, 1996a) followed by its use as a tonic, for colds and for other pulmonary complaints.

The inner bark extract showed activity against respiratory syncytial virus (McCutcheon *et al.*, 1995). It also exhibited good antibacterial and antifungal activity (McCutcheon *et al.*, 1992; 1994). The essential oil of the related species *Oplopanax elatus* has also been reported to have antifungal activity (Mi *et al.*, 1987). Our results showed positive activity against Gram-positive and Gram-negative bacteria and fungi as well as anti-*Mycobacterium* activity (McCutcheon *et al.*, 1992; 1995; 1997).

We isolated two novel polyynes viz., 1,2-dihydro-9-heptadecene-4,6-diyne-3,8-diol (**31**) and 17,18-dihydro-9-octadecene-12,14-diyne-1,11,16-triol-1-acetate (**33**) with three

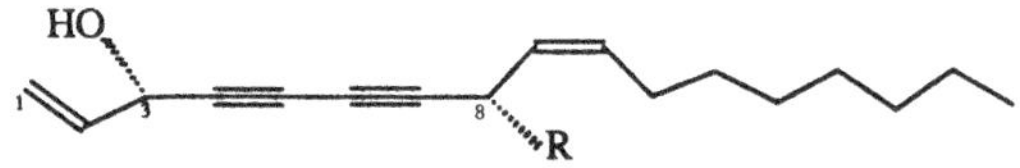

29 R = H; 9-heptadecene-4,6-diyne-3-ol
30 R = OH; 9-heptadecene-4,6-diyne-3,8-diol
31 R = OH; 1,2 dihydro; 1,2-dihydro-9-heptadecene-4,6-diyne-3,8-diol

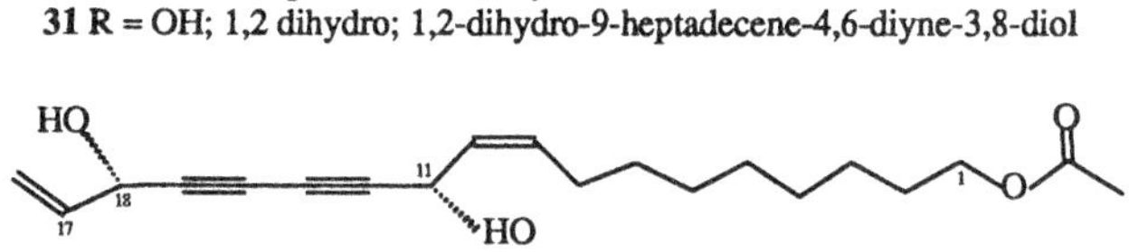

32 9,17-octadecadiene-12,14-diyne-1,11,16-triol-1-acetate
33 17,18 dihydro; 17,18-dihydro-9-octadecene-12,14-diyne-1,11,16-triol-1-acetate

known polyynes (**29**, **30** and **32**) with antimycobacterial and antifungal activity which were reported for the first time from this plant. Compound **29** was found most active against *S. aureus and B. subtilis* with an of MIC 3.1 μg/ml while against *P. aeruginosa* the MIC was 12.5 μg/ml. Against *E. coli* and *C. albicans* the MIC was 6.25 μg/ml. In comparison to **29**, compound **30** and **32** had higher MIC values. The novel compound **31** was active at a level of 6.25 μg/ml and 12.5 μg/ml against *S. aureus*, *B. subtilis*, *E. coli* and *C. albicans*, respectively while against *P. aeruginosa* it was 50.0 μg/ml. The activity of compound **33** was poor. All the isolated compounds, **29–33** were active against *Mycobacterium tuberculosis* and *Mycobacterium avium* at 100 μg/disc. Complete inhibition was noticed at lower concentrations with compound **29** and **33** showing activity at 10 μg/disc (Kobaisy *et al.*, 1997). Our finding for anti-mycobacterial activity supports the traditional usage of *O. horridus* to treat tuberculosis.

Rhus glabra L. (Anacardiaceae) – Smooth Sumac

The genus *Rhus* is most renowned for its species which elicit allergic eczematous contact dermatitis. *Rhus glabra* was used as a medicinal plant by First Nations peoples throughout North America and is reputed to be the most potent of five *Rhus* species which were used (Burgess, 1881). The traditional uses of *Rhus glabra* by First Nations included the treatment of bacterial diseases, such as syphilis, gonorrhea, dysentery and gangrene (Erichsen-Brown, 1989). Branches were used for gonorrhea, kidney trouble, bowel trouble, tonsilitis, sores and to stop children's bedwetting (Banks 1953). Decoctions were used for pyorrhea, canker sores, ease aching teeth, sore throat gargle and to ease colds and coughs (Weslager, 1973).

Amongst the Okanagan-Colville people of British Columbia, sumac fruit decoctions were drunk during childbirth. It was also drunk or used as a wash for persons suffering from frost-bitten limbs and for itchy scalp. A bark decoction was drunk and applied externally in a poultice for a tight chest. The milky latex was used directly as a salve for sores and poison ivy rashes while the leaves were chewed and held in the mouth to heal sore gums and lips (Turner *et al.*, 1980).

Amongst the Thompson people, a sumac stem decoction was reputed to be a powerful remedy for syphilis but thought to be a dangerous poison if made too strong. The root was chewed for sore mouth or tongue and a decoction drunk for stomach ulcers and internal wounds which would not heal (Turner *et al.*, 1990).

First Nations peoples in the neighboring Washington state used sumac for eye problems, tuberculosis, venereal disease and for kidney ailments (Hunn, 1990).

We isolated the methylester of 3,4,5-trihydroxy benzoic acid (**34**), 4-methoxy-3,5-dihydroxy benzoic acid (**35**) and gallic acid (**36**) (Saxena *et al.*, 1994) as well as 3-oxo-oleanolic acid (**37**) and oleanolic acid (**38**) (unpublished). Compounds **37** and **38**, which have not been isolated previously from *Rhus glabra*, showed activity against *Pseudomonas aeruginosa* (MIC 12.5 and 22 μg/ml respectively) and *Escherichia coli* (25 and 50 μg/ml respectively). Compound **37** was also active against *Bacillus subtilis* (MIC 12.5 μg/ml) (unpublished).

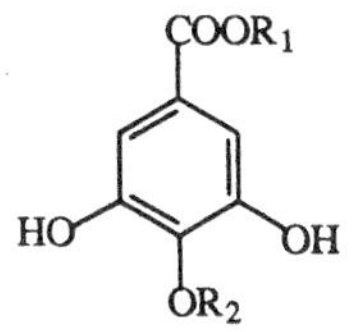

34 R_1 = Me, R_2 = H; Methyl gallate
35 R_1 = H, R_2 = Me; 4-methoxy gallic acid
36 R_1 = R_2 = H; Gallic acid

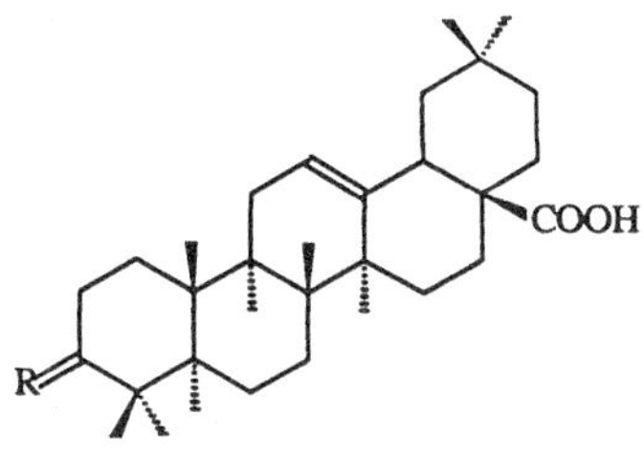

37 R = O; 3-oxo-oleanolic acid
38 R = H, OH; Oleanolic acid

CONCLUSIONS

These screenings not only validate the efficacy of First Nations traditional medicines and provide new leads for clinical therapeutics, but they also give insights into some of the factors which may influence pharmacological activity and hence the means to improve the effectiveness of future screenings. The most promising leads identified by the screenings were followed up with chemical isolation work which has been summarized above.

ACKNOWLEDGMENTS

We thank the Canadian Bacterial Diseases Network for support of this research.

REFERENCES

Ambrogi, V., Artini, D., de Carneri, I., Castellino, S., Dradi, E., Logemann, W., Meinardi, G., di Somma, M.D. and Tosolini, G. (1970) Studies on the antibacterial and antifungal properties of 1,4-naphthoquinones. *Brit. J. Pharmacol.*, **40**, 871–880.

Arnason, T., Hebda, R.J. and Johns, T. (1981) Use of plants for food and medicine by Native Peoples of eastern Canada. *Can. J. Bot.*, **59**, 2189–2325.

Arisawa, M., Kinghorn, A.D., Cordell, G.A. and Farnsworth, N.R. (1984) Ipomopsin, a new biscoumarin from *Ipomopsis aggregata*. *J. Nat. Prod.*, **47**, 106–112.

Arisawa, M., Pezzuto, J.M., Kinghorn, A.D., Cordell, G.A., Phoebe, C.H. and Farnsworth, N.R. (1985) Plant anticancer agents XXX: Cucurbitacins from *Ipomopsis aggregata* (Polemoneaceae). *J. Pharm. Sci.*, **73**, 411–413.

Balza, F., Lopez, I., Rodriguez, E. and Towers, G.H.N. (1989) Dithiacyclohexadienes and thiophenes from *Ambrosia chamissonis*. *Phytochemistry*, **28**, 3523–3524.

Banks, W.H. Jr. (1953) *Ethnobotany of the Cherokee Indians.* M.Sc. Thesis, University of Tennessee.

Bohlmann, F., Misra, L.N., Jakupovic, J., King R.M. and Robinson, H. (1985) Guaianiolides, heliangolides, diterpenes and cycloartenol derivatives from *Balsamorrhiza sagittata*. *Phytochemistry*, **24**, 2029–2036.

Bohm, B.A., Choy, J.B. and Lee, A.Y-M. (1989) Flavonoids of *Balsamorrhiza* and *Wyethia*. *Phytochemistry*, **28**, 501–503.

Burgess, T. (1881) On the beneficient toxical effects of the various species of *Rhus*. *Can. Pharmacol. J.*, **14**, 161–168.

Cardellina, J.H. and Vanwagenen, B.C. (1985) Antifungal agents from *Lomatium dissectum*. *Abstracts of the International Research Congress on Natural Products*. University of North Carolina, Chapel Hill, N.C. Abstract 211.

Compton, B. (1993) *Upper North Wakashan and Southern Tsimshian Ethnobotany*. Ph.D. Thesis, University of British Columbia.

Constabel, C.P. and G.H.N. Towers (1989) The complex nature of the mechanism of toxicity of antibiotic dithiacyclohexadiene polyines (thiarubrines) from the Asteraceae. *Planta Med.*, **55**, 35–37.

Craig, A.R., Das, K.C., Farmer, W.J., Lin,Yu-Yin, Woo,Wai-Kuan and Weinstein, B., (1971) Constituents of the leaves and rootbark of *Ceonothus velutinus*. *Phytochemistry*, **10**, 908.

Das, K.C. and Farmer,W.J., Weinstein, B. (1970) Phytochemical studies. IX. A new flavone, velutin. *J. Org. Chem.*, **35**, 3989–3990.

De Clercq, E. (1988) Recent advances in the search for selective antiviral agents. *Adv. Drug Res.*, **17**, 1–59.

De Clercq, E. (1993) Antiviral agents: characteristic activity spectrum depending on the molecular target with which they interact. *Adv. Virus Res.*, **43**, 1–55.

Erichsen-Brown, C. (1989) *Medicinal and other use of North American Plants: A historical survey with special reference to the Eastern Indian Tribes*. Dover Publications Inc., New York.

Galasso, G.J. (1988) Promises to keep: clinical use of antiviral drugs. In E. De Clercq, (ed.), *Clinical Use of Antiviral Drugs*, Martinus Nijhoff Publishing, Boston.

Georgopapadakou, N.H. and Walsh, T.J. (1994) Human mucoses: drugs and targets for emerging pathogens. *Science*, **264**, 371–371.

Hart, J. (1974) *Plant Taxonomy of the Salish and Kootenai Indians of Western Montana*. M.A. Thesis, University of Montana.

Hausen, B.M. and Schiedermair, I. (1988) The sensitizing capacity of Chimaphilin, a naturally-occuring quinone. *Contact Dermatitis*, **19**, 180–183.

Hudson, J.B. (1990) *Antiviral Compounds from Plants*, CRC Press, Boca Raton, FL.

Hunn, E.S. (1990) *Nch'i-Wana "The Big River"*, University of Washington Press, Seattle.

Kari, P.R. (1977) *Dena'ina K'et'una. Tanaina Plantlore*, Adult Literacy Labroratory, Anchorage, Alaska.

Kari, P.R. (1987) *Tanaina Plantlore. Dena'ina K'et'un, Second Edition*, National Park Service, Alaska Region, Anchorage, Alaska.

Kobaisy, M., Abramowski, Z., Farmer, S., Lermer, L., Saxena, G., Hancock, R.E.W. and Towers, G.H.N. (1997) Antimicrobial polyynes of devil's club (*Oplopanax horridus*), a North American native medicinal plant. *J. Nat. Prod.*, **60**, 1210–1213.

Kosuge, T., Yokota, M., Sugiyama, K., Mure, T., Yamazawa, H. and Yamamoto, T. (1985) Studies on Bioactive Substances in Crude Drugs Used for Arthritic Diseases in Traditional Chinese Medicine. III. Isolation and Identification of Anti-inflammatory and Analgesic Principal for the whole Herb of *Pyrola rotundifolia* L. *Chem. Pharm. Bull.*, **33**, 5355–5357.

Lee, M., Tanaka, J., Nonaka, G. and Nishiaka, I. (1992) Hirsunin, an ellagitanin with a diarylheptanoid moiety, from *Alnus hirsuta* var. *microphylla*. *Phytochemistry*, **31**, 967–970.

Leighton, A.L. (1985) *Wild Plant Use by the Woods Cree (Nihithawak) of East-Central Saskatchewan*. Canadian Ethnology Service Paper No. 101, National Museum of Man Mercury Series, National Museums of Canada, Ottawa.

MacKinnon, A., Pojar, J. and Coupe, E. (eds.) (1992) *Plants of Northern British Columbia*. Lone Pine Publishing, Vancouver, B.C.

Matsuura, H., Saxena, G., Farmer, S.W., Hancock, R.E.W. and Towers, G.H.N. (1995a) Antibacterial and antifungal compounds from *Empetrum nigrum*. *Planta Med.*, **61**, 580.

Matsuura, H., Saxena, G., Farmer, S.W., Hancock, R.E.W. and Towers, G.H.N. (1996a) An antibacterial thiophene from *Balsamorhiza sagittata* (Pursh) Nuttall. *Planta Med.*, **62**, 65–66.

Matsuura, H., Saxena, G., Farmer, S.W., Hancock, R.E.W. and Towers, G.H.N. (1996b) Antibacterial and antifungal polyine compounds from *Glehnia littoralis* ssp. *leiocarpa*. *Planta Med.*, **62**, 256–259.

McCutcheon, A.R. (1996) *Ethnopharmacology of Western North American Plants*, Ph.D. Thesis, University of British Columbia, Vancouver, B.C.

McCutcheon, A.R., Ellis, S.M., Hancock, R.E.W. and Towers, G.H.N. (1992) Antibiotic screening of medicinal plants of the British Columbian native peoples. *J. Ethnopharmacol.*, **37**, 213–233.

McCutcheon, A.R., Ellis, S.M., Hancock, R.E.W. and Towers, G.H.N. (1994) Antifungal screening of medicinal plants of the British Columbian native peoples. *J. Ethnopharmacol.*, **44**, 157–169.

McCutcheon, A.R., Roberts, T.E., Gibbons, E., Ellis, S.M., Hancock, R.E.W. and Towers, G.H.N. (1995) Antiviral screening of British Columbian medicinal plants. *J. Ethnopharmacol.*, **49**, 101–110.

McCutcheon, A.R., Stokes, R.W., Thorson, L.M., Ellis, S.M., Hancock, R.E.W. and Towers, G.H.N. (1997) Anti-mycobacterial screening of British Columbian medicinal plants. *Int. J. Pharmacog.*, **35**, 1–7.

Mi, H.M., Li, C.G., Su, Z.W., Wang, N.P., Zhao, J.X. and Jiang, Y.G. (1987) Studies on the chemical constituents and antifungal activities of essential oil from *Oplopanax elatus* Nakai. *Acta Pharm. Sinica*, **22**, 549–552.

Moerman, D.E. (1986) *Medicinal Plants of Native America*. Research Reports in Ethnobotany, Contribution 2, University of Michigan Museum of Anthropology, Technical Reports, Number 19. Ann Arbor, Mich.

Muir, A.D., Cole, A.L.J. and Walker, J.R.A. (1982) Antibiotic compounds from New Zealand plants. I. Falcarindiol, an anti-dermatophyte agent from *Schefflera digitata*. *Planta Med.*, **44**, 128–133.

Norton, R.A., Finlayson, A.J. and Towers, G.H.N. (1985) Two dithiacyclohexadiene polyacetylenes from *Chaenactis douglasii* and *Eriophyllum lanatum*. *Phytochemistry*, **24**, 356–357.

Pojar, J. and MacKinnon, A. (eds.) (1994) *Plants of Coastal British Columbia including Washington, Oregon and Alaska*. Lone Pine Publishing, Vancouver, B.C.

Sachdev, K. and Kulshreshtha, D.K. (1986) Phenolic constituents of *Coelogyne ovalis*. *Phytochemistry*, **25**, 499–502.

Saxena, G., McCutcheon, A.R., Farmer, S., Towers, G.H.N. and Hancock, R.E.W. (1994) Antimicrobial constituents from *Rhus glabra*. *J. Ethnopharmacol.*, **5**, 196–199.

Saxena, G., Farmer, S., Hancock, R.E.W. and Towers, G.H.N. (1995a) Antimicrobial compounds from *Alnus rubra* Bong. *Int. J. Pharmacog.*, **33**, 33–36.

Saxena, G., Farmer, S., Towers, G.H.N. and Hancock, R.E.W. (1995b) Use of specific dyes in the detection of antimicrobial compounds from crude plant extracts using a thin layer chromatography agar overlay technique. *Phytochem. Anal.*, **6**, 125–129.

Saxena, G., Hancock, R.E.W. and Towers, G.H.N. (1996a) Antimicrobial metabolites from British Columbian medicinal plants: ethnopharmacology and phytochemistry. In S.K. Jain (ed.), *Ethnobiology in Human Welfare*. Deep Publications, New Delhi, pp. 87–90.

Saxena, G., Farmer, S.W., Hancock, R.E.W. and Towers, G.H.N. (1996b) Chlorochimaphilin: a new antibiotic from *Moneses uniflora*. *J. Nat. Prod.*, **59**, 62–65.

Sheth, K., Catalfomo, P. and Sciuchetti, L.A. (1967) Phytochemical investigation of the leaves of *Chimaphila umbellata* var. *occidentalis*. *J. Nat. Prod.*, **30**, 78–83.

Sternberg, S. (1994) The emerging fungal threat. *Science*, **266**, 1632–1634.

Taylor, R.L. and MacBride, B. (1977) *Vascular Plants of British Columbia: a descriptive resource inventory*. University of British Columbia Press.

Towers, G.H.N., Abramowski, Z., Finlayson, A.J. and Zucconi, A. (1985) Antibiotic properties of thiarubrine A, a naturally occurring dithiacyclohexadiene polyine. *Planta Med.*, **51**, 225–229.

Turner, N.J. (1973) The ethnobotany of the Bella Coola Indians of British Columbia. *Syesis*, **6**, 193–220.

Turner, N.J. (1975) *Food Plants of the British Columbian Indians: Coastal Peoples*. British Columbia Provincial Museum Handbook No. 34, Royal British Columbian Provincial Museum, Victoria, British Columbia.

Turner, N.J. (1978) *Food Plants of the British Columbian Indians: Interior Peoples*. British Columbia Provincial Museum Handbook No. 36, Royal British Columbian Provincial Museum, Victoria, British Columbia.

Turner, N.J. (1982) Traditional use of Devil's-club (*Oplopanax horridus*; Araliaceae) by native people in western North America. *J. Ethnobiology*, **2**, 17–38.

Turner, N.J. (1988) Ethnobotany of coniferous trees in Thompson and Lillooet Interior Salish of British Columbia. *Econ. Bot.*, **42**, 176–194.

Turner, N.J. and Bell, M.A.M. (1971) Ethnobotany of the coast Salish Indians of Vancouver Island. *Econ. Bot.*, **25**, 257–310.

Turner, N.J. and Bell, M.A.M. (1973) Ethnobotany of the Southern Kwakiutl Indians of British Columbia. *Econ. Bot.*, **27**, 257–310.

Turner, N.J. and Efrat, B.S. (1982) *Ethnobotany of the Hesquiat Indians of Vancouver Island*. Cultural Recovery Paper No. 2, Ethnobotanical Contribution No. 1 of the Hesquiat Cultural Committee, British Columbia Provincial Museum, Victoria, B.C.

Turner, N.J. and Hebda, R.J. (1990) Contemporary use of bark for medicine by two Salishan native elders of southeast Vancouver Island, Canada. *J. Ethnopharmacol.*, **29**, 59–72.

Turner, N.J., Bouchard, R. and Kennedy, D.I.D. (1980) *Ethnobotany of the Okanagan-Colville Indians of British Columbia and Washington*. British Columbia Provincial Museum No. 21, Occasional Papers Series, British Columbia Provincial Museum, Victoria, British Columbia.

Turner, N.J., Thomas, J., Carlson, B.F. and Ogilvie, R.T. (1983) *Ethnobotany of the Nitinaht Indians of Vancouver Island*. Occasional Papers of the British Columbia Provincial Museum No. 24, Published by the British Columbia Provincial Museum, Victoria and Parks Canada (Government of Canada), Ottawa.

Turner, N.J., Thompson, L.E., Thompson, M.T. and York, A.Z. (1990) *Thompson Ethnobotany: Knowledge and Uses of Plants by the Thompson Indians*. British Columbia Provincial Museum Memoir No. 25, British Columbia Provincial Museum, Victoria, British Columbia.

Vanwagenen, B.C. and Cardellina, J.H. (1986) Native American food and medicinal plants 7. Antimicrobial tetronic acids. *Tetrahedron*, **42**, 1117–1122.

Vesikari, T., Isolauri, E., Ruuska, T., Delem, A. and Andre, F.E. (1988) Rotavirus: new vaccine and vaccination. In E. Kurstak, R.G. Marusyk, F.A. Murphy and M.H.V. Regenmortel (eds.), *Applied Antiviral Research* (*Volume 1*): *New Vaccines and Chemotherapy*, Plenum Medical Book Co., New York.

Weslager, C.A. (1973) Magic Medicines of the Indians. The Middle Atlantic Press, Somerset, New Jersey.

8. BIOACTIVE COMPOUNDS FROM PANAMANIAN PLANTS

MAHABIR P. GUPTA[1], ANDREW MARSTON[2] and KURT HOSTETTMANN[2]

[1]*International Coordinator, CYTED, Centre for Pharmacognostic Research on Panamanian Flora, College of Pharmacy, University of Panama, and Smithsonian Tropical Research Institute, Panama, Republic of Panama*
[2]*Institut de Pharmacognosie et Phytochimie, Université de Lausanne, CH-1015 Lausanne, Switzerland*

INTRODUCTION

The flora of Panama is one of the richest in the Neotropical area. It is estimated that in Panama there are between 8,000 and 10,000 vascular plant species, of which 17.3% are endemic. There is also a large number of non-vascular species but they have not been investigated fully. In addition, Panama has over 687 species of ferns and about 1,000 species of orchids, of which 50% are endemic. Epiphytes, climbers and lianas are a major component of the Panamanian tropical forest. Mosses abound in moist cloud forests as well as other parts of the country. Panama, with its biogeographic situation, plays an important role as a biological bridge between the northern and southern lobes of the continent. Indeed, three of the four principal routes of bird migration converge in its territories (Cobos-Moran 1992).

Natural products are one of our most promising sources of medicinals for the future (Clark 1986; Speedie 1992; Farnsworth 1985). It is estimated that only 5,000 plant species have been studied extensively for medicinal applications out of the total of 250,000–300,000 species (Abelson 1990). The distinguished Brazilian scientist Otto Gottlieb (1981) wrote "Nothing at all is known about the chemical composition of 99.6% of our flora". Hence, plants represent a valuable supply of new agents, while phytotherapy and herbal medicine provide a fund of exploitable knowledge (Rouhi 1997). At the present time, the interest in botanical medicine is increasing steadily (Srivastava *et al.* 1996; Weil 1989; Plotkin 1988). It is widely accepted that the tropical rainforests are the richest source of biomass on Earth. Tropical rainforests presently cover an area of about 9 million square kilometres. Of these, 5.1 million square kilometres are in tropical America. The value of biodiversity as a raw material for pharmaceutical and biotechnology industries is only a portion of its value to society. Biodiversity is the natural biological capital of the Earth and presents important opportunities for all nations (Soejarto and Farnsworth 1989). It makes good economic sense to study and conserve biodiversity (Watson *et al.* 1995). The number of species of seed plants occuring in the tropics is estimated at 155,000 (Prance 1977), of which approximately 120,000 occur in moist tropical forests alone (Myers 1984, 1986). Of these, roughly 60% occur in tropical America. This biotic richness is due to the high biological diversity of this tropical ecosystem. Many Latinoamerican countries are thus considered to be megadiverse countries. For example, Brazil, Costa Rica, Colombia and Panama have more plant species per square kilometre than the United States, Great Britain and New Zealand combined (Gentry 1986). The value of undiscovered pharmaceuticals in tropical forests has been estimated at US $ 147 billion (Mendelsson and Balick 1995).

Panama is a unique terrestrial bridge of extreme biological importance. Presently, throughout the isthmus, there is an amalgam of biotas of the north and of the south. A consequence of this mixing is a vast plant and animal diversity. This diversity is supported by a unique collection of environmental factors, which influence and allow the coexistence of so many species (Cobos-Moran 1993). Therefore, its integrity as a biological bridge is necessary for the continued existence and evolution of the species of this important region of the planet.

Panama is one of the 25 most plant-rich countries in the world. Flowering plant richness and biodiversity of Panama ranks fourth in North and Central America (World Conservation Monitoring Center 1992). According to Conservation International, Panama is one of the hotspot countries in the world because of the high number of species and there is a great ecological risk (Anonymous 1997). Therefore, there is an urgent need to intensify our efforts to study the Panamanian flora as a source of novel bioactive molecules.

One way of achieving this is through cooperation, such as the interuniversity link between the Institute of Pharmacognosy and Phytochemistry of the University of Lausanne, and the Pharmacognostic Research Center of Panamanian Flora of the University of Panama. The results of this collaborative research program which have been obtained over the last 12 years are presented here.

Selection of plant material has been based on ethnomedical claims, chemotaxonomic considerations, field observations and random collections.

BIOLOGICAL SCREENING OF PLANT MATERIAL

Following collection of plant material, the procedure adopted in the collaborative research program between Panama and Lausanne has been to screen extracts of the plants, both biologically (Hostettmann *et al.* 1995) and chemically (Hostettmann *et al.* 1997). The latter methodology has been discussed elsewhere in this book. Interesting leads have been pursued and the constituents of the extracts isolated.

For example, screening for antifungal activity has been performed against the plant pathogenic fungus *Cladosporium cucumerinum* and the human pathogenic yeast *Candida albicans* (Rahalison *et al.* 1993). For *C. cucumerinum*, direct bioautography was performed on aluminium-backed TLC sheets, while activity against *C. albicans* was assessed in an agar overlay assay. The plant material was extracted successively with chloroform, methanol and water and each extract was submitted to testing. In total, 153 crude extracts from 28 species (representing 21 plant families) were screened. When tested at 100 μg in the TLC bioassays, 26 (17%) and 25 (16%) of the extracts gave inhibition zones against *C. albicans* and *C. cucumerinum*, respectively. Most of these were chloroform extracts (Rahalison *et al.* 1993).

Similarly, chloroform, methanol and water extracts of Panamanian plants have been investigated for their molluscicidal activity against *Biomphalaria glabrata* snails, intermediate hosts of the parasite *Schistosoma mansoni* (Marston *et al.* 1996). This is one of the parasites responsible for the incidence of schistosomiasis in the tropics, a disease which affects more than 250 million people in over 70 countries. A measure of control of the disease may be possible by killing the aquatic snail vectors in infested water sources. Thus, active plant extracts might provide an important alternative method of achieving snail mortality, instead of synthetic products. Of the Panamanian plant extracts tested, approximately 10% killed *B. glabrata* snails within 48 h in a laboratory assay at 400 ppm or less (Marston *et al.* 1996).

The Panamanian plants collected have also been tested for their larvicidal activity and brine shrimp toxicity (Cepleanu *et al.* 1994). The larvicidal test involves larvae of the mosquito *Aedes aegypti*, which is the vector of yellow fever and dengue hemorrhagic fever. It is important to find new insecticidal or larvicidal compounds against *A. aegypti* since resistance quickly develops to presently used chemicals such as dieldrin. The test involves incubating *A. aegypti* eggs in water, to produce second instar larvae. These are then placed in contact with the plant extracts in tap water and mortality is measured with the naked eye (Cepleanu *et al.* 1994).

ISOLATION OF COMPOUNDS FROM PANAMANIAN PLANTS

Phytochemical research on candidate plants chosen after biological and chemical screening has led to the isolation of numerous new compounds, many of which have very valuable biological activities. Some of these have previously been described (Gupta 1995) so this review will concentrate on other representatives, with the emphasis on more recent results.

Molluscicidal Saponins from *Swartzia simplex* (Leguminosae)

In the course of systematic screening studies for molluscicidal constituents of plants, a series of very active saponins was isolated from fruits of the African tree *Swartzia madagascariensis* (Leguminosae) (Borel and Hostettmann 1987). Furthermore, through field trials carried out in Africa, this plant was shown to be a potential candidate as a natural molluscicide (Suter *et al.* 1986). The importance of these results led to the investigation of a plant from the same genus, *S. simplex*, found in Panama. A methanol extract of the leaves was active against *Biomphalaria glabrata* snails at 400 ppm after 24 h. The extract was fractionated by a combination of silica gel column chromatography and low-pressure liquid chromatography (LPLC) on RP-18, to give seven saponins (**1**–**7**) (Borel *et al.* 1987). These were glycosides of oleanolic acid (**1**, **3**–**6**), gypsogenin (**2**) and gypsogenic acid (**7**).

Saponin **1** was active against the snails (100% of snails killed at 3 ppm) but the others were all inactive up to 100 ppm. Thus, as is often observed with the saponins, the

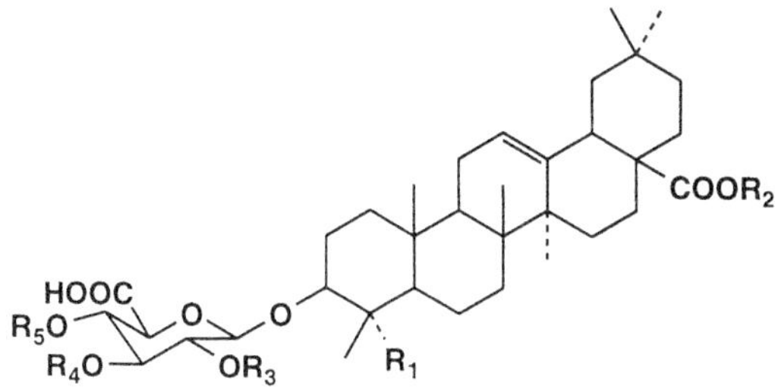

1 R_1=CH_3, R_2=H, R_3=H, R_4=Rha, R_5=H
2 R_1=CHO, R_2=Glc, R_3=H, R_4=H, R_5=H
3 R_1=CH_3, R_2=Glc, R_3=H, R_4=H, R_5=H
4 R_1=CH_3, R_2=Glc, R_3=H, R_4=Rha, R_5=H
5 R_1=CH_3, R_2=Glc, R_3=H, R_4=H, R_5=Glc
6 R_1=CH_3, R_2=Glc, R_3=Xyl, R_4=Rha, R_5=H
7 R_1=COOH, R_2=Glc, R_3=H, R_4=H, R_5=H

monodesmosidic glycoside was responsible for the molluscicidal activity. The potency of **1** is high and is close to the value obtained for the synthetic molluscicide Bayluscide (0.3 ppm) (Cepleanu *et al.* 1994). Basic hydrolysis of the glucosyl moiety at C-28 gave the corresponding monodesmosides of **2–7**; these all had activity against the snails (Borel *et al.* 1987).

Larvicidal Acetogenins from *Annona purpurea* (Annonaceae)

In Panama, the leaves of *A. purpurea* are used in a decoction for the cleansing of skin eruptions (Morton 1981). As a dichloromethane extract of the leaves gave pronounced activity against larvae of the mosquito *Aedes aegypti*, bioactivity-guided fractionation of the extract was undertaken, resulting in the isolation of 6 acetogenins (**8–13**) (Cepleanu *et al.* 1993). Compounds **8–11** were identified as the known acetogenins rolliniastatin 1, bullatacin, cherimoline and sylvaticin, respectively. The structures of the new compounds purpureacin 1 (**12**) and purpureacin 2 (**13**) were determined from UV, NMR and mass spectral data. However, the limited amounts of substance isolated precluded determination of the configuration.

Testing of compounds **8–13** for brine shrimp toxicity, larvicidal activity and antimicrobial activity is shown in Table 1. In common with other acetogenins from the Annonaceae (Fang *et al.* 1993), the compounds isolated from *Annona purpurea* display some potent bioactivities. Of note is the very high brine shrimp toxicity of rolliniastatin 1 (**8**) and the antifungal activities of **11–13**.

Antifungal Compounds from *Baccharis pedunculata* (Asteraceae)

A dichloromethane extract of the leaves of *B. pedunculata* was active against both *Cladosporium cucumerinum* and *Candida albicans*. Fractionation of this extract by a combination of flash chromatography, silica gel column chromatography and gel filtration afforded two flavonoids (**14, 15**), an acetylenic lactone (**16**) and a coumarin (**17**) (Rahalison *et al.* 1995).

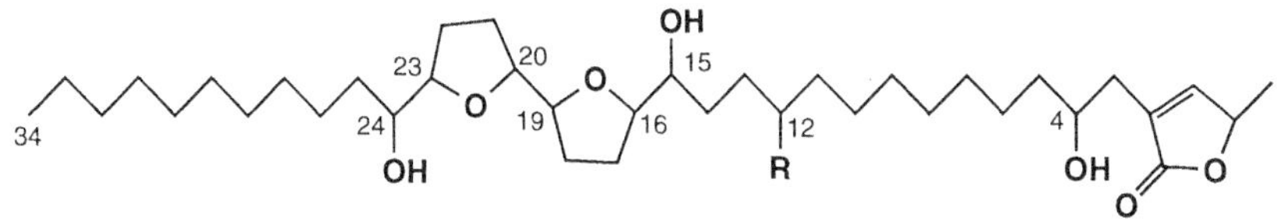

	R	C-15/C-16	C-16/C-19	C-19/C-20	C-20/C-23	C-23/C-24
8 Rolliniastatin 1	H	*threo*	*cis*	*threo*	*cis*	*erythro*
9 Bullatacin	H	*threo*	*trans*	*threo*	*trans*	*erythro*
13 Purpureacin 2	OH					

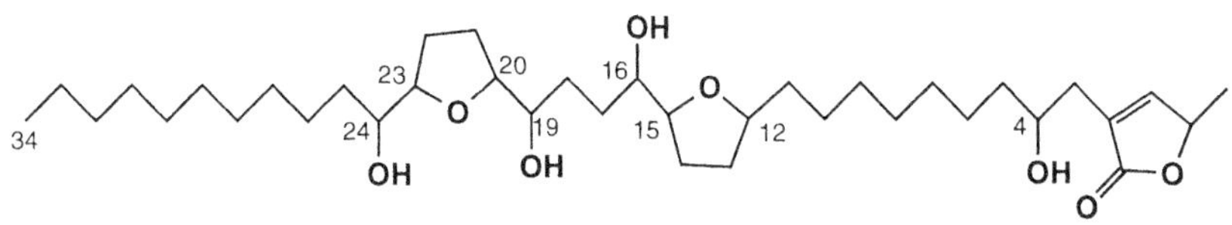

	C-15/C-16	C-19/C-20	C-20/C-23	C-23/C-24
10 Cherimoline	*threo*	*threo*	*trans*	*erythro*
11 Sylvaticin	*threo*	*erythro*	*trans*	*threo*
12 Purpureacin 1				

Table 1 Biological activities of acetogenins from *Annona purpurea* (Annonaceae)

	Brine shrimp toxicity[a] (μg/ml)	*Larvicidal activity*[b] (μg/ml)	*Antifungal activity*[c] (μg)	*Antibacterial activity*[d] (μg)
Extract	3.9	10	100	inactive
8	0.0049	0.2	inactive	inactive
9	0.091	0.3	inactive	inactive
10	0.40	2.0	20	inactive
11	0.57	1.0	0.5	inactive
12	0.53	2.0	0.05	20
13	0.38	1.0	1	inactive
References[e]	5.0	0.1	0.001	0.001

[a]Toxicity towards *Artemia salina*, expressed as $LC_{50}/24\,h$.
[b]Activity against larvae of *Aedes aegypti*, expressed as $LC_{100}/24\,h$.
[c]Antifungal activity against *Candida albicans*; minimal amount inhibiting fungal growth on TLC plate.
[d]Antibacterial activity against *Bacillus subtilis*; minimal amount inhibiting bacterial growth on TLC plate.
[e]Reference compounds: taxol for brine shrimp toxicity; diazion for larvicidal activity; miconazole for antifungal activity; chloramphenicol for antibacterial activity.

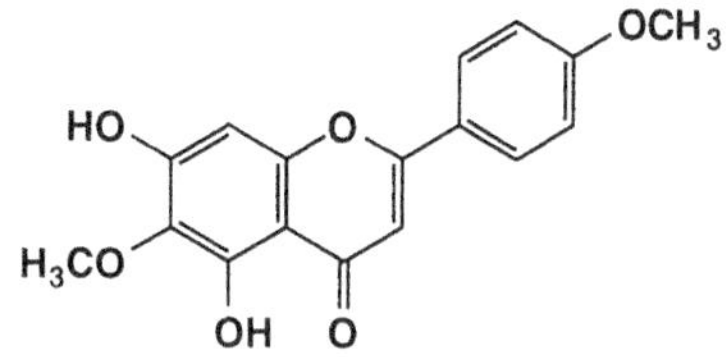

14

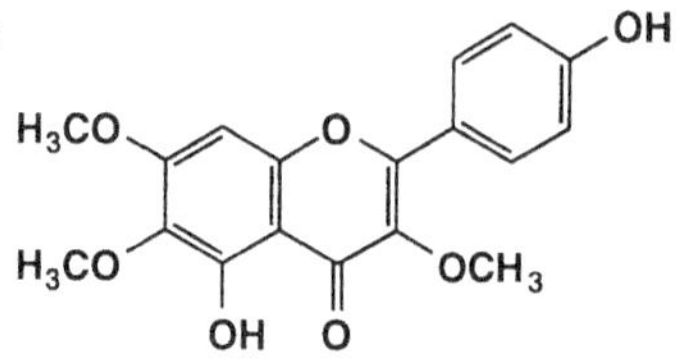

15

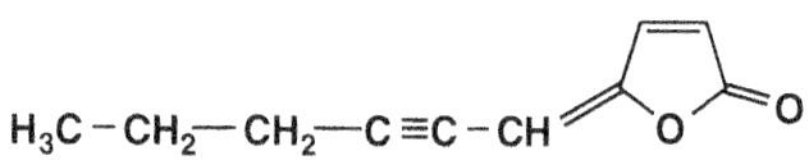

16

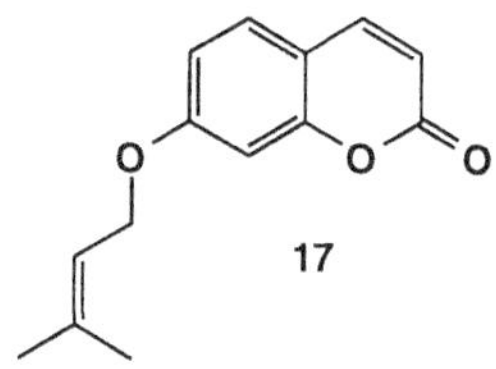

17

Table 2 Antifungal activities of compounds **16** and **17** from *Baccharis pedunculata* (Asteraceae)

Organism	*Antifungal activity* (μg/ml)		
	16	**17**	*Miconazole*
Candida albicans	>20	inactive	0.1
Pityrosporum ovale	20	inactive	10
Aspergillus fumigatus	>20	inactive	10
Epidermophyton floccosum	10	10	10
Microsporum canis	inactive	inactive	10
Microsporum gypseum	10	>20	10
Trichophyton mentagrophytes	inactive	inactive	10

Flavone **14** did not inhibit fungal growth and while the 3-methyl ether **15** was active against *C. albicans* and *C. cucumerinum* in the bioautographic assays, the activity was not high enough to justify testing by a dilution method. Lachnophyllum lactone (**16**), the major component of the dichloromethane extract, exhibited strong activity against some dermatophytes (*Epidermophyton floccosum*, *Microsporum gypseum*, Table 2) and also against *C. albicans* in dilution assays. However, when tested on human epidermal keratinocytes, it produced a higher cytotoxicity (LD_{50} 2 μg/ml) than antifungal therapeutic agents such as econazole (LD_{50} 14 μg/ml); potential therapeutic effects are thus limited.

Coumarin 17 (ulopterol) inhibited the growth of the human pathogenic fungus *Epidermophyton floccosum* at 10 μg/ml and gave reduced cytotoxicity to human keratinocytes (LD_{50} 15 μg/ml). On the other hand, it was not active against the yeasts and the other dermatophytes.

The synthetic antifungal agent miconazole was used as positive control for the antifungal tests. It inhibited fungal growth at 10 μg/ml, except for *C. albicans* and *C. cucumerinum*, which were inhibited at 0.1 and 1 μg/ml, respectively (Rahalison *et al.* 1995).

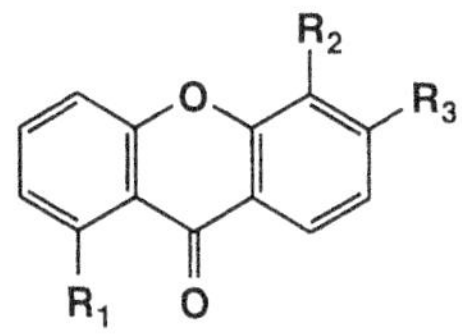

18 R_1=OCH_3, R_2=OH, R_3=H
19 R_1=OH, R_2=OH, R_3=H
20 R_1=OH, R_2=OCH_3, R_3=OH

Antifungal Xanthones from *Marila laxiflora* (Guttiferae)

The shrub *Marila laxiflora* is distributed in lower altitude areas of Panama, Bolivia and Guatemala. It belongs to a genus which has not yet been studied phytochemically. In common with many plants belonging to the Guttiferae, HPLC/UV analysis showed *M. laxiflora* to contain xanthones (Bennett and Lee 1989). In addition, a dichloromethane extract of the roots produced antifungal activity against *Cladosporium cucumerinum* in the TLC autobiographic assay.

The dichloromethane extract was separated into three fractions (1–3) by centrifugal partition chromatography (CPC), using the upper phase of the solvent system petrol ether–ethyl acetate–methanol–water (1 : 1 : 1 : 1) as mobile phase. Diol column chromatography of fraction 2 with petrol ether–chloroform (1 : 1) gave xanthones **19** and **20**. Gel filtration of fraction 3 on Sephadex LH-20 with chloroform–methanol 1 : 1 yielded xanthone **18**. Betulinic acid was isolated from fraction 1. The methanol extract, which was inactive in the antifungal bioassay, was also fractionated. A combination of medium pressure LC (MPLC) on silica gel and octadecyl silica gel, low pressure LC (LPLC) on octadecyl silica gel and gel filtration gave 2-(3,3-dimethylallyl)-1,3,5,6-tetrahydroxyxanthone, rhamnetin and 3,4-dihydroxybenzoic acid (Ioset *et al.* 1998a).

Xanthones **18–20** gave inhibition of spore growth at minimum quantities of 5 μg each in the TLC bioassay with *C. cucumerinum*. The reference compound, amphotericin B, was active at 0.1 μg in the same test. In dilution tests with *C. cucumerinum*, minimum inhibition concentration (MIC) values of 25 and 50 μg/ml were found for **19** and **20**, respectively. Compound **18** was inactive up to 100 μg/ml, while amphotericin B was active at 1.25 μg/ml (Ioset *et al.* 1998a).

Antifungal and Larvicidal Naphthoquinones and a Naphthoxirene from *Cordia linnaei* (Boraginaceae)

After biological screening of several *Cordia* species (family: Boraginaceae) from Panama, a dichloromethane extract of roots from *C. linnaei* was chosen for further study because it showed activity against the fungus *Cladosporium cucumerinum*, the yeast *Candida albicans* and the larvae of the *Aedes aegypti* mosquito.

Five compounds (**21–25**) were isolated from the dichloromethane extract after a combination of silica gel column chromatography, gel filtration on Sephadex LH-20 and LPLC on diol (Ioset *et al.* 1998b). Structure elucidation of compounds **21–24** showed them all to be 6-substituted naphthoquinone derivatives, containing a cyclohexylethyl moiety. Compound **25** was a naphthoxirene since it possessed a 2,3-epoxy moiety. The terpenoid part of this naphthoxirene was identical to that found in cordiaquinone C from *C. corymbosa* (Bieber *et al.* 1994).

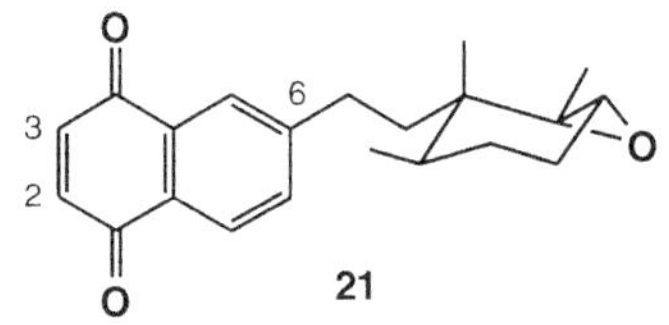

21

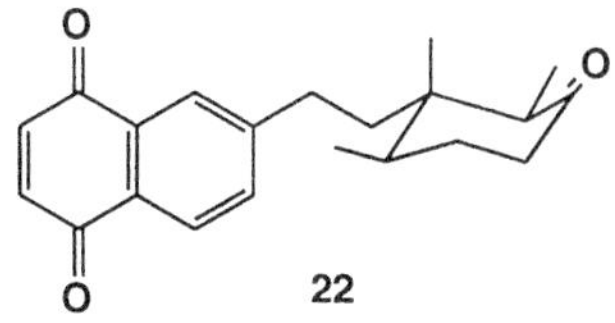

22

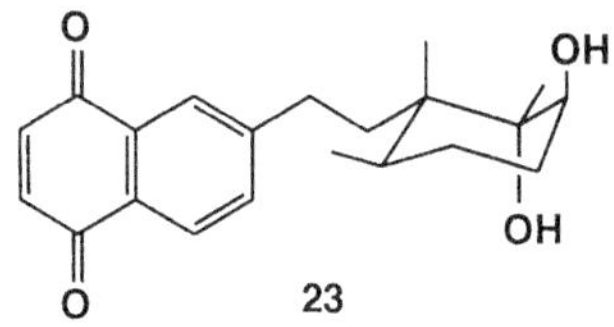

23

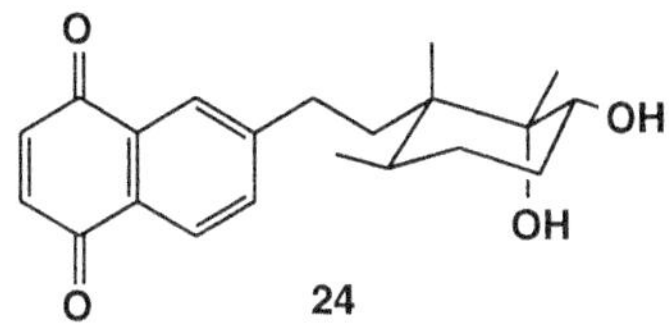

24

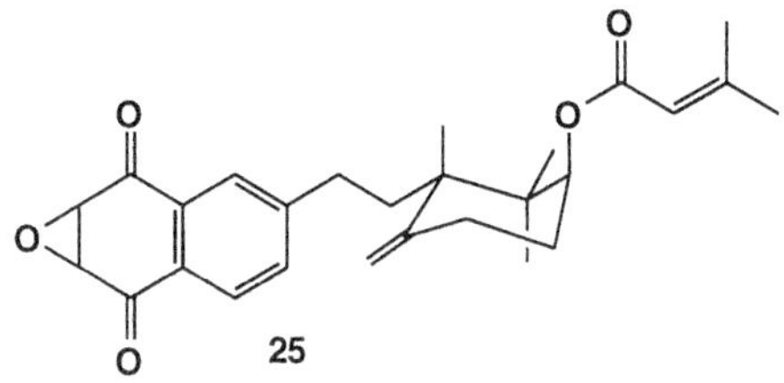

25

Table 3 Antifungal and larvicidal activities of naphthoquinones from *Cordia linnaei* (Boraginacceae)

	Cladosporium cucumerinum[a] (μg)	*Cladosporium cucumerinum*[b] (μg/ml)	*Candida albicans*[a] (μg)	*Candida albicans*[b] (μg/ml)	*Aedes aegypti*[c] (ppm)
21	1	3	1	6	12.5
22	0.5	3	0.5	3	25
23	0.5	1.5	1	6	50
24	1	3	1	6	25
25	>50	>50	>50	>50	not tested
Nystatin	0.2	1	0.1	1	not tested
Plumbagin	not tested	not tested	not tested	not tested	6.25
Diazion	not tested	not tested	not tested	not tested	0.1

[a]Minimum amount of compound required to inhibit growth on a silica gel TLC plate.
[b]Minimal inhibition concentration (MIC) of compound in an agar dilution assay.
[c]Minimum concentration of compound required to kill all larvae after 24 h.

Antifungal and larvicidal potencies of the individual naphthoquinones are reported in Table 3. The activities of the naphthoquines **21–24** against *C. cucumerinum* were very potent and similar to those of the reference compound nystatin. A similar situation was observed for *C. albicans*. Concerning larvicidal activity against *A. aegypti*, the same four naphthoquinones were toxic to the larvae at levels approaching the toxicity of the naphthoquinone plumbagin. However, they were considerably less active than the synthetic reference compound diazion. Naphthoxirene **25** was inactive at the highest tested concentrations against *C. cucumerinum* and *C. albicans*. Thus the presence of the epoxide ring on the naphthoquinone ring would seem to hinder any effect on these microorganisms (Ioset *et al.* 1998b).

Antifungal Benzoic Acid Derivatives from *Piper dilatatum* (Piperaceae)

The Piperaceae family is known to be rich in bioactive compounds (Sengupta and Ray 1987). As *Piper dilatatum* is used by the Kuna Indians of Panama in tonic baths and as no previous phytochemical investigation had been performed on this plant, it was chosen for study within the context of the Panama-Lausanne collaborative venture.

The leaves were found to give a dichloromethane extract which was active against *C. cucumerinum* in the bioautographic TLC assay. Consequently, this extract was fractionated, first by open-column silica gel chromatography and then by various chromatographic procedures (involving gel filtration on Sephadex LH-20, silica gel column chromatography, vacuum liquid chromatography over RP-18), to give six prenylated benzoic acid derivatives (**26**–**31**) and three chalcones (**32**–**34**). Compound **26** was obtained for the first time from a plant source (Terreaux *et al.* 1998).

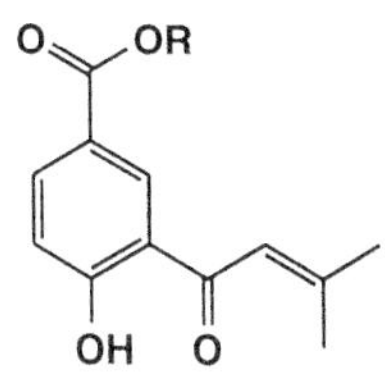

26 R=H
27 R=CH_3

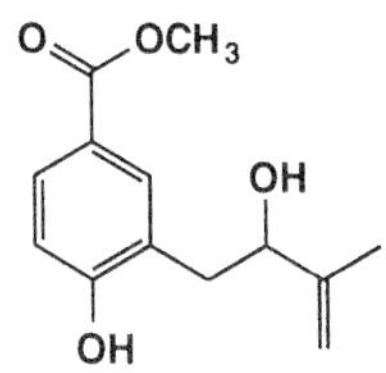

28

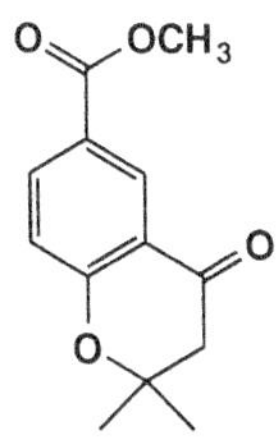

29

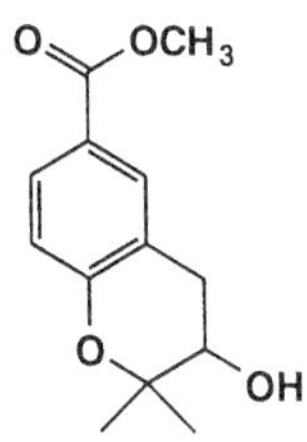

30

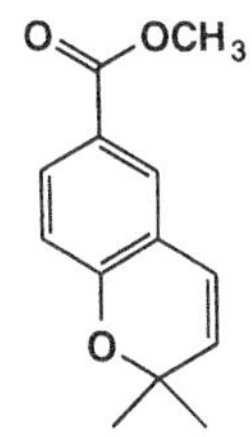

31

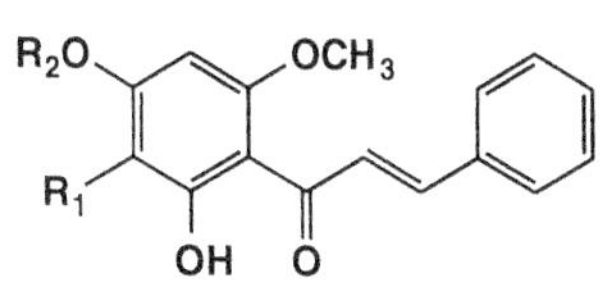

32 R_1=H, R_2=CH_3
33 R_1=R_2=H
34 R_1=OCH_3, R_2=CH_3

Compounds **26** (taboganic acid), **27** (taboganic acid methyl ester), **29** and **30** displayed antifungal activity against *C. cucumerinum* in the TLC bioautographic test. The chroman-4-one **29** was the most active, with only 1 μg required to inhibit fungal growth on the TLC plate. This was the same value as that given by the reference compound miconazole. The MIC value of **29** in an agar dilution assay was 40 μg/ml, while miconazole gave a MIC value of 10 μg/ml. For chroman **30**, 3 μg was required for inhibition in the TLC bioassay, while 5 μg was required for **26** and **27**. The constituents of *P. dilatatum* were also tested against *C. albicans* but no significant activity could be detected (Terreaux *et al.* 1998).

CONCLUSIONS

From the results presented, it is immediately obvious that the Panamanian flora is a valuable source of new bioactive natural products. More work on the phytochemical investigation of Panamanian plants can be found in a review article published in 1995 (Gupta 1995).

The collaboration between the University of Lausanne and the Pharmacognostic Research Center of Panamanian Flora of the University of Panama (CIFLORPAN) has resulted in the isolation of approximately 80 compounds, including 29 new compounds (Table 4). These belong to very diverse compound classes: naphthoquinones, flavonoids, biphenyls, coumarins, xanthones, benzoic acid derivatives, acetogenins, monoterpenes, triterpenes, saponins etc. Of the isolated compounds, 19 had reproducible biological activity. A total of 14 plants have been studied in detail – from the aspect of isolation and biological activity. Certain of these have been described in this chapter, to give an idea of the types of constituents and of their biological activities. For example, several extracts with fungicidal activity have been investigated (*Baccharis pedunculata*, *Marila laxiflora*, *Cordia linnaei*, *Piper dilatatum*), a plant with molluscicidal activity (*Swartzia simplex*) has been studied and larvicidal compounds have been isolated from a member of the Annonaceae family (*Annona purpurea*).

Limited screening of Panamanian plants (Cepleanu *et al.* 1994; Marston *et al.* 1996; Rahalison *et al.* 1993) has indicated that many more species have interesting biological activities and that much work remains to be done on the flora of this country. The potential for discovering new lead compounds is enormous.

ACKNOWLEDGEMENTS

This work has been financed by the following bodies: Swiss National Science Foundation; Organization of American States (UNIPAN-IDB Research Grant No. VIP 3/11.1.2.10.94); EEC (Grant CI. 1.0505 ES (JR)). Thanks are due to Prof. Mireya Correa, Director of the Herbarium of the University of Panama for plant identification.

Table 4 Compounds isolated from Panamanian flora

Plant (Family)	*New compounds*	*Known compounds*	*Biological activity*	*References*
Annona purpurea L. (Annonaceae)	*Acetogenins* Purpureacin 1, Purpureacin 2	Rolliniastatin, Bullatacin, Cherimoline, Sylvaticin	Purpureacin 1: brine shrimp toxicity, antifungal Purpureacin 2: brine shrimp toxicity, larvicidal, antifungal, antibacterial	Cepleanu *et al.* (1993)
Baccharis pedunculata (Mill.) Cabr. (Asteraceae)		Lachnophyllum lactone (acetylenic lactone), flavone, prenylated coumarin and 3-methylether flavone	Lachnophyllum lactone: antifungal, cytotoxic 3-Methylether flavone: antifungal Prenylated coumarin: antifungal, cytotoxic	Rahalison *et al.* (1995)
Cordia linnaei Stearn (Boraginaceae)	Naphthoquinones Cordiaquinone E, Cordiaquinone F, Cordiaquinone G, Naphthoxirene	Cordiaquinone B	Cordiaquinones B, E-G: antifungal, larvicidal	Ioset *et al.* (1998b)
Coutoubea spicata Aubl. (Gentianaceae)	Quercetin 3-*O*-rhamnopyranosyl-(1 → 6)-(4″-*trans*-p-coumaroyl)-*β*-D-galactopyranoside 7-*O*-*α*-L-rhamnopyranoside (4″-*trans*-p-coumaroyl clovin)	Clovin, Gentiopicrin, Swertiamarin		Schaufelberger *et al.* (1987)
Dorstenia contrajerva L. (Leguminosae)		Bergapten, 4-{[3-(4,5-Dihydro-5,5-dimethyl-4-oxo-2-furanyl)-butyl]oxy}-7H-furo [3,2 g] [1]benzopyran-7-one		Terreaux *et al.* (1995a)

Table 4 *(continued)*

Plant (Family)	New compounds	Known compounds	Biological activity	References
Erythrina berteroana Urb. (Leguminosae)	5,7-Dihydroxy-3-[5-hydroxy-4 methoxy-3-(3-methyl-2-butenyl) phenyl]-2,3-dihydro-4H-1-benzopyran-4-one, Sigmoidin β 2′-methylether		Prenylated flavanone: antifungal	Maillard *et al.* (1987, 1989)
Lisianthius jefensis Robyns et Elias (Gentianaceae)	Lisianthioside			Hamburger *et al.* (1990)
Marila laxiflora Rusby (Guttiferae)		Xanthones, rhamnetin, betulinic acid, 3,4-dihydroxybenzoic acid	Xanthones: antifungal	Ioset *et al.* (1998a)
Monnina sylvatica Schlecht & Cham. (Polygalaceae)	3′-Hydroxy-4′,5-dimethoxy-3,4-methylenedioxy-biphenyl, 3′,4′,5-Trimethoxy-3,4-methylenedioxybiphenyl, 3′-Hydroxy-5-methoxy-3,4-methylenedioxybiphenyl, 3′-Hydroxy-5,5′-dimethoxy-3,4-methylenedioxy-biphenyl, Kaempferol 3-*O*-β-D-glucosyl-(1 $\rightarrow$ 2)-*O*-[α-L-rhamnosyl-(1 $\rightarrow$ 6)]-β-D-galactoside	1,5,-Dihydroxy-2,3-dimethoxyxanthone, Kaempferol 3-*O*-robinobioside, Kaempferol 3-*O*-β-D-apio-D-furanosyl-(1 $\rightarrow$ 2)-β-D-galactoside, Kaempferol β-D-apio-D-furanosyl-(1 $\rightarrow$ 2)-*O*-[α-L-rhamnopyranosyl-(1 $\rightarrow$ 6)]-β-D-galactoside	Biphenyls: antifungal, brine shrimp toxicity	Bashir *et al.* (1991a,b; 1992)
Paradrymonia macrophylla Wiehl (Gesneriaceae)	2α, 3β, 19α-Trihydroxy-24-oxo-urs-12-en-28-oic-acid (24-*epi*-pinfaensic acid), 2α, 3β, 19α-Trihydroxy-24-	Glucosyl-2α, 3β, 19α, 24-tetrahydroxyrus-12-en-28-oate (24-hydroxytormentic acid ester glucoside),		Terreaux *et al.* (1996)

	oxo-urs-12-en-28-oic acid diacetate (24-*epi*-pinfaensin acid diacetate), Glucosyl-2α, 3β, 19a-trihydroxy-24-oxo-urs-12-en-28-oate (24-*epi*-pinfaensin), Glucosyl-2α, 3b, 19a-trihydroxy-24-oxo-olean-12-en-28-oate (paradrymonin), Glucosyl-2α, 3β, 19β, 24-trihydroxyolean-12-en-28-oate (sericoside)	2α, 3β, 19α, 24-Tetra-hydroxy-urs-12-en-28-oic acid (24-hydroxytormentic acid), 2α, 3β, 19α-Trihydroxyrus-12-en-28-oic acid diacetate (tormentic acid acetate)		
Piper dilatatum L.C. Rich (Piperaceae)	Taboganic acid, Taboganic acid methyl ester	Chalcones, Prenylated benzoic acid derivatives	Taboganic acid and taboganic acid methyl ester: antifungal	Terreaux *et al.* (1998)
Polygala paniculata L. (Polygalaceae)	7-Methoxy-8-(1,4-dihydroxy-3-methyl-2-butenyl)-coumarin	Aurapten, Phebalosin, Murrangatin		Hamburger *et al.* (1985b)
Securidaca diversifolia S.F. Blake (Polygalaceae)	Quercetin 3-(2″-β-apio-furanosyl-β-D-gluco-pyranoside), Quercetin 3-(2″-β-D-apio-furanosyl-β-D-galactoside), Quercetin 3-(2″-β-D-apio-furanosyl-α-L-arabinopyranoside), Quercetin 3-(2″-β-D-apio-furanosyl-β-D-xylo-pyranoside)	Quercetin 3-sophoroside, Hyperoside, Isoquercitrin, Guaijaverin, Reynoutrin, Kaempferol-3-α-L-arabino-pyranoside, Quercitrin, Kaempferol 3-*O*-(2″-*O*-β-D-apiofuranosyl-β-D-gluco-pyranoside)		Hamburger *et al.* (1985a)

Table 4 *(continued)*

Plant (Family)	New compounds	Known compounds	Biological activity	References
Schultesia lisianthoides (Griseb.) Benth & Hook (Gentianaceae)	5-Hydroxy-1,2,3-methoxyxanthone, 5-Hydroxy-1,2,3,-methoxyxanthone acetate	1,5,6-Trihydroxy-3-methoxyxanthone, 1,5-Dihydroxy-2,3-dimethoxyxanthone, 1,3,8-Trihydroxy-7-methoxyxanthone, 1,3,8-Trihydroxy-5-methoxyxanthone, 1-Hydroxy-3,7,8-trimethoxyxanthone, 1,8-Dihydroxy-3,7-dimethoxyxanthone		Terreaux *et al.* (1995b)
Swartzia simplex Desv. (Leguminosae)	β-D-Glucopyranosyl-[β-D-glucuronopyranosyl-(1 → 3)]-3-β-hydroxyolean-12-en-23-al-28-oate, β-D-Glucopyranosyl-[β-D-glucopyranosyl-(1→ 4)-[β-D-glucopyranosyl-(1 → 3)]-3-β-hydroxyolean-12-en-28-oate, β-D-Glucopyranosyl-[(β-D-xylopyranosyl-(1 → 2))(α-L-rhamnopyranosyl-(1 → 3))-β-D-glucuronopyranosyl (1 → 3)]-3β-hydroxyolean-12-en-28-oate	[α-L-Rhamnopyranosyl-(1 → 3)-β-D-glucuronopyranosyl-(1 → 3)]-3β-hydroxyolean-12-en-28-oic acid, β-D-Glucopyranosyl-[β-D-glucopyranosyl-(1 → 3)]-3-β-hydroxyolean-12-en-28-oate, β-D-Glucopyranosyl-[α-L-rhamnopyranosyl-(1 → 3)-β-D-glucuronopyranosyl-(1 → 3)]-3β-hydroxyolean-12-en-28-oate, β-D-Glucopyranosyl-[β-D-glucuronopyranosyl-(1 → 3)]-3β-hydroxyolean-12-en-23-oic-28-oate	Monodesmosidic saponins: molluscicidal	Borel *et al.* (1987)

REFERENCES

Abelson, P.H. (1990) Medicine from plants. *Science*, **247**, 513.

Anonymous (1997) New map of "biodiversity hotspots" aids targeting of conservation efforts. *Diversity*, **13**, 27–29.

Bashir, A., Hamburger, M., Rahalison, L., Monod, M., Gupta, M.P., Solis, P. and Hostettmann, K. (1991a) Antifungal biphenyls from *Monnina sylvatica*. *Planta Med.*, **57**, 192–193.

Bashir, A., Hamburger, M., Gupta, M.P., Solis, P.N. and Hostettmann, K. (1991b) Flavonol glycosides from *Monnina sylvatica*. *Phytochemistry*, **30**, 3781–3784.

Bashir, A., Hamburger, M., Gupta, M.P., Solis, P. and Hostettmann, K. (1992) Biphenyls and a xanthone from *Monnina sylvatica*. *Phytochemistry*, **31**, 3203–3205.

Bennett, G.J. and Lee, H.-H. (1989) Xanthones from Guttiferae. *Phytochemistry*, **28**, 967–998.

Bieber, L.W., Krebs, H.C. and Schäfer, W. (1994) Further meroterpenoid naphthoquinones from *Cordia corymbosa*. *Phytochemistry*, **35**, 1027–1028.

Borel, C. and Hostettmann, K. (1987) Molluscicidal saponins from *Swartzia madagascariensis* Desvaux. *Helv. Chim. Acta*, **70**, 570–576.

Borel, C., Gupta, M.P. and Hostettmann, K. (1987) Molluscicidal saponins from *Swartzia simplex*. *Phytochemistry*, **26**, 2685–2689.

Cepleanu, F., Ohtani, K., Hamburger, M., Gupta, M.P., Solis, P. and Hostettmann, K. (1993) Novel acetogenins from the leaves of *Annona purpurea*. *Helv. Chim. Acta*, **76**, 1379–1388.

Cepleanu, F., Hamburger, M.O., Sordat, B., Msonthi, J.D., Gupta, M.P., Saadou, M. and Hostettmann, K. (1994) Screening of tropical medicinal plants for molluscicidal, larvicidal, fungicidal and cytotoxic activities and brine shrimp toxicity. *Int. J. Pharmacog.*, **32**, 294–307.

Clark, A.M. (1986) Natural products as a resource for new drugs. *Pharm. Res.*, **13**, 1133–1141.

Cobos-Moran, J.A. (1992) *Los Recursos Naturales de Panama*. Instituto Nacional de Recursos Naturales Renovables, Ministerio de Desarrollo Agropecuario, Panama, p. 25.

Cobos-Moran, J.A. (1993) *Informe Nacional de la Situacion Forestal*. Instituto Nacional de Recursos Naturales Renovables, Ministerio de Desarrollo Agropecuario, Panama, p. 28.

Fang, X., Rieser, M.J., Gu, Z., Zhao, G. and McLaughlin, J.L. (1993) Annonaceous acetogenins: an updated review. *Phytochem. Anal.*, **4**, 27–48.

Farnsworth, N.R., Akerele, D., Bingel, D.S., Soejarto, D.D. and Guo, Z. (1985) Medicinal plants in therapy. *Bull. W.H.O.* **63**, 965–981.

Gentry, A.H. (1986) Endemism in tropical versus temperate plant communities. In M.E. Soulé (ed.), *Conservation Biology*, Sinauer, Sunderland, Mass.

Gottlieb, O. (1981) New and under-utilized plants in the Americas: solution to problems of inventory through systematics. *Interciencia*, **6**, 22–29.

Gupta, M.P. (1995) Panamanian flora: source of bioactive compounds. In K. Hostettmann, A. Marston, M. Maillard and M. Hamburger (eds.), *Phytochemistry of Plants Used in Traditional Medicine*, Clarendon Press, Oxford, pp. 359–398.

Hamburger, M., Gupta, M. and Hostettmann, K. (1985a) Flavonol glycosides from *Securidaca diversifolia*. *Phytochemistry*, **24**, 2689–2692.

Hamburger, M., Gupta, M. and Hostettmann, K. (1985b) Coumarins from *Polygala paniculata*. *Planta Med.*, **51**, 215–217.

Hamburger, M., Hostettmann, M., Stoeckli-Evans, H., Solis, P.N., Gupta, M.P. and Hostettmann, K. (1990) A novel type of dimeric secoiridoid glycoside from *Lisianthius jefensis*. *Helv. Chim. Acta*, **73**, 1845–1852.

Hostettmann, K., Marston, A. and Wolfender, J.-L. (1995) Strategy in the search for new biologically active plant constituents. In K. Hostettmann, A. Marston, M. Maillard and M. Hamburger (eds.), *Phytochemistry of Plants Used in Traditional Medicine*, Clarendon Press, Oxford, pp. 17–45.

Hostettmann, K., Wolfender, J.-L. and Rodriguez, S. (1997) Rapid detection and subsequent isolation of bioactive constituents of crude plant extracts. *Planta Med.*, **63**, 2–10.

Ioset, J.-R., Marston, A., Gupta, M.P. and Hostettmann, K. (1998a) Antifungal xanthones from roots of *Marila laxiflora*. *Int. J. Pharmacog.*, in press.

Ioset, J.-R., Marston, A., Gupta, M.P. and Hostettmann, K. (1998b) Antifungal and larvicidal meroterpenoid naphthoquinones and a naphthoxirene from the roots of *Cordia linnaei*. *Phytochemistry*, **47**, 729–734.

Maillard, M., Gupta, M.P. and Hostettmann, K. (1987) A new antifungal prenylated flavanone from *Erythrina berteroana*. *Planta Med.*, **53**, 563–564.

Maillard, M., Hamburger, M., Gupta, M.P. and Hostettmann, K. (1989) An antifungal isoflavanone and a structure revision of a flavanone from *Erythrina berteroana*. *Planta Med.*, **55**, 281–282.

Marston, A., Dudan, G., Gupta, M.P., Solis, P.N., Correa, M.D. and Hostettmann, K. (1996) Screening of Panamanian plants for molluscicidal activity. *Int. J. Pharmacog.*, **34**, 15–18.

Mendelsson, R., Balick and M.J. (1995) The value of undiscovered pharmaceuticals in tropical forests. *Econ. Bot.*, **49**, 223–228.

Morton, J.F. (1981) *Atlas of Medicinal Plants of Middle America*, C.C. Thomas, Springfield, Illinois.

Myers, N. (1984) *The Primary Source*, Norton, New York.

Myers, N. (1986) Tropical deforestation and a mega-extinction spasm. In Soulé, M.E. (ed.), *Conservation Biology*, Sinauer, Sunderland, Mass.

Plotkin, M.J. (1988) Conservation, ethnobotany and the search for new jungle medicines: pharmacognosy comes of age … again. *Pharmacotherapy*, **8**, 257–262.

Prance, G.T. (1977) Floristic inventory of the tropics: where do we stand? *Ann. Missouri Bot. Garden*, **64**, 659–684.

Rahalison, L., Hamburger, M., Hostettmann, K., Monod, M., Frenk, E., Gupta, M.P., Santana, A.I., Correa, M.D. and Gonzalez, A.G. (1993) Screening for antifungal activity of Panamanian plants. *Int. J. Pharmacog.*, **31**, 68–76.

Rahalison, L., Benathan, M., Monod, M., Frenk, E., Gupta, M.P., Solis, P.N., Fuzzati, N. and Hostettmann, K. (1995) Antifungal principles of *Baccharis pedunculata*. *Planta Med.*, **61**, 360–362.

Rouhi, A.M. (1997) Seeking drugs in natural products. *Chem. Eng. News*, April 7, pp. 14–29.

Schaufelberger, D., Gupta, M.P. and Hostettmann, K. (1987) Flavonol and secoiridoid glycosides from *Coutoubea spicata*. *Phytochemistry*, **26**, 2377–2379.

Sengupta, S. and Ray, A.B. (1987) The chemistry of *Piper* species: a review. *Fitoterapia*, **58**, 147–166.

Soejarto, D.D. and Farnsworth, N.R. (1989) Tropical rain forests: potential sources of new drugs. *Perspectives in Biology and Medicine*, **32**, 244–256.

Speedie, M.K. (1992) Trends in scholarship in medicinal chemistry/pharmacognosy: adventures in drug design and discovery. *Am. J. Pharm. Educ.*, **56**, 414–417.

Srivastava, J., Lambert, J. and Vietmeyer, N. (1996) *Medicinal Plants: An Expanding Role in Development*. World Bank Technical Paper No. 320, The World Bank, Washington D.C., p. 22.

Suter, R., Tanner, M., Borel, C., Hostettmann, K. and Freyvogel, T.A. (1986) Laboratory and field trials at Ifakara (Kilombero District, Tanzania) on the plant molluscicide *Swartzia madagascariensis*. *Acta Tropica*, **43**, 69–83.

Terreaux, C., Maillard, M., Stoeckli-Evans, H., Gupta, M.P., Downum, K.R., Quirke, J.M.E. and Hostettmann, K. (1995a) Structure revision of a furanocoumarin from *Dorstenia contrajerva*. *Phytochemistry*, **39**, 645–647.

Terreaux, C., Maillard, M., Gupta, M.P. and Hostettmann, K. (1995b) Xanthones from *Schultesia lisianthoides*. *Phytochemistry*, **40**, 1791–1795.

Terreaux, C., Maillard, M.P., Gupta, M.P. and Hostettmann, K. (1996) Triterpenes and triterpene glycosides from *Paradrymonia macrophylla*. *Phytochemistry*, **42**, 495–499.

Terreaux, C., Gupta, M.P. and Hostettmann, K. (1998) Antifungal benzoic acid derivatives from *Piper dilatatum*. *Phytochemistry*, in press.

Watson, R.T. (1995) *Global Biodiversity Assessment. Summary for Policy Makers*. Cambridge University Press, Cambridge, p. 46.

Weil, A. (1989) A new look at botanical medicine. *Whole Earth Review*, no. 64, pp. 3–8.

World Conservation Monitoring Center (1992) *Global Biodiversity – Status of the Earth's Living Resources*. Chapman and Hall, London, pp. 84–86.

9. BIOACTIVE NATURAL PRODUCTS OF MEDICINAL AND AGROCHEMICAL INTEREST FROM SELECTED MEXICAN MEDICINAL PLANTS

RACHEL MATA[1], J. FAUSTO RIVERO-CRUZ[1] and ALEJANDRA ROJAS[2]

[1]*Facultad de Química, Universidad Nacional Autónoma de México, Coyoacán, 04510, México DF, México*
[2]*Facultad de Química, Universidad Autónoma de Querétaro, Centro Universitario, Querétaro, México*

INTRODUCTION

México, with its floristic diversity, ethnic variety, long history of human civilization and centuries old record of the use of plants as medicine provides a rich and valuable reservoir of new bioactive substances useful for the development of new drugs and pesticide agents.

The richness of medicinal plants in México can be estimated from the number of species utilized. Since the first contact with Spaniards, efforts have been made to provide a basic listing of plant species employed for healing purposes. Such endeavors, using different sources and conducted on diverse scales, continue today. The most recent effort to produce systematically an inventory of Mexican medicinal plants was carried out by Instituto Nacional Indigenista (INI) using native participants as well as bibliographic and herbarium sources. The bibliography consists of 2049 references on medicinal plants used in México and contains chemical studies on 394 species; chemical and pharmacological investigations on 280 species; and chemical, pharmacological and active principal studies on 88 species. Combining these sources, 3103 species in 183 families are currently used in México (Bye *et al.* 1995 and references cited therein). This information clearly indicates that regardless of the ethnobotanical richness of México only a small percentage of its medicinal flora has been investigated as a source of biologically active compounds.

In the past several factors have limited the investigation of plant natural products as a source of drugs and pesticide agents in México. However, many local scientific groups are currently exploring Mexican medicinal flora from the biological and chemical point of view. Such efforts have led to the isolation of several compounds with interesting biological properties (Mata 1993; Pereda-Miranda 1995; Calera *et al.* 1996; Meckes *et al.* 1996; Martínez *et al.* 1996; Navarrete and Hong 1996; Reynolds *et al.* 1995; Delgado 1996; Alvarez *et al.* 1996, *inter alia*).

In this context, we have established a rationally designed, multidisciplinary research program to obtain bioactive phytochemicals for the development of medicinal and agrochemical agents. Thus, the purpose of this chapter is to focus on some of the findings of our recent investigations on some selected Mexican medicinal plants.

SMOOTH MUSCLE RELAXING CONSTITUENTS FROM *DODONAEA VISCOSA* (SAPINDACEAE) AND *CONYZA FILAGINOIDES* (ASTERACEAE)

Dodonaea viscosa (L.) Jacq. (Sapindaceae) is a shrub widely distributed in México and reputed to possess several medicinal properties. The leaves are used against several kinds of aches, rheumatism, skin infections, fevers, swellings and "gastrointestinal disorders" including diarrhea. Some popular names of this plant are "chapuliztle", "granadina", "cuerno de cabra", "jirumi", "hierba de la cucaracha" and "ocotillo" (Martínez 1989). The chemistry of this species has been recently reviewed (Mata 1993). From the aerial parts several flavonoids, diterpenoids, condensed tannins, triterpenoids and the essential oil were isolated and characterized. Additionally, a novel *p*-coumaric acid ester of L-*myo*-inositol [1-L-1-*O*-methyl-2-acetyl-3-*p*-coumaryl-*myo*-inositol] has been obtained from the aerial parts of the plant. Finally, two biologically active saponins, dodonosides A and B, were obtained from the seeds (Mata 1993 and references therein).

Conyza filaginoides (DC) Hieron (Asteraceae) is a short grayish green annual herb with a bitter taste, commonly known in México as "simonillo". It grows in open, disturbed habitats of temperate pine-oak forests and grasslands from northern México south to Guatemala. Known in the Aztec language as "zacachichic" (bitter herb), it has been prized since prehispanic times to treat gastrointestinal ailments such as flatulence, colic and upset stomachs and to stimulate the appetite. Minor uses include its employment as an emetic, to wash infected eyes and alleviate coughs. Today it is considered as "fresh" herb which is drunk as a tea before meals to alleviate stomachaches, liver pain, and nausea. In central México, it is also a common aperitif. (Mata *et al.* 1997 and references therein).

Recently, in the course of a search for naturally occurring spasmolytic agents from Mexican medicinal plants it was demonstrated that the methanol extracts of *D. viscosa* and *C. filaginoides* exerted a significant concentration-dependent inhibition of the spontaneous contractions of the isolated rat ileum (Rojas *et al.* 1995). Therefore, in order to establish the nature of the active constituents we decided to carry out the phytochemical study of these species using a biological activity-driven procurement of antispasmodic principles. The bioassay employed for monitoring the smooth muscle relaxing activity involved the use of an isolated ileum preparation (rat or guinea pig). In this section the isolation of the spasmolytic principles from *D. viscosa* and *C. filaginoides* is described. In addition, the results of a preliminary pharmacological study designed to characterize the mechanism of action of the active constituents from *D. viscosa* will be discussed.

Initially, the chloroform–methanol (1 : 1) extract of *D. viscosa* showed a concentration-dependent inhibition of the tone and the amplitude of the spontaneous contractions of the guinea-pig ileum. The extract of *D. viscosa* ($IC_{50} = 79.5\ \mu g/ml$) resulted to be 20 times less potent than the methanolic extract of *Datura lanosa* ($IC_{50} = 4.4\ \mu g/ml$), which was used as a positive control because of its high content of scopolamine and atropine [Table 1] (Rojas *et al.* 1996). These results confirmed the spasmolytic activity previously demonstrated for the chloroform–methanol (1 : 1) extract of *D. viscosa*, employing the rat ileum model (Rojas *et al.* 1995).

Table 1 Inhibition of the spontaneous contractions of isolated guinea-pig ileum induced by the extract and the isolated compounds from *D. viscosa*

Substance	*Emax**	IC_{50}(μg/ml)	γ
D. lanosa extract	91.08 ± 1.13	4.54 ± 0.53	0.58 ± 0.03
D. viscosa extract	87.09 ± 2.63	79.51 ± 17.11	0.60 ± 0.06
Sakuranetin	84.74 ± 0.82	15.98 ± 0.82	1.07 ± 0.05
6-hydroxykaempferyl-3,7-dimethyl ether	93.67 ± 1.19	15.89 ± 1.58	0.68 ± 0.04
Ent-15,16-epoxy-9αH-labda-13(16) 14-diene-3β,8αH-diol	92.17 ± 1.47	6.17 ± 0.57	0.78 ± 0.06
Hautriwaic acid	82.95 ± 2.86	8.59 ± 2.33	0.63 ± 0.08

Vaules are means ± S.E., *n* = 6.
*E max indicates the percentage of maximum inhibition.

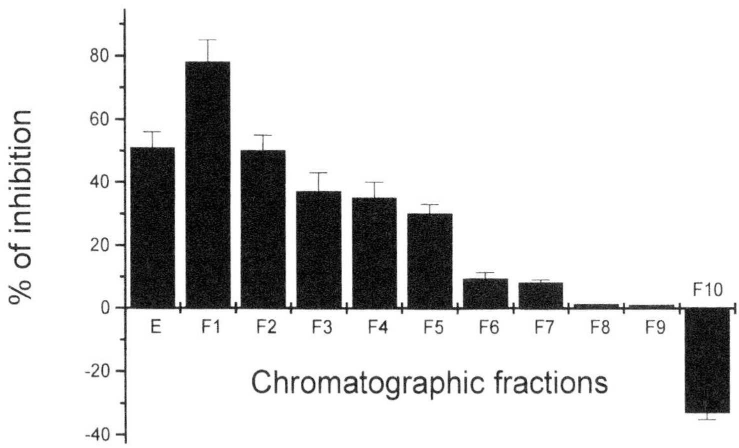

Figure 1 Effect of the chromatographic fractions from the chloroform–methanol (1 : 1) extract of *Dodonaea viscosa* on the spontaneous contractions of the guinea-pig ileum. Values are expressed as the percentages of inhibition of contractile response obtained at the IC_{50} of the original extract calculated as the mean from six data ± S. E. [From Rojas, A., Cruz, S., Ponce-Monter, H. and Mata, R. (1996) *Planta Medica*, 62, 154–159, with permission]

Column chromatography fractionation of the active extract of *D. viscosa* yielded ten primary fractions. Five of these fractions (F1–F5) decreased the spontaneous contractions of the isolated guinea pig ileum when tested at the IC_{50} of the original extract (Figure 1). Intensive chromatography of the five antispasmodic fractions F1 to F5 yielded sakuranetin (**1**), 6-hydroxy-kaempferyl-3,7-dimethyl ether (**2**), hautriwaic acid (**3**) and *ent*-15,16-epoxy-9αH-labda-13(16),14-diene-3β, 8α-diol (**4**) as the major active principles (Figure 2). The isolated metabolites induced a concentration-dependent

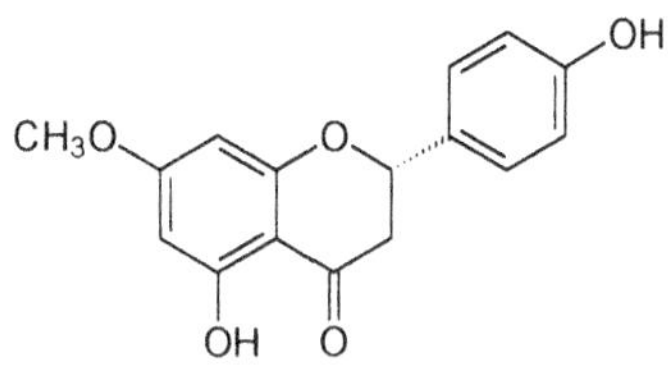

(1)

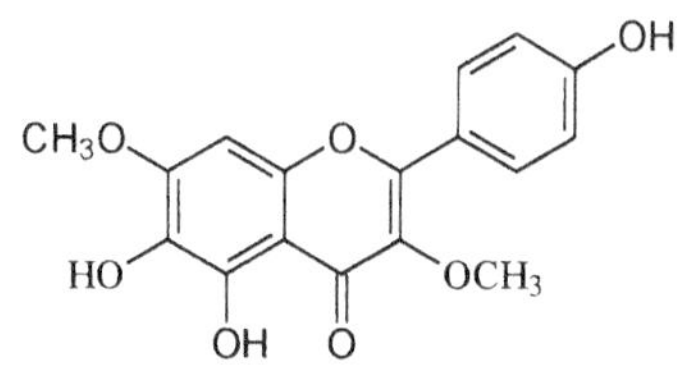

(2)

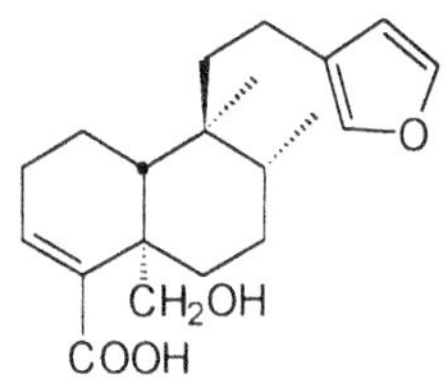

(3)

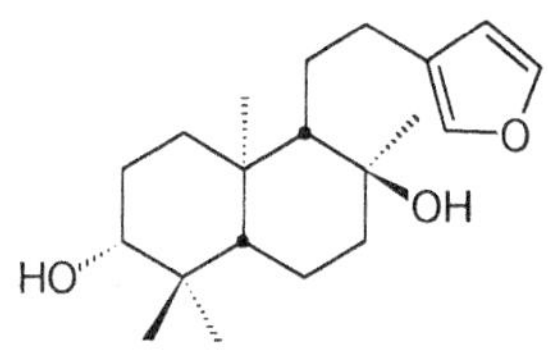

(4)

Figure 2 Smooth muscle relaxing constituents from *Dodonaea viscosa* (Sapindaceae)

inhibition of the spontaneous contractions of the guinea-pig ileum. Figure 3 shows the concentration-response curves (CRC) for the active principles and Table 1 indicates their respective IC_{50} and Emax values (Rojas *et al.* 1996).

The four active principles from *D. viscosa* also inhibited the ileum contractions evoked by electrical stimulation (Figure 4) in a concentration-dependent manner. In these experiments, the *ent*-labdane **4** ($IC_{50} = 19.37\ \mu g/ml$), sakuranetin **1** ($IC_{50} = 4.69\ \mu g/ml$) and the flavonol **2** ($IC_{50} = 37.87\ \mu g/ml$) were more potent than the original extract ($IC_{50} = 106.19\ \mu g/ml$). However, the potency of all compounds was much lower than that exhibited by the anticholinergic drug atropine ($IC_{50} = 0.005\ \mu g/ml$).

In another set of experiments, sakuranetin and the *ent*-labdane significantly inhibited the contractions induced by Ach, histamine and $BaCl_2$ (Figure 5). Since these spasmogens have different modes of action, the antagonism elicited by the flavanone and the *ent*-labdane indicated non-specific affinity for acetylcholine and histamine receptors. Furthermore, these results suggested that the metabolites might be acting at a common step in the contraction mechanism elicited by these agonists, for example interfering with Ca^{2+} mobilization into the smooth muscle cell.

To visualize if metabolites **1** and **4** induced a Ca^{2+} entry blockade, their effect on the contractions elicited by increasing concentrations of Ca^{2+} on the isolated depolarized (K^+, 30 mM) rat uterus was assessed. We took into account that when uterus is exposed to a high concentration of KCl (a depolarizing medium) contractions are the result of an increased Ca^{2+}-influx through voltage-stimulated type-L Ca^{2+} channels. If under these conditions the tissue is bathed in a Ca^{2+} free solution, the uterus relaxes. Subsequent addition of cumulative concentrations of Ca^{2+}, causes a gradual increase in tonic contractions, which are dependent of extracellular Ca^{2+} influx. These contractions are inhibited by Ca^{2+} entry blockers like nifedipine and gallopamil (Granger *et al.* 1986; Varagic *et al.* 1984). In our study, sakuranetin at concentrations of 1, 3 and 5.8 μM and the *ent*-labdane at concentrations of 3, 10 and 18 μM, produced a rightward shift of

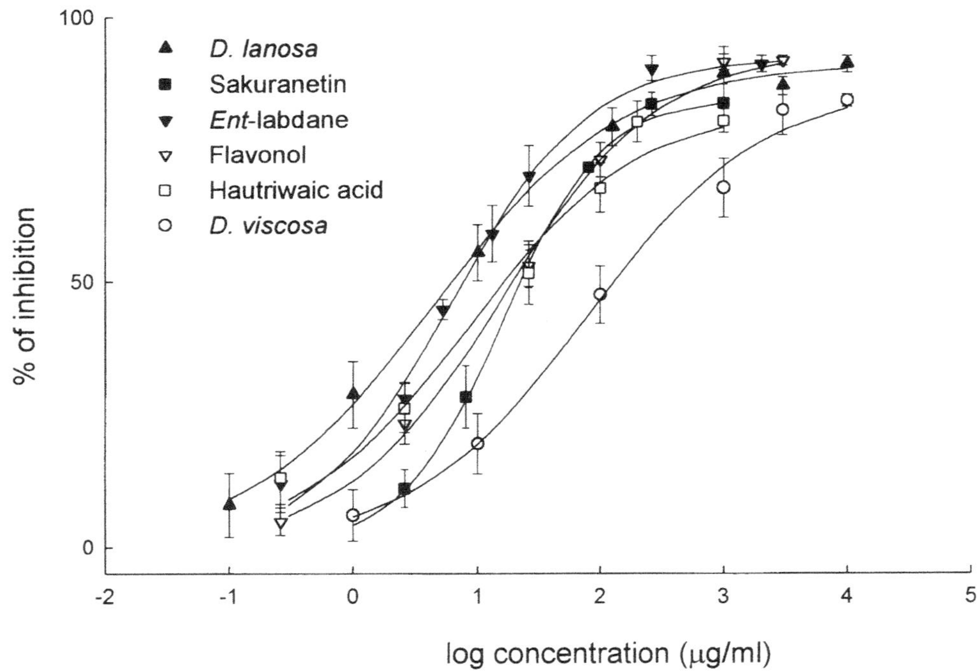

Figure 3 Concentration-response curves for the inhibitory effect of the extract and isolated compounds from *Dodonaea viscosa* on the spontaneous contractions of isolated guinea-pig ileum. Values are expressed as the percentages of inhibition of contractile response calculated as the mean from six data ± S.E. [From Rojas, A., Cruz, S., Ponce-Monter, H. and Mata, R. (1996) *Planta Medica*, **62**, 154–159, with permission]

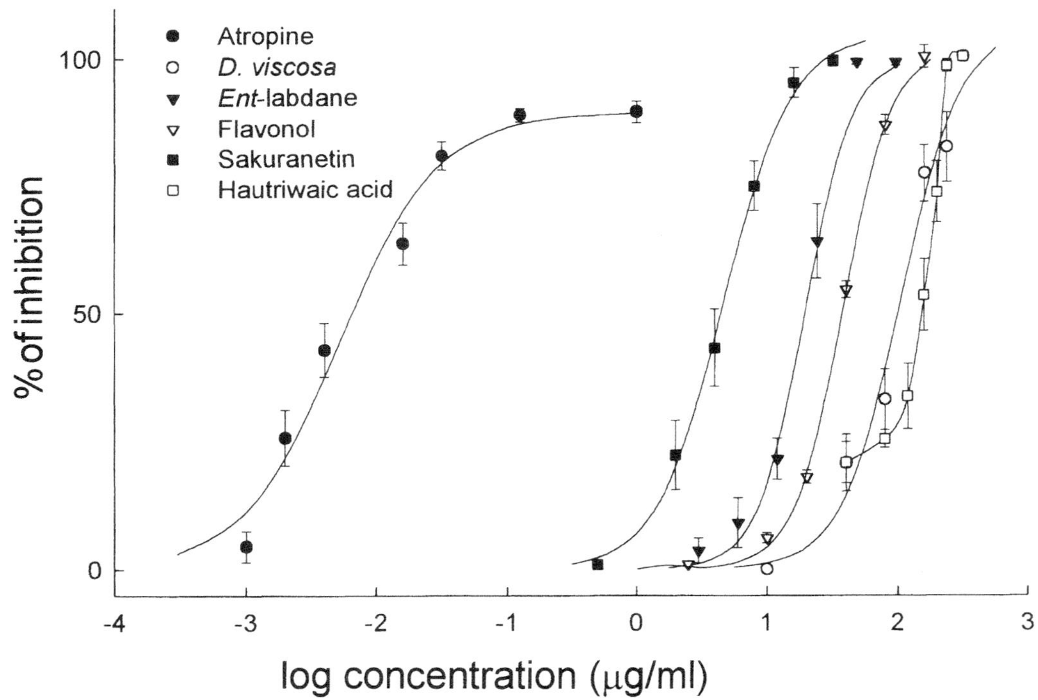

Figure 4 Concentration-response curves for the inhibitory effect of the extract and isolated compounds from *Dodonaea viscosa* on the electrically induced contractions of isolated guinea-pig ileum. Values are expressed as the percentages of inhibition of contractile response calculated as the mean from six data ± S.E. [From Rojas, A., Cruz, S., Ponce-Monter, H. and Mata, R. (1996) *Planta Medica*, **62**, 154–159, with permission]

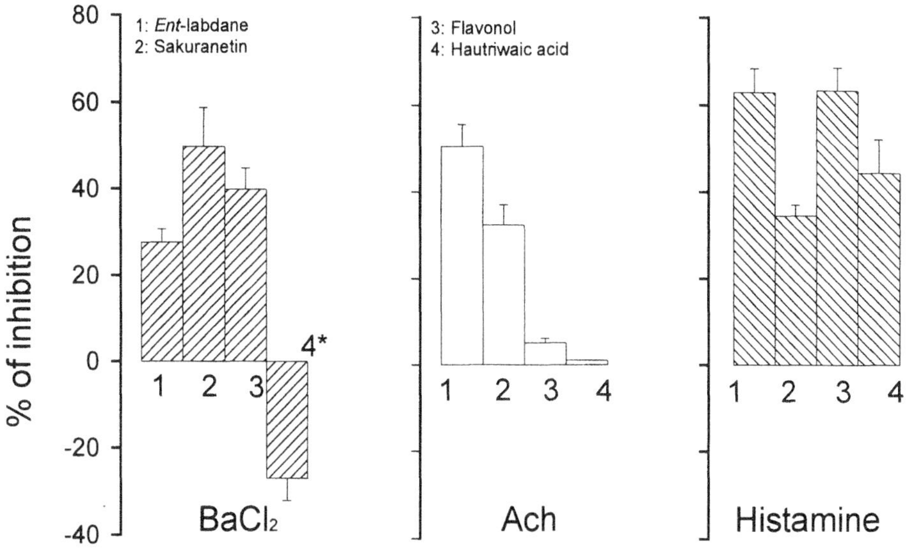

Figure 5 Effect of the isolated compounds from *Dodonaea viscosa* on the contractions induced by Ach, Histamine and Barium chloride on the guinea-pig ileum. In the tests with $BaCl_2$, substances were tested at their IC_{50} for the inhibition of spontaneous contraction of the ileum. In the experiments with Ach and Histamine, the compounds were tested at the IC_{50} for the inhibition of the electrically induced contractions. Values are expressed as the percentages of the mean responses when the agonists alone were added to the bath calculated as the mean from six data ± S.E. [From Rojas, A., Cruz, S., Ponce-Monter, H. and Mata, R. (1996) *Planta Medica*, **62**, 154–159, with permission]

the Ca^{2+}-effect control curves (Figure 6) and a reduction in the maximum effect, without significantly affecting the IC_{50} values. The same trend was observed with nifedipine (10 nM), the positive control. Thus, the shift to the right of calcium-contraction curves produced by sakuranetin and the *ent*-labdane suggested that both compounds might be inhibiting Ca^{2+} influx into the smooth muscle cells.

6-hydroxy-kaempferyl-3,7-dimethyl ether (2) inhibited the contractions evoked by histamine and barium chloride. However, this flavonol did not affect the response to Ach (Figure 5). This lack of antagonism to Ach is not in agreement with the results previously found for other flavonols which usually inhibit Ach-induced contractions of guinea-pig ileum (Fanning *et al.* 1983; Capasso *et al.* 1991; Havsteen 1983).

Finally, hautriwaic acid produced an inhibition of the histamine-induced contractions and an increase of the contractions evoked by barium chloride, without modifying the Ach-induced contractions (Figure 5). It is difficult to explain the variable behavior exhibited by hautriwaic acid on the basis of a simple mechanism of action and obviously more detailed pharmacological investigation is necessary to characterize the mode of action of this diterpene.

To isolate the smooth muscle relaxing principles from *C. filaginoides*, the active extract was fractionated by column chromatography on silica-gel to yield eight primary fractions (F1–F8). F1–F8 decreased the spontaneous contractions of isolated rat ileum

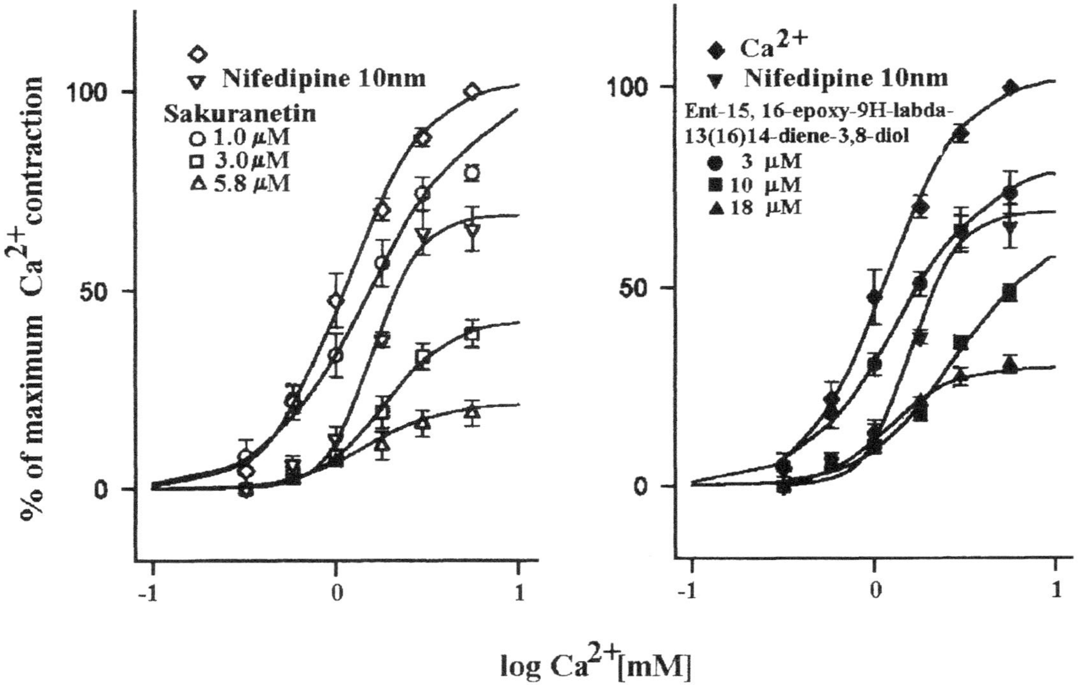

Figure 6 Calcium concentration-response curves of rat uterus in the presence of nifedipine (10 nM), sakuranetin (1, 3 and 5.8 mM) and ent-15, 16-epoxy-9αH-labda-13(16)14-diene3β,8α-diol (3, 10 and 18 mM). Values are expressed as the percentages of the maximum response to Ca^{2+} calculated as the mean from six data ± S.E. [From Rojas, A., Cruz, S., Ponce-Monter, H. and Mata, R. (1996) *Planta Medica*, **62**, 154–159, with permission]

when tested at the IC_{50} of the original extract. Repeated chromatography of the most active fractions (F-3, F-4 and F-8) allowed the isolation of four flavonoids [pinostrobin (**5**), rutin (**6**), 3-*O*-β-D-galactopyranosyl-quercetin (**7**) and 3-*O*-β-D-glucopyranosyl-quercetin (**8**)]; two triterpenoids [erythrodiol (**9**) and 3-β-tridecanoiloxy-olean-12 en-28-ol (**10**)]; β-caryophyllene-4,5-epoxide (**11**) and α-spinasterol (**12**) [Figure 7]. Compound **10** turned out to be a new natural product and its structure was determined by spectral means (Mata *et al.* 1997).

Compounds **5–12** induced a concentration-dependent inhibition of the spontaneous contractions of the rat ileum. Figure 8 shows the CRC for these phytochemicals. In general all the substances were more potent than the crude extract. Among the tested flavonoids, rutin (**6**) and 3-*O*-β-D-glucopyranosyl-quercetin (**8**) possessed a significant high inhibitory activity on the spontaneous contractions of rat ileum, while pinostrobin (**5**), exerted a rather weak effect. Flavonoids **6** and **8** were, respectively, 8.76 and 7 times more potent than quercetin but 18.75 and 15 less potent than atropine. The results obtained with the flavonoids were not unexpected since rutin (**6**), has been reported to induce smooth muscle relaxation in various *in vitro* preparations, including guinea-pig colon and rat duodenum, at the same level of concentrations employed in this investigation. In addition, this flavonoid is a non-competitive inhibitor of angiotensin II and prostaglandin E_2 on the guinea-pig ileum, interferes with arachidonic acid metabolism and inhibits cyclic AMP-phosphodiesterase (Mata *et al.* 1997 and references therein).

Of the remaining compounds, the new triterpenoid **10** and β-caryophyllene-4,5-epoxide (**11**) exhibited a moderate relaxant activity while the sterol **12** and erythrodiol (**9**) proved to be the least potent phytochemicals. The larger activity observed for **10** in comparison to **9**, may result from the ester aliphatic chain. The antispasmodic effect observed for the caryophyllene oxide has been attributed to its calcium current inhibitor property, recently demonstrated by voltage clamp experiments (Sensch *et al.* 1993).

In conclusion, the pharmacological action shown by the extracts and isolated compounds from *D. viscosa* and *C. filaginoides* accounts for the ethnomedical use of both species in Mexican folk medicine. The results of the preliminary pharmacological characterization of sakuranetin and the *ent*-labdane **4**, isolated from *D. viscosa* suggest that the mode of action of both compounds involves an interference with calcium influx into the smooth muscle cells. This type of mechanism has not been previously described for diterpenes possessing a labdane nucleus.

PESTICIDAL LIMONOIDS FROM *SWIETENIA HUMILIS* AND *CEDRELA SALVADORENSIS* (MELIACEAE)

The report of the marked insect antifeedant and growth regulating activity of azadirachtin and toosendanin has led to an increase in interest in the limonoids as green herbicides (Chiu 1995). A recent review by Champagne *et al.* (1992) summarized the effects of 78 naturally occurring limonoids on insect feeding, growth and development. It must be emphasized that this type of compounds are highly valued as pesticides for their low mammalian toxicity, non-neurotoxic mode of action and low persistance.

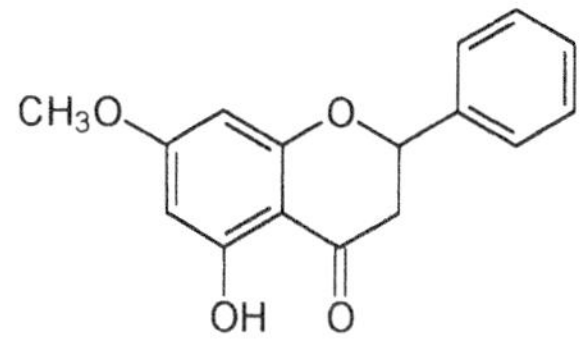

(5)

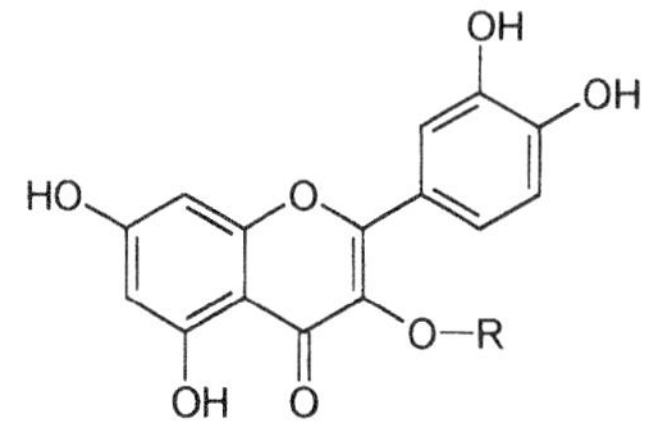

R= Rutinosyl (**6**)
R= β-D-Glucopyranosyl (**7**)
R= β-D-Galactopyranosyl (**8**)

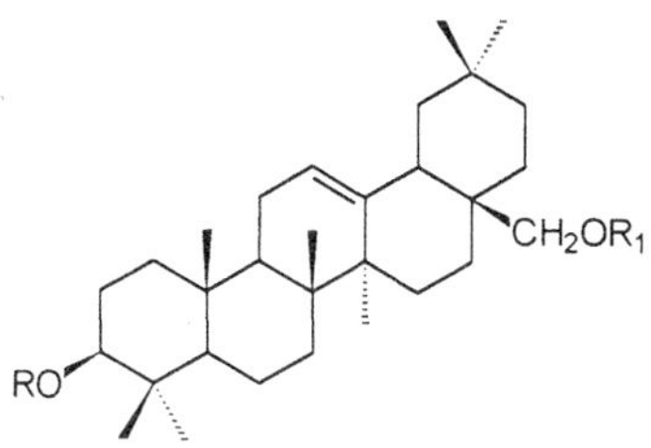

R = R_1 = H (**9**)
R = O=C-$(CH_2)_{11}$-CH_3 (**10**)

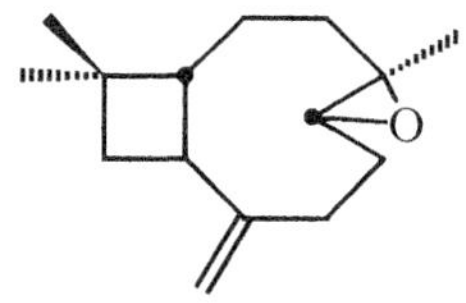

(**11**)

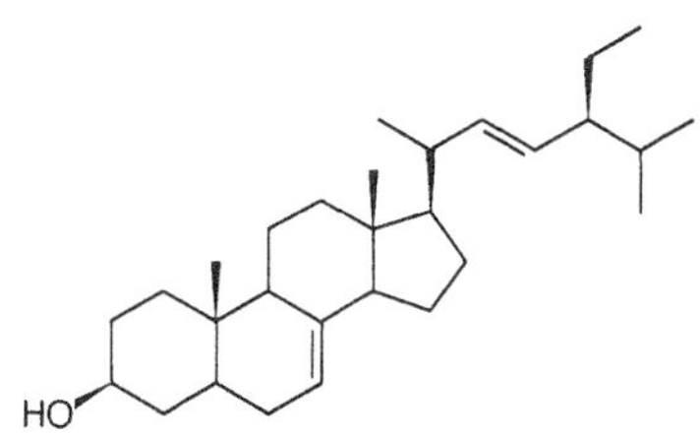

(**12**)

Figure 7 Smooth muscle relaxing constituents from *Conyza filaginoides* (Asteraceae)

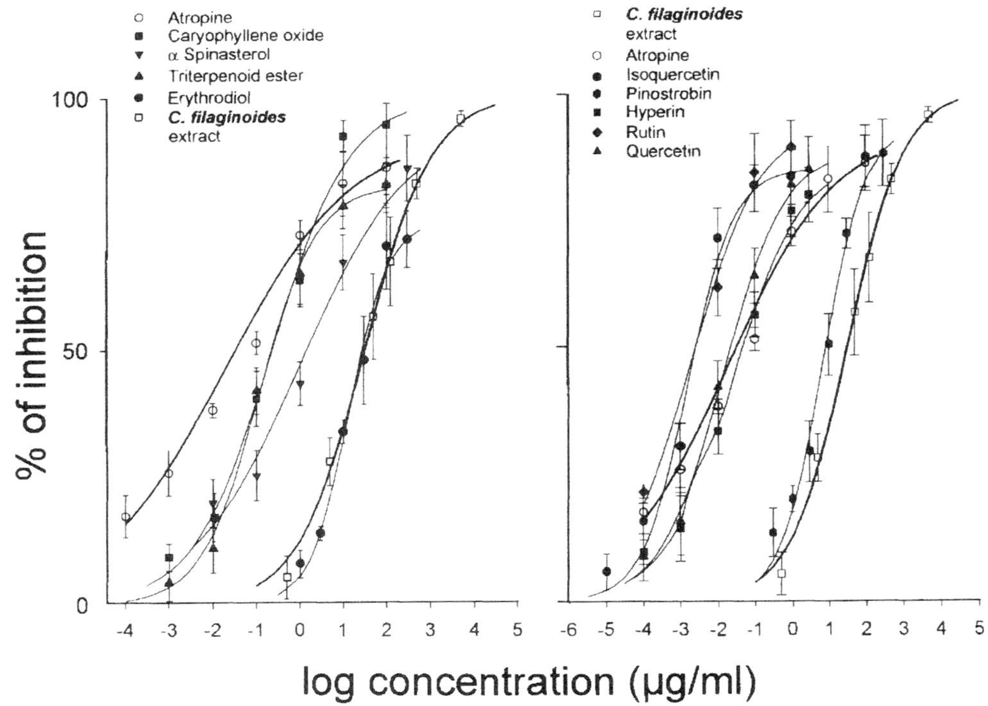

Figure 8 Concentration-response curves for the inhibitory effect of the extract and isolated compounds from *Conyza filaginoides* on the spontaneous contractions of isolated rat ileum. Values are expressed as the percentages of inhibition of contractile response calculated as the mean from six data ± S.E. [Mata, R., Rojas, A., Estrada, S., Acevedo, L., Rojas, I., Calzada, F., Linares, E. and Bye, R. (1997) *Planta Medica*, **63**, 31–35, with permission]

In an effort to find practical insecticides from Central América and México, Arnason and coworkers (1993) have screened extracts of the Meliaceae for effects against Lepidopteran pests such as the European Corn Borer (ECB) *Ostrinia nubilalis* Hubner, a highly polyphagus species which attack important economic crops such as corn, sweet pepper, beet, bean, potato, oat, soybean and wheat. The significantly active extracts included *Swietenia humilis* Zucc. and *Cedrela salvadorensis* Standley, among others.

S. humilis, locally known as "zopilote", "cóbano", "flor de venadillo" and "caoba", is a tree up to 20 m high which grows in the tropical areas of México. Decoctions or infusions of the ground seeds of this plant are highly valued for the treatment of worms, amebiasis, cancer, chest pains and coughs (Soto and Sousa 1995). Previous chemical studies conducted with the seeds led to the isolation and characterization of three limonoids, namely, humilin B, 2-methyl-2-hydroxy-3-isobutyrylmeliac-8(30)-enate, and 2-methylhydroxy-3-tlgloyloxymeliac-8(30)-enate (Okorie and Taylor 1970). *C. salvadorensis* is a small tree of the dry Pacific slope ranging from the state of Jalisco to Chiapas in México, through Central América to the North of Panamá. The stem bark of the plant is considered as febrifuge.

Chromatographic fractionation of the MeOH extract prepared from dry seeds of *S. humilis* allowed the isolation of four new tetranortriterpenoids, humilinolides A–D [**13–16**] (Segura-Correa *et al.* 1993). The identification of compounds **13–16** (Figure 9) was accomplished by conventional spectroscopic methods. The structures of humilinolides A and D were confirmed by single crystal X-ray diffraction studies (Segura-Correa *et al.* 1993; Soriano-García *et al.*, 1996). In this investigation it was also demonstrated that humilinolides A and C inhibited radicle growth of seedlings of *Amaranthus hypochondriacus* L. with IC_{50} of 99.6 μg/ml and 163.0 μg/ml, respectively. *Echinochloa crusgalli* (L.) Beauv was less sensitive to compounds **13** and **15** with IC_{50} of 199.0 μg/ml and 215.0 μg/ml, respectively. Humilinolides B and D were not inhibitory against both seedlings up to 1000 μg/ml.

From the $CHCl_3$ extract of the stem bark of *C. salvadorensis* a new rearranged limonoid, cedrelanolide I [**17**], was isolated (Figure 9). The structure was established by spectroscopic and X-ray diffraction analyses. Cedrelanolide might be biogenetically derived from a methyl angolensate type of precursor (Segura *et al.* 1994).

More recently, limonoids **13–17** were evaluated for their effect on the growth and developmental parameters of European corn borer in comparison with toosendanin, a commercial insecticide (Jiménez *et al.* 1997a). Humilinolides A–D and cedrelanolide I caused significant growth reduction in ECB larvae when incorporated at 50 ppm into diets fed to larvae beginning at the neonate stage. The highest growth reduction was elicited by humilinolide C, which produces an effect similar to that of toosendanin. However, at 5 ppm only humilinolide C, cedrelanolide I and toodendanin showed significant inhibition. Toosendanin and cedrelanolide I at both 5 and 50 ppm induced only moderate larval mortality (< 37%), but the humilinolides produced higher mortality (> 50%). The percentage of larvae which reached pupation decreased compared to the control group for all the test compounds; the most important effect was observed with humilinolides D and C which at 50 ppm resulted in only 10 and 13% of pupation, respectively. Other effects observed in many of the limonoids treated groups included reduction of adult emergence and significant delay in time to pupation and adult emergence.

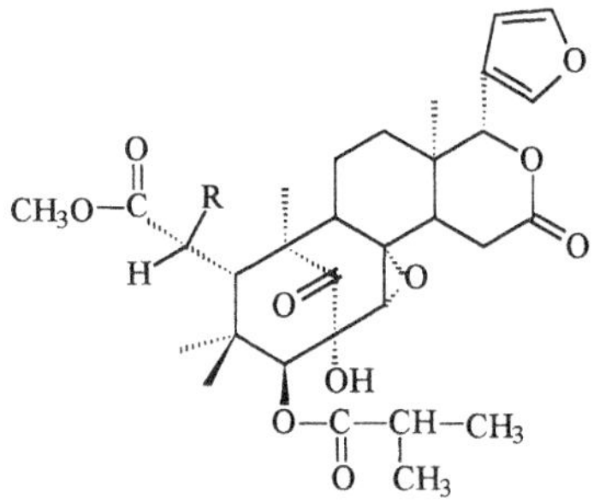

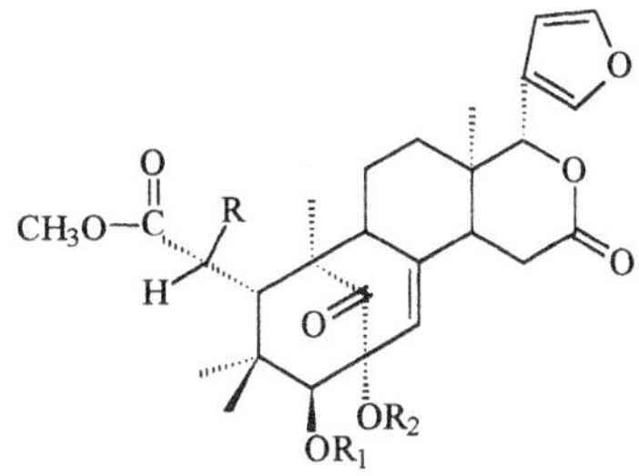

R = OH (**13**)

R = OAc (**14**)

R = H, R_1 = COC(CH_3)=CHCH_3, R_2 = Ac (**15**)

R = OAc, R_1 = Ac, R_2 = H (**16**)

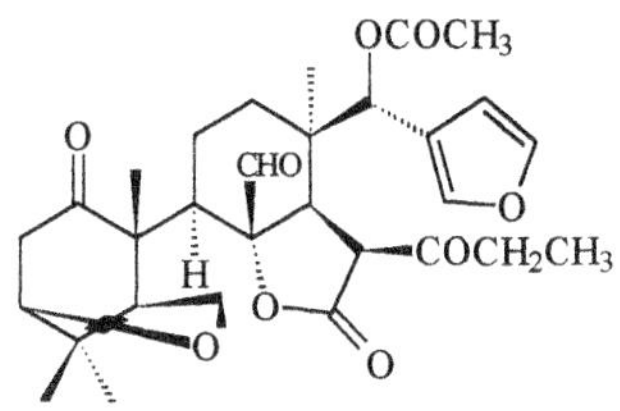

(**17**)

Figure 9 Pesticidal limonoids from *Swietenia humilis* and *Cedrela salvadorensis* (Meliaceae)

The use of the seeds of *S. humilis* for the treatment of cancer in folk medicine prompted us to evaluate the cytotoxic effect of the extract and isolated compounds against three human solid tumor cell lines, lung carcinoma (A-549), breast carcinoma (MCF-7) and colon adenocarcinoma (HT-29). In all the cases the ED_{50} values were higher than 20 μg/ml.

In summary, the effect demonstrated by humilinolides A-D and cedrelanolide I on reducing insect growth, incresing development time and mortality of ECB clearly indicates the potential of these compounds as green insecticides.

CHARACTERIZATION OF *METOPIUM BROWNEI* URUSHIOL COMPONENTS

Metopium brownei (Jacq.) Urban (Anacardiaceae) is a medicinal tree (8–25 m high) which grows exclusively in the Gulf of México, from Veracruz to Quintana Roo. The plant is locally known by the names of "chechém negro", "box chechém", "kabal chechém" and "madera negra venenosa". Several medicinal properties such as antiviral, cathartic, diaphoretic, antirheumatic and sedative are attributed to this species. The beautiful luster, color and close graining of its wood makes it one of the most attractive timber tree species of the region for cabinet work (Martínez 1989; Souza *et al.*, 1992). Previous phytochemical work with the wood of this species led to the detection of eight flavonoids by paper chromatography (Young 1975; 1979). *M. brownei*, like poison ivy and many other plants of the Anacardiaceae family, causes irritation, inflammation, and blistering of the skin of sensitive individuals. These observations suggested the presence of *n*-alk(en)ylcatechol type of allergenes in the plant.

The presence of *n*-alk(en)ylcatechols in the stem bark was established by GC-MS analysis of their corresponding TMSi ether derivatives, which were obtained by treatment of the crude acetone extract with Sigma Sil-A (Rivero-Cruz *et al.* 1997). According to the results summarized in Table 2 the sample from the stem bark consisted largely of 3-*n*-pentadec(en)ylcatechols (99.11%), while the homologous *n*-C_{17}-substituted catechols were present in trace levels (0.89%). With respect to the relative proportions of each alk(en)ylcatechol, the mono- and diolefinic congeners of 3-*n*-pentadecylcatechol were predominant. The acetone extracts derived from the leaves and wood were devoid of *n*-alk(en)ylcatechols.

The acetone extract from the fresh stem bark exhibited antifungal property against *Fusarium oxysporum* and *Helminthosporium longirostratum*. The extract also displayed significant toxicity to brine shrimp larvae (BST); the LC_{50} was 123.93 µg/ml. The active extract was fractionated by flash column chromatography on silica gel, using the BST and the bioautographic antifungal test at each step for activity directed fractionation, to yield only one active fraction. This fraction was toxic to *Artemia salina* (BST $LC_{50} = 89.13$ µg/ml) and also showed antifungal activity against *H. longirostratum* and *F. oxysporum* (Figure 10) at a concentration range between 150 and 700 µg/ml; the inhibition of the radial growth of the phytopathogenic fungi was comparable to that induced by Captan, a commercial fungicide employed as a positive control. According to the GC-MS analysis the active fraction contained the three *n*-C_{15}-alk(en)yl catechols detected in the crude extract. HPLC separation of the methylated catechol active mixture allowed the isolation of compounds **18**–**20** (Figure 11) as their methyl derivatives and their structures were determined by spectroscopic means (Rivero-Cruz *et al.* 1997). Compound **20** is a new natural product, while compounds **18** and **19** have been previously detected in *Rhus vernicifera* and *Gluta renghas* (Yamauchi *et al.* 1981; Du *et al.*

Table 2 Composition of urushiols from the fresh stem bark of *Metopium brownei*

Type of alk(en)ylcatechols	R_t*(min)*	*% of each alk(en)ylcatechol in the mixture*
bis-TMSi-n-C_{15}-catechols		
Saturated	24.23	21.16
Monoene	24.54	38.52
Diene	24.69	39.43
bis-TMSi-*n*-C_{17}-catechols		
Saturated	26.22	0.30
Monoene	26.31	0.59

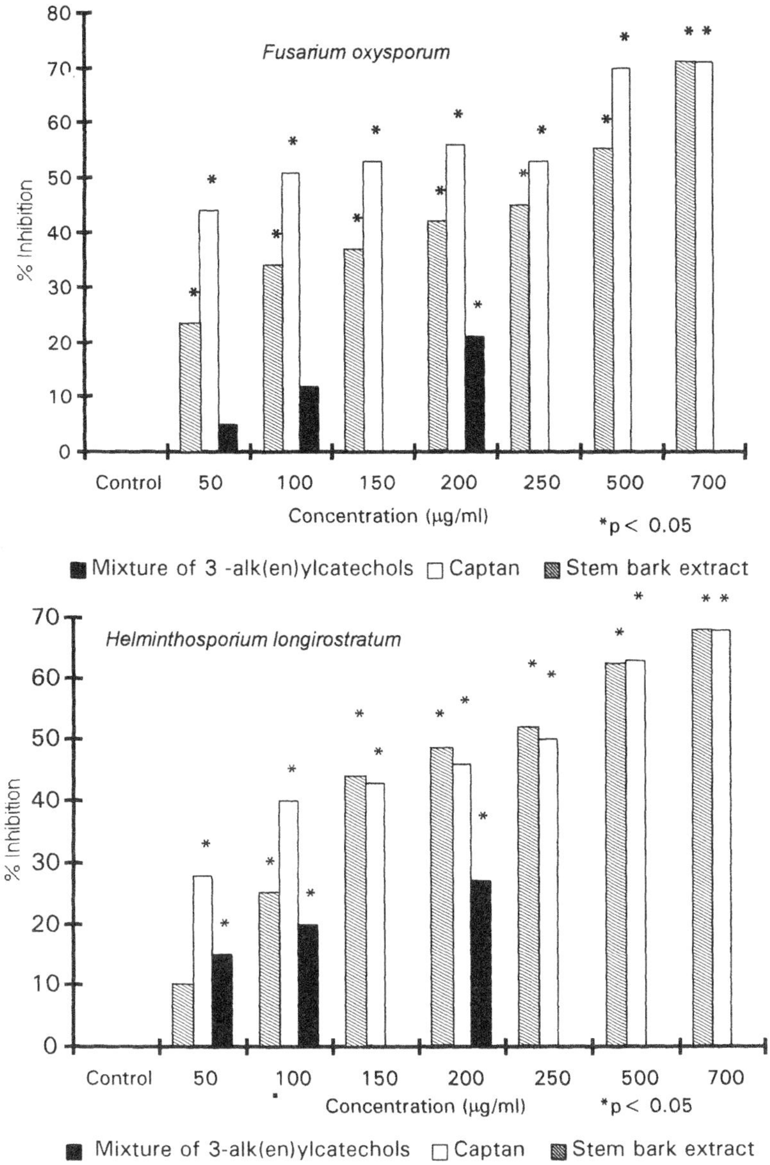

Figure 10 Antifungal activity of the acetone extract and urushiol from *Metopium brownei* against *Helminthosporium longistratum* and *Fusarium oxysporum*

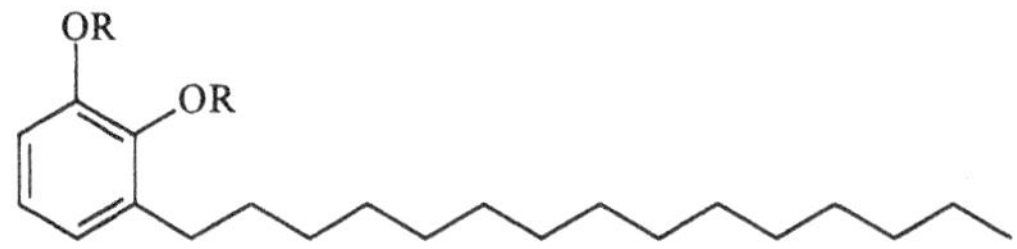

(**18**) R =H
(**18a**) R= CH_3

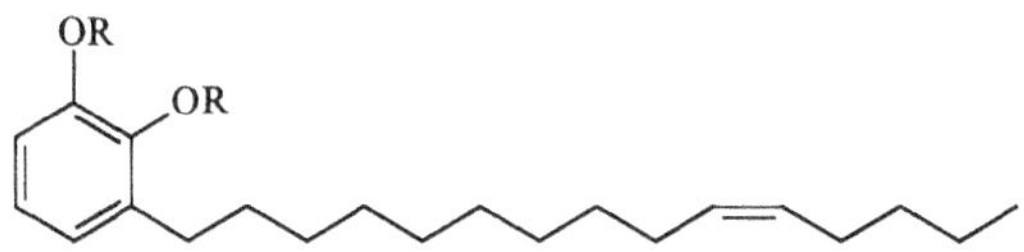

(**19**) R= H
(**19a**) R= CH_3

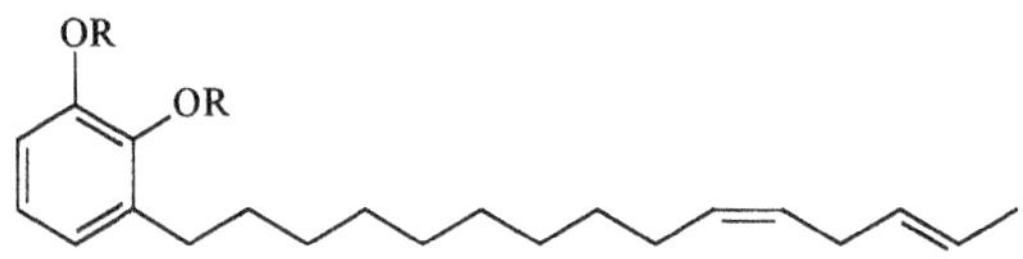

(**20**) R= H
(**20a**) R= CH_3

Figure 11 *Metopium brownei* urushiol components

1984; Jefferson and Wangchareontrakul 1985). It is important to note that the double bond positioning at the alkenyl side chain of compound **20** is unusual.

The urushiol from *M. brownei* exhibited a remarkable antifungal activity against two phytopathogenic fungi; however, the allergenic properties of these compounds likely preclude their development into environmentally safe fungicides. Nevertheless, the antifungal activity elicited by the urushiol could be of importance in the defence mechanism of the plant, preventing the attack of some fungi and bacteria.

The results of our investigation demonstrated that the stem bark of *M. brownei* contains alk(en)ylcatechol-type allergens, possessing 15 carbons unsaturated alkyl side chains, as do most of the allergenic Anacardiaceae species that have been previously studied (Tyman 1991 and Rodríguez 1992). Therefore, the use of this plant as a medicinal agent represents a considerable hazard in Southeastern México, where the species is particularly widespread. It is important to point out that the brine shrimp test turned out to be a simple and efficient method to detect the presence of alk(en)ylcatechols.

PHYTOGROWTH-INHIBITORY COMPOUNDS FROM *HELIANTELLA QUINQUENERVIS* (ASTERACEAE) AND *MALMEA DEPRESSA* (ANNONACEAE)

Helianthella quinquenervis (Hook.) A. Gray (Asteraceae), commonly known as "little sunflower", is a perennial herb with large yellow flowering heads and grows in the sub-alpine coniferous forests of the Rocky Mountains of western United States from Montana to New México. Isolated populations extended into the pine-oak forests of the extreme northern Sierra Madre Occidental and Sierra Madre Oriental of México. It is known to the Tarahumara Indians of Chihuahua, México, as "rarésoa" and its thick, vertical roots are used medicinally. The powdered roots are applied topically to the body of humans and domestic animals to eliminate ectoparasites. The decoction can be employed for the same purpose as well as to improve the quality of the hair. Taken internally, the decoction is said to be good for treating gastrointestinal ailments and ulcers. A previous phytochemical study on the aerial parts of this plant led to the isolation and characterization of one prenylhydroxyphenone and several known acetylchromenes, including demethylencecalin (**21**) and encecalin (**22**) (Castañeda *et al.* 1996 and references therein).

Malmea depressa (Baill.) R. E. Fries (Annonaceae) [syn. *Guatteria leiophylla* (Donn. Sm.) Saff. ex. Standl.], commonly known by the Maya people as "elemuy", "sufricaya", "elemuy-box" and "nazareno prieto", is a tree up to 10 m high. This species is known to occur in Central América and México (from Veracruz to the Península of Yucatán) and grows in the tropical forest as a dominant species. The tree is used by local people as an analgesic agent and for the treatment of several diseases including pelagra, liver and kidney gallstones (Jiménez *et al.* 1996 and references therein).

In our screening subprogram for potential herbicidal agents we found that a methanol extract prepared from the roots of *H. quinquenervis* inhibited radical growth of seedlings of *A. hypochondriacus* (IC_{50} of 25.0 μg/ml) and *E. crusgalli* (IC_{50} of 53.0 μg/ml). Fractionation of the active extract by column chromatography over silica gel, using the bioautographic phytogrowth-inhibitory bioassay at each step for activity directed fractionation, led to the isolation of a new benzofuran, 6-methoxytremetone (**23**) and a new prenylacetophenone, 4-β-D-glucopyranosyloxy-3-[3-methoxy-*trans*-isopenten-(1)-yl]-acetophenone (**24**) (Figure 12). In addition, 6-hydroxy-3-methoxytremetone (**25**), euparin (**26**), demethylencecalin (**21**) and encecalin (**22**) were obtained. Structural assignments of the isolated compounds were based on spectroscopic and spectrometric analyses (Castañeda *et al.* 1996).

The six compounds were evaluated for their ability to inhibit the radicle growth of *A. hypochondriacus* and *E. crusgalli*. According to the results summarized in Table 3, these products reduced the radicle growth of both target species. The two chromenes were the most active and they showed comparable phytotoxicity against both seedlings. Since it has been described that some chromenes are phototoxic, we decided to repeat the dish plate bioassay with the isolated compounds using the same conditions but irradiating first during 2 h with UV or visible light. The results obtained in the three sets of experiments were comparable, therefore, we ruled out any phototoxic effect, and thus concluded that the phytotoxic compounds from *H. quinquenervis* inhibited plant growth by a different mechanism.

More recently, we decided to further explore the herbicidal potential of compounds **21**, **22** and **26** by evaluating their effect on some photosynthetic activities associated with the light reaction phase on freshly lysed spinach chloroplasts. (Castañeda *et al.* 1997). The results indicated that encecalin (**22**) and demethylencecalin (**21**) inhibited proton uptake, ATP synthesis and electron flow (basal, phosphorylating and uncoupled) in a concentration dependent manner, therefore acting as Hill reaction inhibitors.

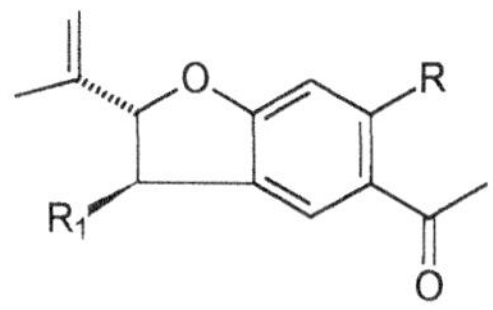

R = OCH_3, R = H (**23**)
R = OH, R_1 = OCH_3 (**25**)

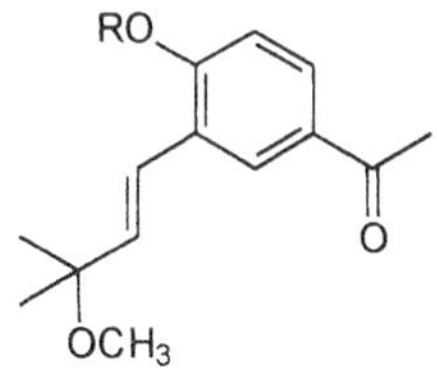

R = β-D-Glucopyranosyl (**24**)

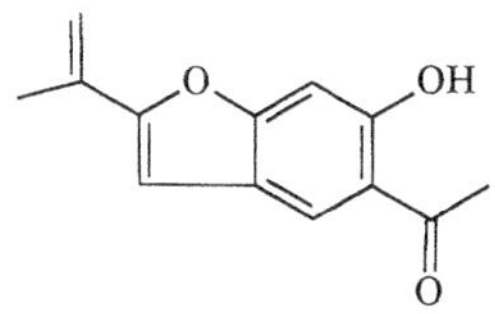

(**26**)

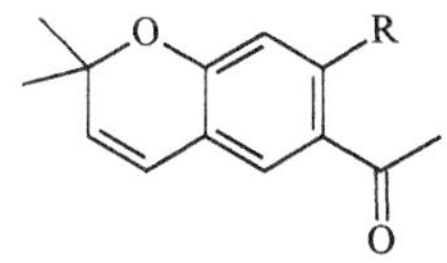

R = OH (**21**)
R = = OCH_3 (**22**)

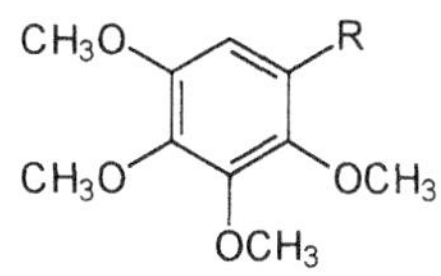

R = CH_2-CH=CH_2 (**27**)
R = CH=CH-CHO (**28**)
R = CH_2-CH=CH_3 (**29**)
R = CH=CH-CH_2-OH (**30**)
R = CHO (**31**)

Figure 12 Phytogrowth-Inhibitory compounds from *Heliantella quinquenervis* (Asteraceae) and *Malmea depressa* (Annonaceae)

Table 3 Phytogrowth-Inhibitory activity of the MeOH extract and compounds **21–26** from *H. quinquenervis*

	Seedling growth IC_{50} (μg/ml)	
Compound	*E. crusgalli*	*A. hypochondriacus*
21	79	27
22	6	7
23	68	11
24	231	148
25	323	61
26	252	111
MeOH extract	53	25

On the other hand, euparin (**26**) blocked ATP synthesis, proton uptake, basal and phosphorylating electron flow without any effect on uncoupled electron flow. This compound also inhibited the light membrane activated Mg^{2+}-ATPase. Thus euparin acts as an energy-transfer inhibitor (Castañeda *et al.* 1997).

The cytotoxic activity of the extract and compounds **21–26** was also evaluated against three human solid tumor cell lines, using adriamycin as the positive control. According to the results summarized in Table 4 demethylencecalin showed cytotoxicity against MCF-7 (breast) and HT-29 (colon), while 6-methoxytremetone (**23**) was only active against A-549 human (lung) solid tumor cells. In all the cases the cytotoxic effects were weak (Castañeda *et al.* 1996).

Compounds **21** and **26** showed also antifungal activity against *T. mentagrophytes* with MIC values of 58 and 29 μg/mL, respectively. Griseofulvin (MIC of 2.6 μg/mL) was used as the reference antibiotic. None of the isolated compounds was active against *Candida albicans* and *Aspergillus niger* (Castañeda *et al.* 1996).

It has been demonstrated that encecalin (**22**) and related chromenes possess insecticidal and repellent properties (Proksch and Rodriguez 1983). Thus, the presence of this type of compounds in *H. quinquenervis* might be related to the ethnomedical application of the roots to eliminate ectoparasites in humans and domestic animals.

The initial phytogrowth-inhibitory activity of the $CHCl_3$ extract of the stem bark of *M. depresa* on seedlings of *A. hypochondriacus* and *E. crusgalli* gave IC_{50} values of 134.0 and 457 μg/ml, respectively. The essential oil of the stem bark also inhibited the radicle growth of both species with IC_{50} of 116 and 475.2 μg/ml, respectively. Successive chromatographies of the active extract, guided by phytogrowth-inhibitory bioassays, allowed the isolation of the known phenyl propanoids 1,2,3,4-tetramethoxy-5-(2-propenyl) benzene (**27**), 2,3,4,5-tetramethoxycinnamaldehyde (**28**), *trans*-isomyristicin (**29**), 2,3,4,5-tetramethoxycinnamyl alcohol (**31**) and the novel C_6C_1 derivative 2,3,4,5-tetramethoxy- benzaldehyde (**31**) (Jiménez *et al.* 1996). It should be point out that tetraoxygenation of simple aromatic compounds is rare in nature, and their presence in *M. depressa* could be of chemotaxonomic significance.

Table 4 Cytotoxicity and Brine Shrimp Lethality (BST) of the MeOH extract and compounds **21–26** from *H. quinquenervis*.

	ED_{50}(μg/ml)			LC_{50}(μg/ml)
Compound	*MCF-7*[a]	*HT-29*[a]	*A-549*[a]	*BST*
21	2	2	10	31
22	29	18	25	31
23	33	16	1	8
24	44	61	6	>1000
25	34	16	31	68
26	6	6	22	44
Extract	3	>100	5	9
Adriamycin[b]	2×10^{-2}	4×10^{-2}	2×10^{-3}	ND

[a] Cytotoxicities in human breast (MCF-7), colon (HT-29) and lung (A-549) tumor cell lines. Activity criteria: extracts, $ED_{50} < 20$ μg/ml; pure compounds, $ED_{50} < 4$ μg/ml. [b] Positive standard control.

The active essential oil obtained by water and steam distillation of the stem bark was also investigated by GC-MS analysis, which showed the presence of **27–29**, **31** and an unidentified compound. Identification was made by comparison of the GC mobilities, the interpretation of mass fragmentation patterns, and coelution with compounds **27–29**, and **32** isolated from the $CHCl_3$ extract. According to the GC-MS analysis compound **27** was also the major component of the essential oil (95%). Compounds **28–29**, **31** and the unknown were present in a concentration of 1.25, 1.21, 1.34 and 1.11%, respectively.

Compound **27** produced significant inhibition of radicle growth in *A. hypochondriacus* (IC_{50} = 43.0 μg/ml) but it was less sensitive against *E. crusgalli* (IC_{50} = 810 μg/ml). These results were in agreement with those of Horada and Nakayama, who demonstrated that this compound reduced radicle growth of rice seedlings by 55% at a concentration of 100 μg/ml (Horada and Nakayama 1989). The new natural product **31** and the mixture of phenylpropanoids **28–31** were markedly less active in the phytogrowth-inhibitory bioassays than **27** (Jiménez *et al.* 1996).

The effect of natural product **27** on several photosynthetic activities was also investigated on freshly lysed spinach chloroplasts (Jiménez *et al.* 1998). The results indicated that this compound inhibits proton uptake, ATP synthesis and electron flow (basal, phosphorylating and uncoupled) in a concentration-dependent manner, therefore acting as a Hill's reaction inhibitor. The concentration values producing 50% inhibition (I_{50}) for each electron transport flow (basal, phosphorylating and uncoupled) were 4.80, 3.30 and 4.30 μM, respectively. The I_{50} for ATP-synthesis and H^+-uptake inhibition were 1.40 and 2.30 μM, respectively.

In summary, the results of this study showed that the isolated compounds from *M. depressa* and *H. quinquenervis* possess promising phytogrowth-inhibitory activity. Therefore, they could be developed into environmentally safe herbicide agents.

CONCLUSIONS

Selected Mexican plants from the Anacardiaceae (*M. brownei*), Annonaceae (*Malmea depressa*), Asteraceae (*C. filaginoides* and *H. quinquenervis*), Meliaceae (*C. salvadorensis* and *S. humilis*) and Sapindaceae (*D. viscosa*) families were identified as a source of potential pesticidal or medicinal agents. Bioassay-guided isolation and *in vitro* biological testing yielded several spasmolytic, phytotoxic, insecticidal and antifungal compounds. These bioactive secondary metabolites could be exploited as leads for further chemical synthesis or as commercial products in their own right. In addition, the results of our investigation on *D. viscosa*, *C. filaginoides* and *H. quinquenervis* provide scientific support for the major reputed medicinal properties of these plants.

ACKNOWLEDGEMENTS

This work has been conducted under the auspices of several project grants funded by Consejo Nacional de Ciencia y Tecnología (CONACyT, Convenio 400313-5-2576PM),

Dirección General de Asuntos del Personal Académico, UNAM (DGAPA, Proyectos IN203494, IN206795 and IN205197) and Dirección General de Estudios de Posgrado, UNAM. A. Rojas acknowledges financial support from Dirección de la Facultad de Química, U. A. Q. The work on pesticidal agents have been conducted in collaboration with Dr. J. Thor Arnason, University of Ottawa, Canadá; Dr. Blas Lotina, Departamento de Bioquímica, Facultad de Química, UNAM and Dr. Ana L. Anaya, Instituto de Fisiología Celular, UNAM. The work on *D. viscosa* was accomplished with Dr. Silvia Cruz (Sección de Terapéutica experimental, CINVESTAT, IPN) and Dr. Héctor Ponce-Monter (Centro Medico Nacional Siglo XXI, IMSS). The technical assistance of Isabel Chávez, Beatríz Quiroz, Luis Velasco-Ibarra, Javier Pérez-Flores, Graciela Chávez, Marisela Gutiérrez and Isela Rojas is gratefully acknowledged. R. Mata is grateful to the following graduate students for their valuable work in this research: Ma Adelina Jiménez, Perla Castañeda, Daniel Chávez, Samuel Estrada and Laura Acevedo. Without their continuous effort and enthusiasm most of this work would have not been possible.

REFERENCES

Alvarez, L., Marquina, S., Villareal, M.L., Santos, D., Aranda, E. and Delgado, G. (1996) Biactive polyacetylenes from *Bidens pilosa. Planta Medica*, **62**, 355–357.

Arnason, J.T., MacKinnon, S., Durst, A., Philogéne, B.J.R., Hasbun, C., Sánchez, P., Poveda, L., San Roman, L., Isman, M.B., Satasook, C., Towers, G.H.N., Wiriyachitra, P., and McLauglin, J.L. (1993) In *Phytochemical potential of tropical plants. Recent Advances in Phytochemistry*. Vol. 27 (eds. K.R. Downum, J.T. Romeo and H. A. Stafford), pp. 107–131, Plenum Press, New York.

Bye, R., Linares, E. and Estrada, E. (1995) In *Phytochemistry of medicinal plants. Recent Advances in Phytochemistry*. Vol. 29 (eds. J.T. Arnason, J.T. Romeo and R. Mata), Plenum Press, New York, pp. 65–82.

Calera, M.R., Mata, R., Lotina-Hennsen, B. and Anaya, A.L. (1996) Uncoupling behavior of the 4-phenylcoumarins in spinach chloroplasts: Structure–Activity relationships. *J. Agr. and Food Chem.*, **44**, 2966–2969.

Capasso, A., Pinto, A., Mascolo, N., Autore, G. and Capasso, F. (1991) Reduction of agonist-induced contractions of guinea-pig isolated ileum by flavonoids. *Phytother. Res.*, **5**, 85–87.

Castañeda, P., Goméz, L., Mata, R., Lotina-Hennsen, B., Anaya, A.L. and Bye, Robert (1996) Phytogrowth-Inhibitory and antifungal constituents of *Helliantella quinquenervis. J. Nat. Prod.*, **59**, 323–326.

Castañeda, P., Mata, R. and Lotina-Henssen, B. (1997) unpublished results.

Champagne, D.E., Koul, O., Isman, M.B., Scudder, G.G.E. and Towers, G.H.N. (1992) Biological activity of limonoids from the rutales. *Phytochemistry*, **31**, 377.

Chiu, S.-F. (1995) In *The neem tree Azadirachta indica A. Juss and other meliaceous plants: Sources of unique natural products for integrated pest management, medicine and other purposes* (eds. H. Schmutterer, K.R.S. Ascher, M.B. Isman, M. Jacobson, C. Ketkar, M. Kraus, H. Rembold and R.C. Saxena), GTZ, Eschborn, pp. 661–668.

Delgado, G. (1996) *Bioactive constituents of some Mexican medicinal Compositae*. In P.D.s. Caligari and D.J.N. Hind (eds.). Compositae: Biology & Utilization. Proceedings of the international Compositae conference, Kew, 1994. (D.J.N. Hind, Editor-in-Chief), Vol. 2, pp. 505–515, Royal Botanic Gardens, Kew.

Du, Y., Oshima, R., Yamauchi, Y., Iwatsuki, H. and Kumanotani, S. (1984) High performance liquid chromatography, analysis of urushiol of the lac tree, *Rhus vernicifera*, without derivatization. *J. Chromatography*, **295**, 179–186.

Fanning, M.J., Macander, P., Drzewiecki, G. and Middleton, E. (1983) Quercetin inhibits anaphylactic contraction of guinea pig ileum smooth muscle. *International Archives of Allergy and Applied Immunology*, **71**, 371–373.

Granger, S.E., Hollingworth, M. and Weston, A.H. (1986) Effects of calcium antagonists on tension development and calcium influx in rat uterus. *Brit. J. Pharmacol.*, **87**, 147–156.

Havsteen, B. (1983) Flavonoids, a class of natural products of high pharmacological potency. *Biochem. Pharmacol.*, **32**, 1141–1148.

Horada, J. and Nakayama, M. (1989) Effects of nonterpenic essential oils on the growth of rice seedlings. *Weed Res.*, Japan, **34**, 65–67.

Jefferson, A. and Wangchareontrakul, S. (1985) Long chain phenols urushiol, laccol, thitsiol and phenylalkylcatechol compounds in Burmese lac from *Melanorrhoea usitate*. *J. of Chromatography*, **367**, 145–154.

Jiménez, A. Mata, R., Lotina-Henssen, B., Anaya, A.L. and Velasco-Ibarra, L. (1996) Phytogrowth-Inhibitory compounds from *Malmea depressa*. *J. Nat. Prod.*, **59**, 202–204.

Jiménez, A., Mata, R., Pereda-Miranda, R., Calderón, J., Isman, M.B., Nicol, R. and Arnason, J.T. (1997a): Insecticidal limonoids from *Swietenia humilis* and *Cedrela salvadorensis* (Meliaceae). *J. Chem. Ecol.*, **23**, 1225–1234.

Jiménez, A., Mata, R., Anaya, A.L. and Lotina-Hennsen, B. (1998) Interference of 1,2,3,4-tetra-methoxy-5(2-propenyl)-benzene with photosynthetic electron transport, *Z. Naturforsch*, **53**, 55–59.

Martínez, M. (1989) *Las plantas medicinales de México*. 6th Edition. Editorial Botas, México, D.F., pp. 291, 411, 563.

Martínez-Vázquez, M., Ramírez-Apan, T.O., Aguilar, H. and Bye, R. (1996). Analgesic and antipyretic activities of an aqueous extract and of the flavone Linarin of *Buddleia cordata*. *Planta Medica*, **62**, 137–140.

Mata, R. (1993) In *Phytochemical potential of tropical plants. Recent Advances in Phytochemistry*, Vol. 27 (eds. K.R. Downum, J.T. Romeo and H.A. Stafford), pp. 107–131, Plenum Press, New York.

Mata, R., Rojas, A., Estrada, S., Acevedo, L., Rojas, I., Calzada, F., Linares, E. and Bye, R. (1997) Smooth muscle relaxing flavonoids and terpenoids from *Conyza filaginoides*. *Planta Medica*, **63**, 31–35.

Meckes, M., Calzada, F., Tortoriello, J., Gónzalez, J.L. and Martínez, M. (1996) Terpenoids isolated from *Psidium guajava* hexane extract with depressant activity on central nervous system. *Phytother. Res.*, **10**, 600–603.

Navarrete, A. and Hong, E. (1996) Anthelmintic properties of α-sanshool from *Zanthoxylum liebmanni-aum*. *Planta Medica*, **62**, 250–251.

Okorie D.A. and Taylor D.H.A. (1970) Limonoids from *Swietenia humilis*. *Phytochemistry*, **10**, 469–470.

Pereda-Miranda, R. (1995) In *Phytochemistry of medicinal plants. Recent Advances in Phytochemistry*, Vol. 29 (eds. J.T. Arnason, J.T. Romeo and R. Mata), Plenum Press, New York, pp. 83–112.

Proksch, P. and Rodriguez, E. (1983) Chromenes and benzofurans of the Asteraceae, their chemistry and biological significance. *Phytochemistry*, **22**, 2335–2347.

Reynolds, W.F., Yu, M., Enriquez, R.G., González, H., León, I., Magos, G. and Villareal, M.L. (1995) Isolation and characterization of cytotoxic and antibacterial tetrasaccharide glycosides from *Ipomea stans*. *J. Nat. Prod.*, **58**, 1730–1734.

Rivero-Cruz, J.F., Chávez, D., Hernández-Bautista, B., Anaya, A.L. and Mata, R. (1997) Separation and characterization of *Metopium brownei* urushiol components. *Phytochemistry*, in press.

Rodríguez, E. (1992) *Tropical and temperate plant contact allergens*, Proceedings of the National Conference on Environmental Hazards to the Skin, Washington, DC, October 14–16, pp. 1–13.

Rojas, A., Cruz, S., Ponce-Monter, H. and Mata, R. (1996) Smooth muscle relaxing compounds from *Dodonaea viscosa. Planta Medica*, **62**, 154–159.

Rojas, A., Cruz, S., Rauch, V., Bye, R., Linares, E. and Mata, R. (1995) Spasmolytic potential of some plants used in Mexican traditional medicine for the treatment of gastrointestinal disorders. *Phytomedicine*, **2**, 51–55.

Segura, R., Calderón, J., Toscano, R., Gutiérrez, A. and Mata. R. (1994) Cedrelanolide I, a new limonoid from *Cedrela salvadorensis*, *Tetrahedron Lett.*, **35**, 3437–3440.

Segura-Correa, R., Mata, R., Anaya, A.L., Hernández-Bautista, B., Villena, R., Soriano-Garcia, M., Bye, R. and Linares, E. (1993) New tetranortriterpenoids from *Swietenia humilis. J. Nat. Prod.*, **56**, 1567–1575.

Sensch, O., Vierling, W., Brandt, W. and Reiter, M. (1993) Calcium-channel blocking effect of constituents of clove oil. *Planta Medica*, **59**, Supplement issue, A687.

Soriano-García, M., Segura Correa, R., Jiménez, A. and Mata, R. (1996) Structure of Humilinolide D. *Acta Crystallographica Sect.*, **C52**, 1527–1259.

Soto, J.C. and Sousa, M. (1995) *Plantas medicinales de la Cuenca del Río Balsas. Cuaderno 25, Instituto de Biología, UNAM*, pp. 79.

Souza, M., Tellez, O. and Cabrera-Cano, E. (1992) *Imagenes de la flora quintanarroense. Centro de Investigación Científica de Quintana Roo A.C. and Instituto de Biología UNAM*, pp. 12–17.

Tyman, J.H.P. (1991) The chemistry of non-isoprenoid phenolic lipids. *Studies in Natural Products Chemistry*, **9**, 313–381.

Varagic, V.M., Milovanovic, S.R. y Srkalovic, G. (1984) The effect of calcium-channel-blocking agents on the various types of smooth muscle. Activation of the isolated rat uterus. *Arch. Int. Pharmacodyn.*, **270**, 79–87.

Yamauchi, Y., Murakami, T. and Kumanotani, J. (1981) Separation of urushiol by high performance liquid chromatography on an 8% octadecylsilane chemically bonded silica gel urushiol in the sap of the lac tree japanese (*Rhus vernicifera*) and that in the japanese lac-making process. *J. Chromatography*, **214**, 343–347.

Young, D.A. (1975) Flavonoid chemistry and the phylogenetic relationships of the Julianaceae. *Systematic Bot.*, **7**, 149–162.

Young, D.A. (1979) Heartwood flavonoids and the infrageneric relationships of *Rhus* (Anacardiaceae). *Amer. J. Bot.*, **66**, 502–510.

10. THE BRAZILIAN FOLK MEDICINE PROGRAM TO VALIDATE MEDICINAL PLANTS – A TOPIC IN NEW ANTIHYPERTENSIVE DRUG RESEARCH

ANTONIO JOSÉ LAPA[1,*], MARIA TERESA R. DE LIMA-LANDMAN[1], ROBERTA M. CYSNEIROS[1], ANTONIO CARLOS R. BORGES[1], CADEN SOUCCAR[1], IRINÉIA P. BARETTA[2] and THEREZA CHRISTINA M. DE LIMA[2,†]

[1]*Natural Products Section, Department of Pharmacology, Escola Paulista de Medicina/UNIFESP, 04044-020 São Paulo, SP, Brazil*
[2]*Department of Pharmacology, Universidade Federal de Santa Catarina, 88015-420 Florianópolis, SC, Brazil*

One might say that current pharmaceutical development in all Latin American countries is alike since it shares the same beginning and evolution. Only recently some countries with distinct economical progress have advanced in the production of medicine. In most of them poverty still cohabits with a wealthy and well developed society whose demands are much different from each other. The former seeks urgent improvement in primary health care whereas the latter requires a sophisticated modern medicine. National programs oscillate between one or another strategy, but none of them has ever succeeded in establishing self-sustained development and production of medicine. Since it is not ethically acceptable to assign different medicine standards (quality) for either population, efforts have been directed to standardize medicine according to international definitions. The protocol for the characterization of a medicine (phytomedicine included) is well defined (OMS 1975, WHO 1993a). However, the legal requirements for the acceptance of a phytomedicine vary with the country's legislation and needs. For some the traditional use is sufficient to assure a low toxicity, while for others it is mandatory to make a critical evaluation taking into account the accumulated scientific experience of the last decades. This procedure is known as validation.

The whole process of drug screening and therapeutic trials is expensive and time consuming (Dimasi 1995). Programs and attitudes to fulfill the above needs are envisioned by authorities according to the degree of social development, the immediate need and the local ability to cope with the proposition of validation. Whatever the case, it is foreseen that control and self-determination on health policies in Latin-American countries will occur only when proper conditions to develop new medicines are introduced,

*International Coordinator of Project X-4: "Innovative antihypertensive and/or vasodilator medicines obtained from scientifically validated medicinal plants used in Iberoamerica."

†International Coordinator of RIVAPLAMED: "Iberoamerican Network for the Validation of Medicinal Plants." Subprogram X – Pharmaceutical Fine Chemistry, Program CYTED (Iberoamerican Program of Science and Technology for Development).

which strengthen the competition of the local pharmaceutical industry. Validation of medicinal plants and other reputed phytomedicines seems to be the most appropriate approach for this endeavour because they are naturally valued substrates highly accepted in folk medicine and, therefore, economically competitive to similar agents. Validation will also stimulate training of specialists in the area, creation of university–industry partnerships, effective involvement of health professionals in the research and evaluation of new medicines, a motivated (as yet non-existant) pharmacosurveillance and selection of appropriate medicines for distinct populations in developing countries.

The reasons discussed above have stimulated several cooperative projects supported by international organizations (CYTED, LANBIO, ORSTOM, CONVERGENCE, for instance) and many local projects in each Latin-American country, all of them sharing the same purpose of obtaining new competitive medicines from natural resources. The co-sponsor of the present Symposium (CYTED), for instance, established a much-needed and effective Scientific Program (Subprogram X – Pharmaceutical Fine Chemistry) specifically dedicated to the development of pharmaceuticals from regional medicinal plants. The adequacy of CYTED's Program strategy is well appreciated considering that all 21 Iberoamerican countries are grouped under the common objective of studying medicinal plants using either a chemical approach (RIPRONAMED Network) or pharmacological studies (RIVAPLAMED Network), to provide information for a rational selection of active plants for phytomedicine production (RIPROFITO Network) or active compounds for molecular modeling (SYNTHESIS of ANALOGS project).

CYTED

Main Goal

"SCIENTIFIC AND TECHNOLOGICAL INTEGRATION OF IBEROAMERICAN COMMUNITIES TO ACHIEVE SUSTAINED AND EQUITABLE SOCIOECONOMICAL DEVELOPMENT."

General Strategies

- To establish **complementary partnerships** considering individual potentiality and limitations.
- To mobilize the **endogenous scientific/technological** capabilities engaged in scientific networks and projects
- To develop common scientific/technological policies giving priority **to innovation and competitiveness**
- To establish **bi-directional, cooperative, co-financed projects** based on existing facilities of the associated laboratories and industries.

Brazil is the largest Latin-American country with 8.5×10^6 km^2 and 155×10^6 inhabitants living in a variety of ecosystems. The population is concentrated in the Southeastern States where the major industries are found. These regions also correspond to the higher economic income. The average annual income in the country, however, is no more than US$ 3,000.00 per capita, showing the irregular wealth distribution

(IBGE 1995). In fact since 48% of the total population is under 20 years old (IBGE 1995) it is estimated that only about 30–40% of the remaining people are economically active. It can be assumed from this statistical data that only a small part of the Brazilian population has easy access to pharmaceutical products. The largest part must rely on alternative products among which medicinal plants are the most widely consumed. Traditional medicine in Brazil is not well defined as in other countries. Ethnic related medical practices and selection of natural medicines are restricted to small groups (native Indians, African descendants, "caboclo" countryside remnants) which lack formal ways to preserve information about their culture and traditions. Among them, small groups of immigrants, preserving their own traditions, may be found dispersed in several states. All the groups are cosmopolitans, except some Indian tribes still remaining isolated in the Amazon forest. All of them are rapidly losing their ancestral culture under the massive invasion by industrialized society. Nonetheless, it was the long term use of native and of some foreign plants introduced by immigrants that was at the origin of the present Brazilian folk medicine. Most of the medicinal plants actually used were defined when chemical synthesis was incipient and the international trade was scarce. The major national industries in Brazil did not survive the 60's and the incursion of the multinational corporations. The remaining local factories do not invest in new medicines. These pharmaceutical companies make profits either with medicinal plants and reputed phytomedicines, or by mixing, packing and distributing imported active substances. Not a single modern medicine can be considered as fully developed locally. Yet the potentiality is huge, considering that a quarter of the Earth's plants grow in Brazil.

For this reason, the Central of Medicine (CEME) of the Ministry of Health, sponsored for 15 years the **PPPM** (Research Program on Medicinal Plants), a National Program directed to assess the risk/benefit ratio of plants used in Brazilian folk medicine.

PPPM - CEME

RESEARCH PROJECT ON MEDICINAL PLANTS

Main Goal

TO OBTAIN NATIONAL PHARMACEUTICALS USING BRAZILIAN MEDICINAL PLANTS AS SUBSTRATES

This program has stimulated the pre-clinical pharmacodynamic and toxicological studies of about 70 selected plants. Those with high scientific output were, coincidentally, used for the most common sicknesses like pain and inflammation, gastrointestinal disorders, asthma, cardiovascular diseases and nervous breakdown. These indications correspond also, to the predominant therapeutic categories in the country's pharmaceutical market (50%). It is, therefore, expected that once fully validated, they will have a competitive place in the general medicine market.

BRAZILIAN NATIONAL PROGRAM ON MEDICINAL PLANTS
CENTRAL OF MEDICINES (CEME)
MINISTRY OF HEALTH

General Goals

1. Evaluation of risk/benefit of folk medicines
2. Provision of medicines for primary health care
3. Supply substrates for development and production of new medicines
4. Establishment of a scientific and technological background in the field

The PPPM program was based on a sequence of stages. After collection of native plants and botanical identification, the plant germoplasms were preserved and the rest dried in the shade. The ground dry material was extracted as indicated by folk medicine, usually prepared as 2–5% tea. The extract was freeze-dried and the powder screened for the pharmacological effects which oriented the plant selection. When positive preliminary results were obtained, special plantations were provided and the new material re-evaluated in a wide pharmacological screening. Simultaneously, the material was submitted to toxicological tests according to the protocol suggested by WHO (1975), incorporated into Brazilian Law (Brasil 1978, 1988) and into new legislation specific for phytotherapeutic agents (Brasil 1995). Similar protocols were published by WHO (1993a, 1993b). Those plant extracts with substantiated efficacy and undetected toxicity (a total of 7 plants) were recommended for clinical trials according to the protocols required by the Brazilian legislation (Brasil 1988) and also resumed in a WHO publication (1993b). At the same time chemical purification of the active extracts was undertaken to allow determination of the cellular and molecular mechanisms of the pharmacological activities.

PHASES OF PPPM – CEME – BRAZIL

I	Ethnopharmacological selection
II	Botanical / Pharmacognostic field identification
III	Pharmacological screening
IV	Agronomically controlled plantation
V	Pharmacological / Toxicological preclinical evaluation
VI	Clinical trials
VII	Chemical purification
VIII	Transfer to production

The results obtained with *Cecropia glazioui* Sneth are herein summarized as a positive result of the PPPM – CEME program. The genus *Cecropia* is used in Brazilian folk medicine to decrease "high blood pressure". The species was chosen from 8 other plants selected by an ethnobotanical committee based on folk information of their effectiveness in hypertension and/or kidney diseases.

PHARMACOLOGICAL STUDIES OF *CECROPIA GLAZIOUI* SNETH, A NEW ANTIHYPERTENSIVE PHYTOMEDICINE

Cecropia glazioui Sneth (Moraceae) is popularly used in folk medicine of tropical and subtropical Latin America as a cardiotonic, diuretic and to treat "high blood pressure" (Matta 1913; Braga 1960; Matos 1989).

The leaves of *C. glazioui* were initially collected in the remaining of Brazilian South coast forest, evaluated in preliminary *in vivo* screening and then cultivated on the farm of CPQBA, a pluridisciplinary research center of the University of Campinas. The harvest was supervised by Drs. P.M. Magalhães and I. Montanari who were also responsible for the plant identification and stabilization. A voucher specimen is deposited at the university.

The extract of *C. glazioui* was prepared as described by folk medicine. The ground dried leaves were extracted in hot water (2.5%, 72°C) for 30 min (yield 10%). The aqueous extract (AE) was concentrated under vacuum and freeze-dried.

The repeated daily oral administration of AE by gavage to normotensive rats (AE – 0.15; 0.30 and 0.50 mg/kg/day, *p.o.*) caused a reduction of about 20 mm Hg in the systolic blood pressure measured in conscious rats by the non-invasive tail method. The steady-state level of this hypotensive effect was independent on the doses of AE used but the time to achieve the maximum effect was dose-related (Figure 1). The effect of AE was reversed after withdrawal of the drug. These results confirmed the hypotensive effect popularly attributed to the plant and triggered further pharmacological studies to determine the mechanism of the hypotensive action.

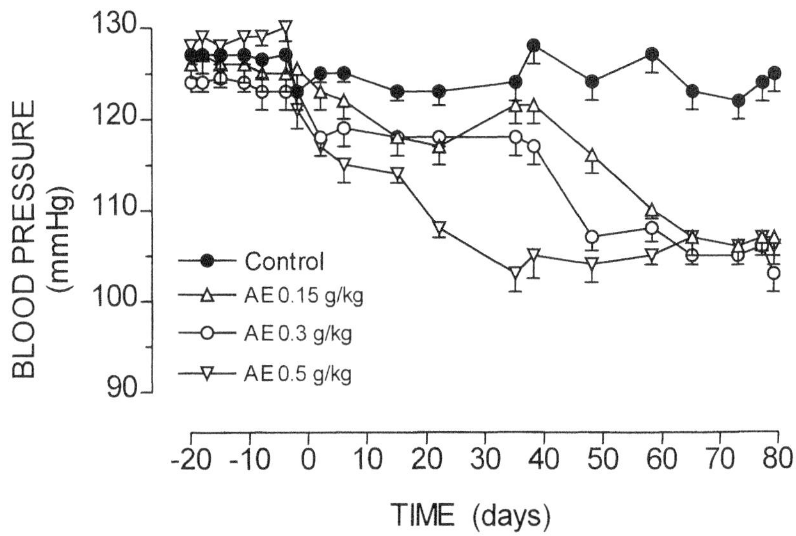

Figure 1 Mean systolic arterial blood pressure measured in conscious rats by the non-invasive tail method. Rats were chronically treated by gavage with the aqueous extract of *C. glazioui* Sneth (AE – 0.15; 0.30 and 0.50 g/kg/day). All groups received water (10 mL/kg/day, *p.o.*) during an initial control period (20 days). Symbols and vertical bars are means ± S.E.M. of 5 animals in each group

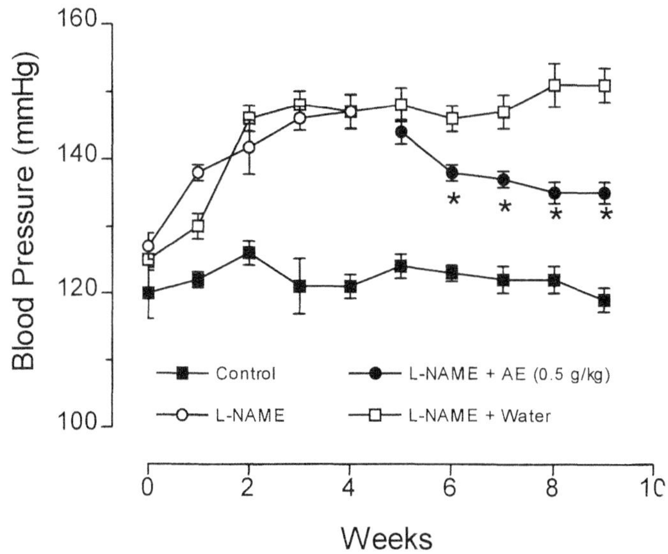

Figure 2 Mean systolic arterial blood pressure measured in normotensive and L-NAME-induced hypertensive rats (10 mg/kg/day, twice a day, *p.o.*). Rats were treated with L-NAME + AE (0.50 g/kg/day, *p.o.*) or L-NAME + water (10 mL/kg/day, *p.o.*). The control group received water (10 mL/kg/day, *p.o.*) during the whole treatment period. Symbols and vertical bars are means ± S.E.M. of 10 animals in each group. *Different from L-NAME + water group ($p < 0.001$, Student "t" test)

Rats treated with the inhibitor of NO synthase, N^G-nitro-L-arginine methyl ester (L-NAME – 10 mg/kg/day, twice a day, *p.o.*) presented gradual hypertension of about 20 mmHg established from 127 ± 2 mmHg in basal conditions to 146 ± 2 mmHg after 3 weeks of treatment, leveling out thereafter. Administration of AE (0.5 g/kg/day, *p.o.*) to these hypertensive animals reduced the systolic blood pressure to 135 ± 2 mmHg within 3 weeks of treatment (Figure 2).

This result indicated that the hypotensive effect induced by the *C. glazioui* aqueous extract was most likely due to an action exerted beyond the membrane receptors and in some step between the excitation–contraction coupling of the vascular smooth muscle and after the release of nitric oxide from endothelial cells (Daniel *et al.* 1995; Rossitti *et al.* 1995).

For *in vitro* studies the AE was chemically purified. Partition with n-butanol yielded an active butanolic fraction (Fbut – yield 10%) that after solvent evaporation was dissolved in Krebs solution and tested in rabbit aortic rings against noradrenaline (NA). This physiological agonist contracted the aortic rings smooth musculature proportionally to the concentration [EC_{50} 5×10^{-6} M – confidence limits (CL) 3 and 8.5×10^{-6} M]. Previous incubation of Fbut (10 and 100 μg/mL) 10 min prior to noradrenaline did not alter the EC_{50} value, but the highest concentration reduced the maximal contraction induced by NA by 22%, indicating a non competitive blockade of the α_1 stimulation pathway (Figure 3).

These results indicated that the hypotensive action of *C. glazioui* was post-junctional and independent on the adrenergic receptor activation. In fact, these observations confirmed the previous *in vivo* indication that Fbut was acting in a late step of the

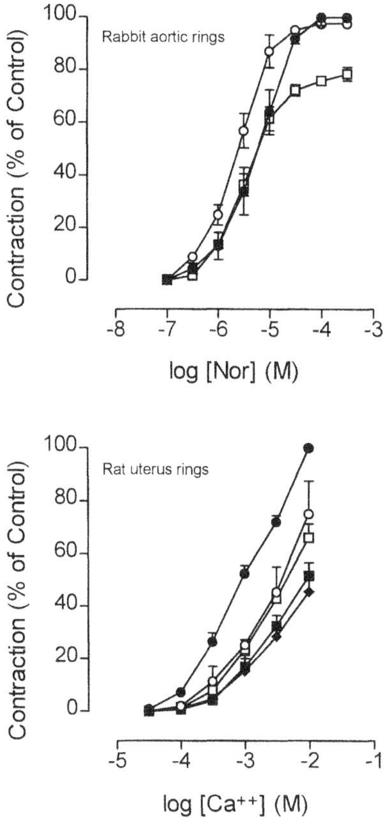

Figure 3 Effects of the butanolic fraction (Fbut) of the extract of *C. glazioui* Sneth in smooth muscle preparations. *Top figure* – Concentration–response curves to noradrenaline in rabbit aortic rings in the absence (○) and in the presence of Fbut (● – 10 μg/mL; □ – 100 μg/mL). *Bottom figure* – Concentration–response curves to $CaCl_2$ in KCl (80 mM) depolarized rat uterus rings in the absence (●) and in the presence of Fbut (○ – 7.5 μg/mL; □ – 15.0 μg/mL; ■ – 60 μg/mL and ◆ – 125 μg/mL). Symbols and vertical bars are means ± S.E.M. (*n* = 3–5 experiments)

excitation–contraction coupling in smooth muscle cells, now postulated to be the Ca^{++} influx or mobilization (Walsh *et al.* 1995).

Concentration–response curves to Ca^{++} were obtained in rat uterus preparations depolarized by high potassium concentration (KCl – 80 mM, $CaCl_2$ – 0 mM). The rat uterus musculature contracted to Ca^{++} proportionally to the concentration (EC_{50} 4.8×10^{-4} M – CL 3.7 and 6.4×10^{-4} M). Incubation of Fbut (7.5; 15 and 30 μg/mL) reduced the maximum contraction by 24, 33 and 40%, respectively, without shifting the Ca^{++} concentration–response curves. Higher concentrations of Fbut (60 and 125 μg/mL) however, reduced the maximal contractile effect by 48 and 54% but also shifted the Ca^{++} curve to the right by 0.8 and 0.9 log units indicating a possible action of Fbut on the voltage-dependent calcium influx (Figure 3).

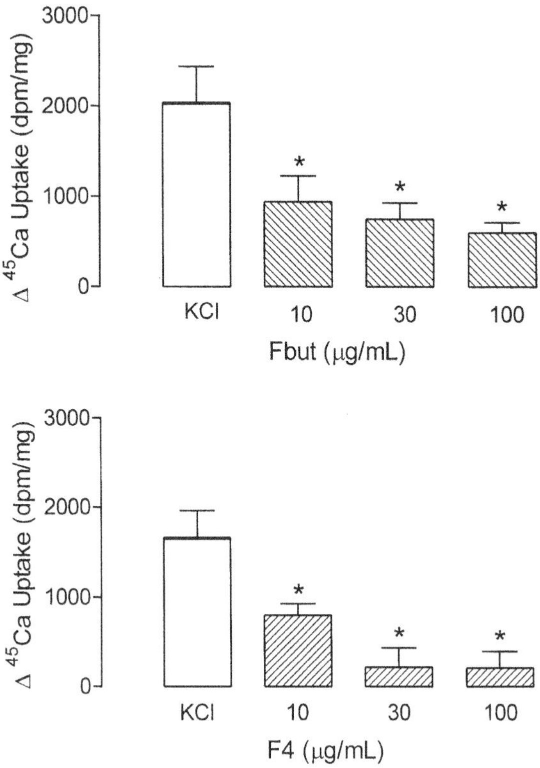

Figure 4 ^{45}Ca uptake in rat uterus rings depolarized by KCl (80 mM) in the presence of 10, 30 and 100 μg/mL of Fbut (*top*) or F4 (*bottom*). Vertical bars are means ± S.E.M. from 3–5 experiments. *Different from KCl group ($p < 0.05$) as indicated by the Student "t" test

^{45}Ca influx was measured in rat uterus rings depolarized by KCl (80 mM). The basal ^{45}Ca influx (3088 ± 400 dpm/mg of tissue) was increased in KCl depolarized preparations by 71 ± 12%. Previous incubation of Fbut (10, 30, and 100 μg/mL) blocked the ^{45}Ca influx stimulated by KCl by 48 ± 10; 60 ± 3 and 67 ± 6%, respectively (Figure 4).

Ca^{++} currents were recorded in chromaffin cells from the PC12 cell line. The voltage-activated Ca^{++} currents were recorded in "whole cell configuration" in the presence of 5 μM TTX and with a pipette containing 10 mM $BaCl_2$. The current/voltage plot yielded a bell shaped curve with the maximum current amplitude around +10 mV membrane potential characteristic of activation of the L-type calcium channels. Incubation of Fbut (100 μg/mL) blocked the peak Ca^{++} current within 5 minutes. This blocking effect was similar to that obtained with the positive control, nifedipine (10^{-9} M), a well known calcium antagonist, reinforcing the putative Ca^{++} channel blockade exerted by Fbut (Figure 5).

Further purification of Fbut by HPLC resulted in 4 subfractions named F1 (RT 5 and 7 min), F2 (RT 7.2 and 11.7 min), F3 (RT 15 and 31 min) and F4 (RT 32 and 39 min) with yields of 7.3, 10.9, 71.6 and 10.2%, respectively (Figure 6).

Incubation of F1, F2, and F3 subfractions (10–100 μg/mL) in rat uterus rings depolarized by KCl (80 mM) did not change the ^{45}Ca influx but F4 (10, 30 and 100 μg/mL)

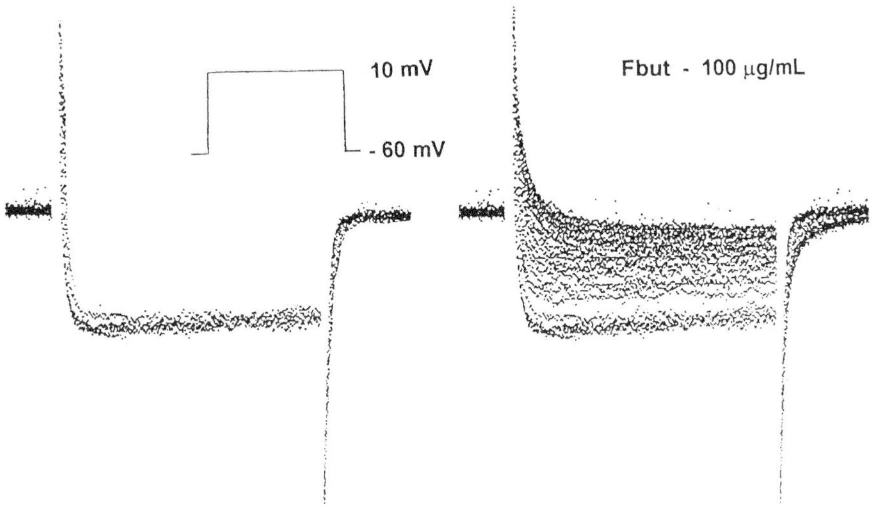

Figure 5 Calcium currents recorded from chromaffin cells of PC12 cell line by the patch-clamp technique in "whole cell configuration", in the absence and presence of Fbut (100 μg/mL)

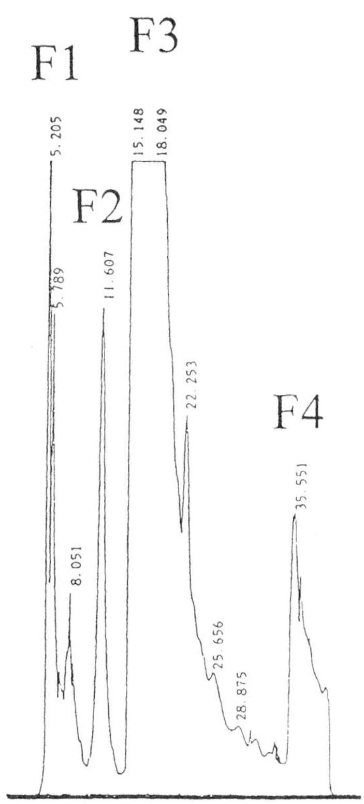

Figure 6 HPLC chromatogram of the Fbut from *Cecropia glazioui* Sneth in a Shimpack-Prep-ODS (25 × 2 cm id) eluted in acetonitrile (2.5 to 100%)

blocked the ^{45}Ca influx by 61.2 ± 4.8; 82.5 ± 5.3 and $95.7 \pm 4.3\%$, respectively (basal influx $= 2316 \pm 43$ dpm/mg of tissue) (Figure 4).

Taken altogether the results indicated that the hypotensive action induced by the AE of *C. glazioui* is due to a blockade of the voltage-gated calcium channels induced by F4. Further purification and chemical identification of the active principle is currently being undertaken.

It was noteworthy that during the *in vivo* recordings of the tail-blood pressure the rats chronically treated with AE were much calmer than control animals and those treated with L-NAME. This observation prompted us to study the effects of *C. glazioui* upon the central nervous system (CNS) since a positive correlation was reported between anxiety/stress and hypertension (Julius and Basset 1987).

Considering that hypertensive animals were shown to present behavioral alterations *per se* (Danysz *et al.* 1983; Soderpalm 1989; Svensson *et al.* 1991; Pare 1992) and that anxiety is a complex response involving the cardiovascular and the central nervous systems (Shekhar *et al.* 1993), the depressant effects induced by treatment with *C. glazioui* was thought to be related to the hypotensive effect and not really to an effect upon the CNS. To address this question the effects of AE were investigated in normotensive and hypertensive rats using the same L-NAME model (10 mg/kg, twice a day, *i.p.*, for 20 days) previously described. Acute administration of AE (0.5 g/kg, *p.o.*) or equivalent volume of tap water 60 min before the experiments, did not modify the animals' behavior in the rota-rod, open field, plus-maze as well as the core temperature in either normotensive rats or L-NAME-induced hypertensive rats. The L-NAME treated group, however, showed an enhancement in locomotor activity (control $= 46.0 \pm 3.3$; L-NAME $= 61.9 \pm 4.5$), as previously described for genetically hypertensive animals (Danysz *et al.* 1983; Soderpalm 1989), besides an increased number of falls on the rota-rod test (control $= 5.8 \pm 0.6$; L-NAME $= 8.3 \pm 0.7$) although not altering the time spent on the rotatory bar.

Animal behavior was also evaluated after 15 days of treatment of the above groups with AE (0.5 g/kg *bid*, *p.o.*). The basal temperature and the catalepsy time measured before and 1 h after the last gavage did not differ between controls or L-NAME-treated animals. The latency for immobility in the forced swimming test though, was longer in the AE treated normotensive group of rats (22.3 ± 7.8 min to 66.8 ± 18.1 min).

Following 15 days treatment with AE (0.5 g/kg, twice a day, *p.o.*) the animals locomotor activity was decreased by 65%. This effect was similar in both normotensive and L-NAME induced hypertensive rats (control $= 34.2 \pm 6.5$; L-NAME $= 35.2 \pm 11.4$). Rearing behavior in the hole board test was also decreased by 56 and 82% respectively in normotensive (13.3 ± 2.3) and L-NAME hypertensive rats (15.5 ± 2.5) (Figure 7), but head-dipping, grooming and defecation were not affected by either treatment.

In the plus-maze test, AE treatment showed a linear trend ($F_{3,19} = 6.288$; $P = 0.0214$) to increase the frequency of entries in open arms (control $= 10.0 \pm 6.8$; AE $= 27.5 \pm 13.2$; L-NAME $= 8.9 \pm 5.3$; L-NAME + AE $= 41.0 \pm 12.1$) without significantly altering the time spent on open arms, closed arms entries, total number of entries, time spent in the central platform, stretch attend postures, head-dippings, rearings, grooming and defecation parameters.

The described effects demonstrated that acute treatment with AE of *C. glazioui* does not have any evident CNS depressant action. Nevertheless, repeated treatment with this

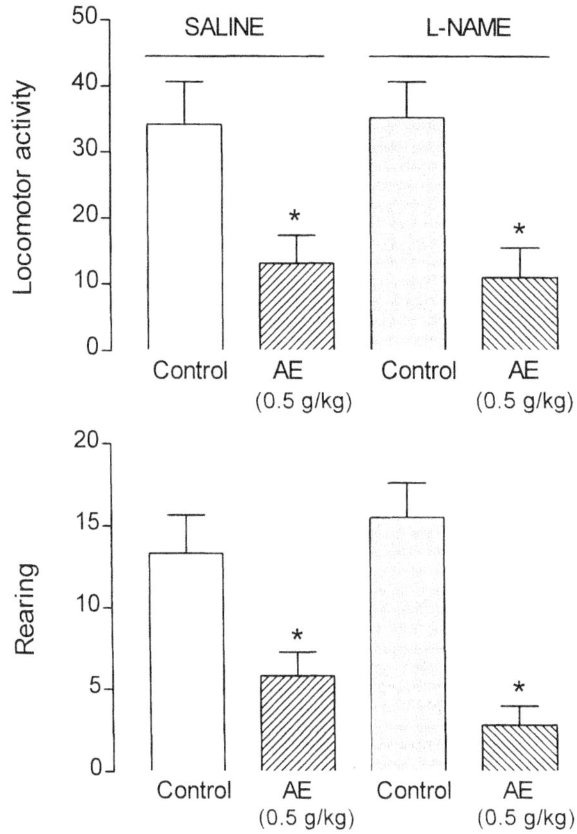

Figure 7 Locomotor activity (*top*) and rearing behavior (*bottom*) in normotensive (saline) and L-NAME-induced hypertensive rats (10 mg/kg, twice a day, *i.p.*) after 15 days of treatment with water (control) or AE of *C. glazioui* Sneth (0.5 g/kg/day, twice a day, *p.o.*). Columns and vertical bars are means ± S.E.M, respectively. ($*p<0.05$, one-way ANOVA, Newman-Keuls' test)

plant species clearly promoted a CNS depression. The results obtained in the hole-board and plus-maze tests suggest an anxiolytic/sedative profile for the *C. glazioui* aqueous extract. A putative antidepressant effect of *C. glazioui* that may be related to calcium channel blockade, as described by other calcium antagonists in the literature, cannot be ruled out (De Vry 1997). Inasmuch as some drugs used in psychiatric clinical practice share both anxiolytic/sedative and antidepressant activity (Handley 1994), the mechanisms underlying these central actions of *C. glazioui* deserve further investigation.

In summary, the on-going Brazilian program of validation of medicinal plants has been presented, along with a pharmacological study of one of the selected plants. The results so far obtained indicate that chronic treatment with *C. glazioui* Sneth extracts as used in Brazilian folk medicine reduces the arterial blood pressure in both normotensive and hypertensive rats, and depresses the central nervous system. Both effects may be related to the blockade of calcium influx through voltage-gated channels. Chemical purification and molecular modeling of the active principle(s) may introduce new leads to pharmaceutical development, notwithstanding the possibility that either the plant extract itself or its purified fractions could be used as new competitive therapeutic agents.

This work was supported by grants from Central de Medicamentos (CEME) and Conselho Nacional de Desenvolvimento Científico e Tecnológico (CNPq).

REFERENCES

Braga, R. (1960) *Plantas do Nordeste especialmente do Ceará.* Imprensa Oficial, Ceará.

Brasil (1978) Resolução Normativa 1/78. Câmara Técnica de Medicamentos. Conselho Nacional de Saúde. Diário Oficial da União, seção 1, 17/10/1978.

Brasil (1988) Resolução número 1. Conselho Nacional de Saúde. Ministério da Saúde. Diário Oficial da União, seção 1, 5/1/1989.

Brasil (1995) Portaria número 6. Secretaria de Vigilância Sanitária. Ministério da Saúde. Diário Oficial da União, seção 1, 6/2/95

Daniel, E.E., Van Breemen, C., Schilling, W.P. and Kwan, C.-Y. (1995) Regulation of vascular tone: cross-talk between sarcoplasmic reticulum and plasmalemma. *Can. J. Physiol. Pharmacol.*, **73**, 551–557.

Danysz, W., Plaznik, A., Pucilowski, O., Plewako, M., Obersztyn, M. and Kostowski, W. (1983) Behavioral studies in spontaneously hypertensive rats. *Behav. Neural. Biol.*, **39**, 22–29.

De Vry, J., Fritze, J. and Post, R.M. (1997) The management of coexisting depression in patients with dementia: potential of calcium channel antagonists. *Clin. Neuropharmacol.*, **20**, 22–35.

Dimasi, J. (1995) Trends in drug development: costs, times and risks. *Drug Information J.*, **29**, 375–384.

Handley, S. (1994) Future prospects for the pharmacological treatment of anxiety. *CNS Drugs*, **2**, 397–414.

IBGE (1995) *Bulletin of the Brazilian Institute of Geography and Statistics*, Rio de Janeiro.

Julius S. and Basset, D.R. (1987) *Handbook of Hypertension,* Elsevier, New York.

Matos, F.J.A. (1989) *Plantas Medicinais: Guia de Seleção e Emprego de Plantas Usadas em Fitoterapia no Nordeste do Brasil.* IOCE, Fortaleza.

Matta, A.U. (1913) *Flora Médica Braziliense.* Seção de Obras da Imprensa Oficial, Manaus.

OMS (1975) *Pautas para la Evaluación de los Medicamentos Destinados al Hombre.* Organización Mundial de la Salud, Ginebra.

Pare, W.P. (1992) The performance of WKY rats on three tests of emotional behavior. *Physiol. Behav.*, **51**, 1051–1056.

Rossitti, S., Frangos, J., Girard, P.R. and Bevan, J. (1995) Regulation of vascular tone. *Can. J. Physiol. Pharmacol.*, **73**, 544–550.

Shekhar, A., Sims, L.S. and Bowsher, R.R. (1993) GABA receptors in the region of the dorsomedial hypothalamus of rats regulates anxiety in the elevated plus-maze test. I. Physiological measures. *Brain Res.*, **627**, 17–24.

Soderpalm, B. (1989) The SHR exhibits less "anxiety" but increased sensitivity to the anticonflict effect of clonidine compared to normotensive controls. *Pharmacol. Toxicol.*, **65**, 381–386.

Svensson, L., Harthon, C. and Linder, B. (1991) Evidence for a dissociation between cardiovascular and behavioral reactivity in the spontaneously hypertensive rat. *Physiol. Behav.*, **49**, 661–665.

Walsh, M.P., Kargacin, G.J., Kendrick-Jones, J. and Lincoln, T.M. (1995) Intracellular mechanisms involved in the regulation of vascular smooth muscle tone. *Can. J. Physiol. Pharmacol.*, **73**, 565–573.

WHO (1993a) *Research Guidelines for Evaluating the Safety and Efficacy of Herbal Medicines.* World Health Organization, Manila.

WHO (1993b) *International Ethical Guidelines for Biomedical Research Involving Human Subjects.* World Health Organization, Geneva.

11. RECENT DEVELOPMENTS IN THE CHEMISTRY AND PHARMACOLOGY OF BOLDO AND BOLDINE

BRUCE K. CASSELS

Department of Chemistry, Facultad de Ciencias, Universidad de Chile, Casilla 653, Santiago, Chile

The leaves of the central Chilean tree known as boldo (*Peumus boldus* Mol., Monimiaceae) are widely used in herbal teas in southern South America, and the leaves, their alcohol extracts and more or less crude alkaloid mixtures are found in commerce beyond this traditional area of distribution, particularly in Europe. The uses for which these products are recommended are quite varied, but the virtues most commonly cited are as digestive aids and tranquilizers. Little evidence is available to support these uses, however, and the frequent claims that the essential oil, the alkaloids or the flavonoids of boldo are the "active principles", although reasonable as hypotheses, lack an adequate scientific basis. This, of course, is just one more example of the all-too-frequent, sometimes disingenuous assumption that whatever has been isolated (or shown to be present) from (or in) a "medicinal" plant must be responsible for whatever activity is claimed (although not necessarily proven).

Since 1990 I have been collaborating with Prof. Hernán Speisky, of the Nutrition and Food Technology Institute (I.N.T.A.) of the University of Chile, on a variety of topics mostly related to boldo and boldine [(*S*)-2,9-dihydroxy-1,10-dimethoxyaporphine, **1**], the major alkaloid of boldo leaves and stem bark. Our interest arose from the simple observation that the structure of boldine incorporates two phenolic functions, each with bulky, electron-donating *ortho*- and *para*-substituents, suggesting that the phenoxy free radicals which could arise as a consequence of free-radical attack upon either of the phenol groups would be stabilized both thermodynamically and kinetically, a feature generally associated with chain-breaking antioxidants. It was not long before we knew that we had a highly effective antioxidant in our hands, and our first results were published a few months later (Speisky *et al.* 1991a). In this early paper we showed that boldine, in low micromolar concentrations, is able to prevent brain homogenate

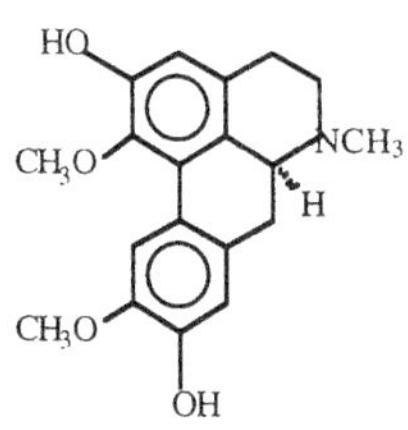

1

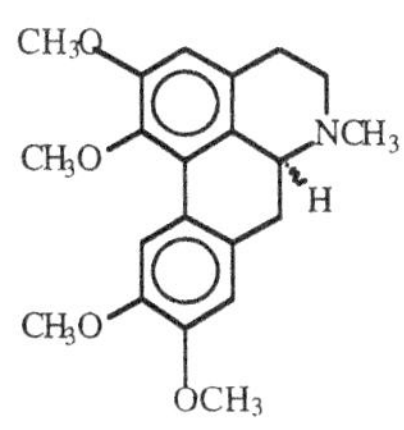

2

autoxidation, the peroxidation of red blood cell plasma membranes induced by 2,2′-azobis(2-amidinopropane) (AAP), and AAP-induced inactivation of lysozyme. We remarked then that its antioxidative potency and its low toxicity might favor its use for therapeutic purposes. In this connection it is worth pointing out that, although boldine appears to be less potent in these models than the reference compound propyl gallate, its antioxidative efficiency is comparable to the widely used flavonoid (+)-3-cyanidanol and exceeds that of the still-popular silybin. The alkaloidal character of boldine is also an unusual asset insofar as it allows the antioxidant to exist in a fat-soluble and a water-soluble form (the free base and its salts, respectively), and is thus able to exert its effects both in aqueous phases (the buffered saline solution containing lysozyme and hydrophilic AAP-derived free radicals) and in lipidic phases (brain or red blood cell membranes). We also tested boldine for its inhibitory effect on spontaneous and metal-induced fish oil oxidation (Valenzuela *et al.* 1991), comparing its behavior with those of ($\pm$)-α-tocopherol (synthetic vitamin E), quercetin, and the commonly used synthetic antioxidants, butylated hydroxytoluene (BHT) and butylated hydroxyanisole (BHA). Boldine, in all assays, showed antioxidative effects which were similar to those of quercetin and even better than those of the synthetic vitamin E, BHT or BHA. These results indicate that boldine can react with water-soluble alkylperoxyl radicals (such as those derived from AAP), lipid radicals, or both. At about the time when we began our studies on boldine, Lanhers *et al.* (1991) had also shown that this alkaloid, in the same concentration range found by us, protects isolated hepatocytes against damage induced by *t*-butyl hydroperoxide, and protects mice *in vivo* against carbon tetrachloride-induced hepatotoxicity.

Prof. Speisky carried out additional studies to determine the effectiveness of boldine in preventing lipid peroxidation of microsomes under a variety of conditions, iron-dependent and iron-independent, enzymatic and non-enzymatic, and involving cytochrome P_{450}-catalyzed activation of carbon tetrachloride (Cederbaum *et al.* 1992). Lipid peroxidation was always inhibited by about 50% using 15 μM boldine, in close agreement with our earlier results with brain and red blood cell membranes and also with aqueous lysozyme solutions. Unlike other well-known antioxidants, concentrations of boldine which almost completely inhibited lipid peroxidation were found to have no effect on microsomal mixed-function oxidase activity. On the other hand, the alkaloid completely protected this enzyme activity against inactivation resulting from lipid peroxidation. Boldine did not react efficiently with superoxide anion radical or with hydrogen peroxide, but it competed much more effectively than ethanol and somewhat more effectively than dimethyl sulfoxide with 2-keto-4-thiomethylbutyric acid as a hydroxyl radical scavenger, suggesting that its reaction rate with the very reactive hydroxyl free radicals approaches diffusion control. Nevertheless, microsomal lipid peroxidation is insensitive to hydroxyl radical scavengers, so boldine must be protecting this subcellular system by some other mechanism. It seems possible that this involves direct reaction with lipid radicals, or that boldine serves to repair the tocopheryl radical as has been postulated for ascorbate, glutathione, and endogenous phenols. These results suggested that boldine might be an ideal lead compound for the development of clinically useful antioxidants, with the additional advantage that it does not seem to stimulate hydroxyl radical production, as is the case for propyl gallate, BHT, BHA, and many other phenolic compounds.

The isolation of fairly large amounts of boldine was necessary for our pharmacological studies, and the stem bark of boldo is an unusually good source as it contains upwards of 6% based on dry weight. We therefore developed appropriate technology to obtain boldine on a scale of tens of grams in my laboratory. As a prerequisite for the optimal preparation of solutions for biological work, water-soluble salts were required. In particular, boldine hydrochloride was found to crystallize nicely, to the extent that macroscopic monocrystals could be separated and subjected to X-ray diffraction analysis (Asencio *et al.* 1996), prompting a long-overdue discussion of the conformational preferences of aporphine molecules. All intramolecular bonds and angles of boldine hydrochloride were found to be within the expected ranges, and the most salient feature of the structure is the non-planarity of the biphenyl system, with a twist angle of 26.3° between the least-squares aromatic ring planes. The variation of the twist angle between different aporphines had been discussed twenty years ago for only three crystal structures, which were insufficient to lend direct experimental support to Shamma's conjecture that the difference in absolute value of the specific rotations of 1,2,9,10- and 1,2,10,11-tetraoxygenated aporphines could be explained on the basis of a greater twist angle in those compounds in which steric compression between C-1 and C-11 substituents leads to greater steric strain. Comparison of the boldine hydrochloride structure with a dozen previously published aporphine X-ray structures led to a discussion which would seem to be more reliable, although somewhat disappointing. The series of structures tabulated by us shows that when only the C-1 or the C-11 position is substituted with either a hydroxyl or a methoxyl group, the twist angle is less than 30° (sometimes, but not usually, considerably less). When both neighboring positions carry oxygen substituents, the twist angle is slightly greater than 30°, but only very slightly so. It therefore seems possible that, in the crystal phase, packing forces tend to smooth out the differences in the twist angles dictated by the sum of effects attributable to the non-planarity of partially saturated ring C and to the mutual repulsion of the atoms or substituent groups on C-1 and C-11. A systematic study of aporphines in solution, using a combination of quantitative theoretical and experimental approaches, would now seem to be necessary to solve this problem.

The concomitant production of considerable quantities of boldine mother liquors made further study of the minor alkaloids an obvious goal and, leaving aside the well-studied less polar substances, coclaurine (3, Asencio *et al.* 1993a) and laurolitsine (norboldine) (4, Asencio *et al.* unpublished results) were isolated and characterized. Coclaurine had not been recognized previously as a boldo constituent, and laurolitsine

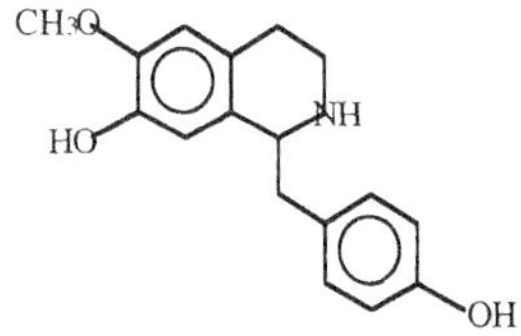

3

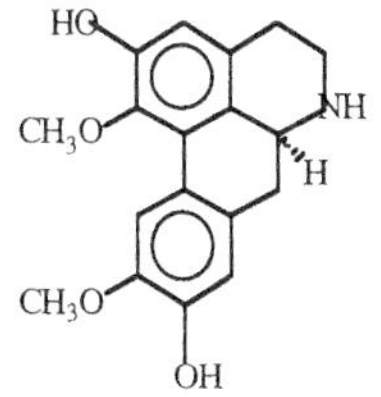

4

had only been found in the wood. The accumulation of coclaurine in boldo bark is of interest considering that it is now regarded as the universal benzyltetrahydroisoquinoline precursor. In fact, it is reasonable to assume that sufficiently sensitive assays should reveal the presence of coclaurine in any benzylisoquinoline alkaloid-producing plant. In line with this, both enantiomers of coclaurine (**3**) were found, with the (*R*)-isomer dominating the mixture. Of the other twelve alkaloids or alkaloid artefacts identified thus far as boldo constituents, (*S*)-reticuline and nine (*S*)-reticuline-derived aporphines are obvious (*S*)-coclaurine metabolites, while the proaporphine pronuciferine and the morphinandienone sinoacutine are presumably derived from (*R*)-coclaurine. Our finding that laurolitsine (**4**) can be readily isolated from boldine mother liquors provided us with a valuable starting material for the preparation of *N*-substituted and enantiomeric boldine analogues which are currently under study.

In 1992 a Spanish team working independently at the University of Valencia had published comparable results for boldine and some other benzylisoquinoline alkaloids using different model systems (Martínez *et al.* 1992). These authors showed that non-phenolic aporphine alkaloids such as glaucine (**2**, *O,O'*-dimethylboldine, which is also found in boldo) and anonaine are also effective antioxidants, and they suggested that the aporphine skeleton was in some way involved in the manifestation of this property. In view of these results, we postulated that non-phenolic aporphines might behave as chain-breaking antioxidants via abstraction of a benzylic hydrogen atom, particularly from C-6a of the aporphine skeleton, which would involve resonance stabilization of the resulting aporphine radical, not only by spin delocalization throughout the highly conjugated biphenyl system but also with participation of the basic nitrogen atom. This hypothesis received support from a structure-antioxidative activity analysis comparing apomorphine, boldine, glaucine, glaucine methiodide, and the benzyltetrahydroisoquinolines coclaurine (**3**) and norarmepavine (Cassels *et al.* 1995). Our results led to the conclusion that the aporphine skeleton, when the nitrogen atom is not quaternized, is itself a good predictor of antioxidative activity, regardless of the presence or absence of phenolic groups. Thus, glaucine can protect lysozyme against AAP-induced inactivation, but this ability is lost to all practical purposes when the nitrogen atom is quaternized to give the *N*-methylglaucinium ion. As we were able to confirm that 1-benzyl-1,2,3,4-tetrahydroisoquinoline alkaloids are only poor antioxidants, even when they possess one or more phenol groups, we took this as evidence that spin delocalization over the biphenyl system is an important feature in the ease with which chain-breaking, persistent free radicals can be formed from aporphines. If substantiated, we expect this hypothesis to support the future development of novel antioxidant molecules which may be of clinical interest.

Another early contribution to the chemistry of boldine was a kinetic study of its methylation with diazomethane. This allowed the transient accumulation of predicentrine (**5**, 9-*O*-methyl-boldine) but not of the recognized boldo constituent *N*-methyllaurotetanine (**6**, 2-*O*-methylboldine), *en route* to the completely *O*-methylated glaucine (**2**), to be explained as a kinetically controlled process (Asencio *et al.* 1993b). It thus appears that the C-9 phenolic function is more reactive than the group at C-2. This observation may be related to the greater acidity of the 9-hydroxyl group, as determined by spectrophotometric titration, and the apparently greater reactivity of *N*-methyllaurotetanine (**6**)

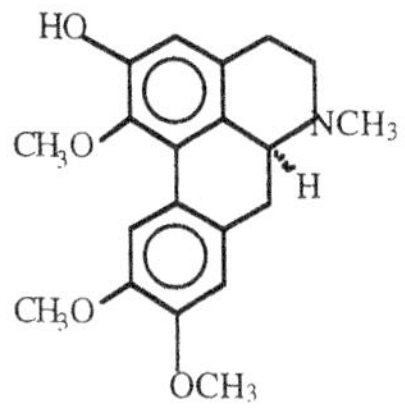

5

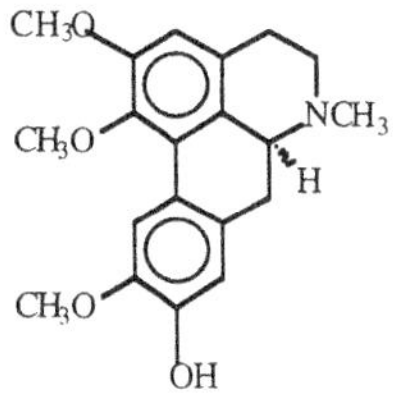

6

towards ABTS [2,2′-azinobis(3-ethylbenzthiazoline-5-sulfonic acid)] cation radicals with regard to predicentrine (5) (Asencio *et al.* unpublished results). An additional feature working in this direction should be the more exposed steric environment of the C-9 hydroxyl due to the preferred (*anti*) coplanarity of the C-10 methoxy group with the aromatic system in contrast to the necessarily perpendicular arrangement of the C-1 methoxy group, as a consequence of steric compression between the C-2 hydroxyl and the C-11 hydrogen atom.

A decade ago it had been shown that several synthetic antioxidants inhibit the respiration and growth of the protozoon *Trypanosoma cruzi*, the causative agent of Chagas' disease, in culture. In view of the well-established antioxidative properties of boldine and our (then unpublished) results showing significant antioxidant activity for glaucine (2) and, to a lesser extent, for 1-benzyl-1,2,3,4-tetrahydroisoquinolines, we tested some of these substances against epimastigote cultures of *T. cruzi*. Approximately 100 mM concentrations of boldine (1), predicentrine (5) and glaucine (2) were shown to be toxic to the parasite cultures, the catecholic apomorphine (which in some systems behaves as a better antioxidant) was slightly more potent, while the phenolic nonaporphines coclaurine and norarmepavine and particularly the nonphenolic nonaporphine codeine were less active (Morello *et al.* 1994). Apomorphine, boldine, predicentrine and glaucine inhibited cell respiration, suggesting the possibility that they might act by blocking mitochondrial electron transport. Perhaps significantly, the trypanocidal potencies of these alkaloids appear to be correlated with their antioxidative activities. Unfortunately, however, such high trypanocidal concentrations are rather unpromising.

Among the almost forgotten indications for the use of boldo in traditional Chilean medicine, headache, earache, toothache, rheumatism and inflammation of the urinary tract suggest the possibility that boldo extracts might possess analgesic and anti-inflammatory activity, and/or antipyretic actions. Anti-inflammatory activity could in fact be demonstrated for a "purified" alcohol extract of boldo leaves (Lanhers *et al.* 1991). As anti-inflammatory and antipyretic actions are often associated with antioxidative activity, it seemed to us that the claimed and demonstrated properties of boldo extracts might be due in part to the presence of boldine, in spite of its very low concentrations in boldo leaf infusions. It should be remembered that, although present-day Chilean folk medicine relies heavily upon aqueous infusions, hydroalcoholic extracts were commonly used a century ago, and native Mapuche medicine made frequent use of concentrated decoctions which were often administered rectally, thus avoiding first-pass metabolism of whatever they contained and could be absorbed. We therefore tested pure boldine, administered orally, in two *in vivo* models of inflammation and fever. Boldine exhibited dose-dependent anti-inflammatory activity in the carrageenan-induced guinea pig paw edema test with an oral ED_{50} of 34 mg/kg (which was compared with sodium naproxen, producing a similar effect at 1.7 mg/kg), and rapidly but briefly reduced bacterial pyrogen-induced hyperthermia in rabbits to an extent varying between 51% and 98% at 60 mg/kg (naproxen only attains similar levels of antipyretic activity at 25 mg/kg, with effects which peak at longer times than for boldine). *In vitro* studies carried out in rat aortal rings showed that boldine inhibits prostaglandin biosynthesis by 53% at 75 μM, an effect which may be mechanistically linked to the *in vivo* results described above (Backhouse *et al.* 1994).

Lanhers *et al.* (1991) had shown that boldine protects isolated liver cells against the oxidative and cytolytic challenge induced by carbon tetrachloride, an effect depending upon cytochrome P_{450} activation of the hepatotoxin. Using *tert*-butyl hydroperoxide as a direct source of lipophilic oxygen-centred free radicals, we were also able to demonstrate the protective effect of boldine in isolated hepatocytes (Bannach *et al.* 1996). In a similar vein, boldine protects red blood cells against the damage induced by azo free radical generators (Jiménez *et al.* 1996). It also exhibits cytoprotective and anti-inflammatory effects in an animal model of acute colitis (Gotteland *et al.* 1997). Our study showed that the intrarectal administration of boldine solution (100 mg/kg), in rats treated intrarectally 24 hours earlier with 4% acetic acid, led to the improvement of morphological, biochemical, and functional parameters of the colonic mucosa. This improvement was similar or greater than that achieved with the same dose of 5-amino-salicylic acid, a reactive oxygen metabolite scavenger commonly used in the treatment of inflammatory bowel disease.

Both boldine and glaucine inhibit phorbol ester-induced down-regulation of gap junctional intercellular communication, a process in which free radicals and peroxidation products are involved as intracellular mediators, and which is associated with tumor formation (Hu *et al.* 1995). Using rat liver epithelial cells, both alkaloids (and the antioxidative hypolipidemic drug probucol) completely inhibited the action of 10 nM 12-*O*-tetradecanoylphorbol-13-acetate (TPA) at 50 μM each. At this concentration, all three antioxidants inhibited TPA-induced translocation of protein kinase C to the particulate fraction of the cells, with concomitant inhibition of TPA-induced hyperphosphorylation of gap junctional connexin and TPA-induced internalization of this protein from the plasma membrane of the cells. In addition, 10 μM boldine and probucol totally inhibited the TPA-induced, allopurinol-independent (and therefore xanthine oxidase-independent) accumulation of intracellular oxidants. These results indicate that all three substances possess antitumor promotive potential in this model of intercellular gap junctional communication, and that this activity may be due in part to the destruction of oxidants within the cells.

A partially purified boldo leaf extract (containing little boldine) was shown to reduce basal intestinal motility in humans, prolonging oro-cecal transit time (Gotteland *et al.* 1995), and confirming early results in animal models. Boldine itself exerts an *in vitro* smooth muscle-relaxing effect on spontaneous and acetylcholine-induced contraction of the rat ileum (Speisky *et al.* 1991b). Although no direct α-adrenergic effect of boldine or glaucine could be demonstrated in rat vas deferens (De Las Heras *et al.* 1990), both alkaloids behave as selective inhibitors at vascular α_1-adrenergic receptors and potential-operated calcium channels (Ivorra *et al.* 1993), and these properties could be partially responsible for the intestinal effects of boldine which in turn may explain some of the traditional uses of the plant.

With these results in hand, the interest of carrying out further *in vivo* studies was obvious. There were some indications that boldine is absorbed from the gut and therefore exhibits systemic (e.g. anti-inflammatory) actions, and also that the alkaloid appears to be able to penetrate the blood-brain barrier to show central (e.g. "neuroleptic-like", antipyretic?) actions. However, the oral activity of boldine was unremarkable and its duration was short, suggesting low bioavailability, and probably rapid metabolic inactivation and/or excretion. Similarly, the reported neuroleptic-like, anticonvulsant and antinociceptive properties could only be demonstrated at relatively high subcutaneous doses. In order to be able to study the pharmacokinetics of boldine an appropriate quantitative assay was required, so we developed a UV-detected HPLC method which allowed the determination of boldine in plasma and other body fluids (Speisky *et al.* 1993). Using this method, we found that the *in vivo* plasma half-life of boldine is only a few minutes, at least partially explaining some of our earlier results. We now know (Liberona *et al.* unpublished results) that boldine is rapidly and extensively glucuronidated in rat liver and excreted in urine as a glucuronide. On the other hand, oxidative demethylation pathways seem to be unimportant, at least in the rat.

The low plasma concentrations achieved after oral administration of boldine, and its short plasma half-life (which, nevertheless, are sufficient to allow some pharmacological effects to be observed), led to further chemical studies aiming to increase the intestinal absorption and reduce the excretion of boldine (and/or possibly active metabolites), or conversely to curtail its absorption in order to maximize its concentration in the intestinal lumen. Thus, conditions were found to allow the efficient preparation of the acetate and pivalate esters of boldine (Asencio *et al.* unpublished results), although the synthesis of more hydrophilic anionic hemiesters has proven unsuccessful so far. Other chemical studies showed that electrophilic halogenation of boldine (with molecular bromine, *N*-chloro-, *N*-bromo-, or *N*-iodosuccinimide), led to rapid substitution at C-3, with slower formation of the 3,8-dihalo derivative in the cases of chlorine and bromine (Sobarzo *et al.* unpublished results). By now, some pharmacological results are available for these derivatives (*vide infra*). Nitrosation (with nitrous acid) also occurred at C-3, allowing the preparation of 3-aminoboldine and oxazole and oxazinone derivatives annellated at C-2/C-3 (Sobarzo *et al.* unpublished results).

Prior to the beginning of our work on boldo and boldine, a paper had been published suggesting that this alkaloid, administered subcutaneously in mice, produces "neuroleptic-like" effects which might be attributed to dopamine receptor antagonism (Zetler 1988). Consequently, boldine was subjected – at the University of Rouen, France (Asencio *et al.* 1997) – to displacement studies with radioligands selective for D_1 ([^{3}H]-SCH 23390) and D_2 receptors ([^{3}H]-raclopride) with the result that this compound binds to both striatal receptor subtypes with $IC_{50} = 0.4–0.5\,\mu M$. In these biochemical models, glaucine is one order of magnitude less potent. Nevertheless, *in vivo* binding studies showed that 40 mg/kg boldine (*i.p.*) did not modify specific striatal [^{3}H]-raclopride binding and only decreased [^{3}H]-SCH 23390 binding by 25%. On the contrary,

40 mg/kg glaucine (*i.p.*) displaced both radioligands by about 50%. Behaviors (climbing, sniffing, grooming) elicited in mice by apomorphine were not modified by boldine at doses up to 40 mg/kg, but were almost completely abolished by 40 mg/kg glaucine. In the apomorphine-induced rat yawning and penile erection model, boldine and glaucine appeared to be similarly effective, inhibiting both behaviours by more than 50% at 40 mg/kg. Boldine and glaucine, injected (*i.p.*) at doses up to 40 mg/kg, were poor modifiers of dopamine metabolism in mouse and rat striatum as well as in other dopaminergic rat brain areas. These results suggest that the unfavorable pharmacokinetics of boldine as compared to glaucine may be an important negative factor in the expression of CNS activities, in spite of its tenfold greater affinity for dopamine receptors. We also consider them as further justification of our interest in the development of more lipophilic and metabolically less labile boldine analogues.

In vitro binding studies using 3-chloro-, 3,8-dibromo- (**8**) and 3-bromoboldine (**7**) indicated some increase in D_1 receptor affinity with regard to unsubstituted boldine (attaining IC_{50} *vs.* [^{3}H]-SCH 23390 = 77 nM for 3-bromoboldine) and a slight reduction in D_2 receptor affinity (*vs.* [^{3}H]-raclopride), and thus a small but probably significant increase in D_1/D_2 selectivity, going from 1.3 for boldine to 5.1–5.4 for 3-chloro- and 3,8-dibromoboldine, respectively, and to 12 for 3-bromoboldine (Asencio *et al.* unpublished results). Further *in vitro* binding studies on halogenated boldine, predicentrine and glaucine derivatives are being carried out, and will be followed by *in vivo* biochemical and behavioral work if improved receptor affinities and/or selectivities seem to warrant it.

Similar *in vitro* binding studies carried out at the University of Valencia, Spain, showed that boldine is a selective inhibitor of [^{3}H]-prazosin-labelled α_{1A} receptors in rat cerebral cortex. The pK_H of boldine (for the high affinity site, assumed to represent α_{1A} receptors) was found to be 8.31, with a selectivity ratio of 70 *vs.* α_{1B} receptors. Comparison of boldine (**1**) with predicentrine (**5**) and glaucine (**2**) in the same model indicated a loss of both affinity and selectivity upon *O*-methylation, particularly of the C-2 oxygen atom, with pK_H of 8.13 for predicentrine and 7.12 for glaucine, and selectivity ratios of 60 and 15, respectively. This was interpreted as suggesting a role of the C-2 hydroxyl group, present in boldine and predicentrine but not in glaucine, in binding to the α_{1A} receptor subtype (Madrero *et al.* 1996). The smooth muscle-relaxant action of boldine in rat uterus and aorta, although reflecting its affinity for α_1 receptors, may also be attributed in part to the blockage of Ca^{2+} entry via the benzothiazepine ([^{3}H]-diltiazem-labelled) receptor in spite of its low affinity (IC_{50} = 33 μM) for this site (Ivorra *et al.* 1993).

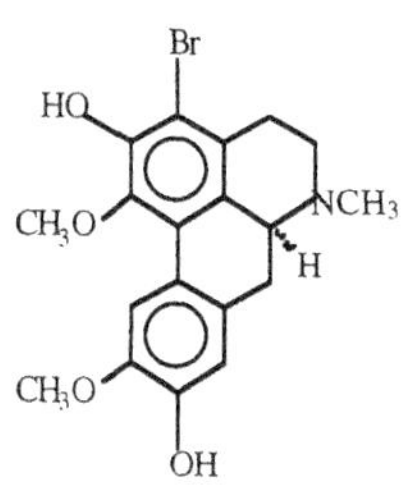

7

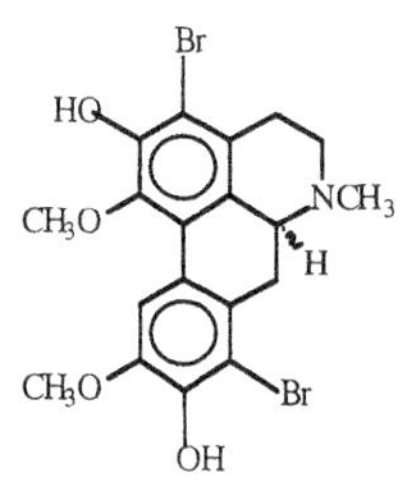

8

In contrast to the results for the *O*-methylated boldine derivatives, affinity and selectivity for the α_{1A} receptor subtype (*vs.* the α_{1B} subtype) increases slightly upon bromination, at least at C-3, going from pK_H (for the high affinity [^{3}H]-prazosin site) = 8.31 and a selectivity ratio of 70 for boldine to 8.79 and 80 for 3-bromoboldine (**7**) and 8.85 and 85 for 3,8-dibromoboldine (**8**) (Madrero *et al.* 1997). Other halogenated boldine derivatives have now been sent to the laboratory in Valencia for similar studies which should clarify the role of C-3 halogen substitution.

The first results of our research on boldo and boldine (up to 1994), together with an extensive discussion of the earlier literature, were summarized by us in a review several years ago (Speisky and Cassels 1994). In view of the sustained interest in antioxidants as evidenced by the number of publications appearing in the literature, and considering our continuing activity in this area, in which we have not only shown previously unrecorded biological activities, but also made significant advances in the synthesis of boldine analogues and observed promising effects of some of these derivatives in monoaminergic systems, we believe that an update is warranted. This chapter is a first step in this direction.

REFERENCES

Asencio, M., Cassels, B.K., Speisky, H. and Valenzuela, A. (1993a) (R)- and (*S*)-coclaurine from the bark of *Peumus boldus*. *Fitoterapia*, **64**, 455–458.

Asencio, M., Cassels, B.K. and Speisky, H. (1993b) The methylation of boldine with diazomethane. *Bol. Soc. Chil. Quím.*, **38**, 331–334.

Asencio, M., Cassels, B.K., Manríquez and V., Boys, D. (1996) Structure of (*S*)-1,10-dimethoxy-2,9-dihydroxyaporphinium chloride (boldine hydrochloride). *Acta Cryst.*, **C52**, 1581–1583.

Asencio, M., Delaquerrière, B., Cassels, B.K., Speisky, H., Comoy, E. and Protais, P. (1997) Biochemical and behavioral effects of boldine and glaucine. *Pharmacol. Biochem. Behav.*, submitted.

Backhouse, N., Delporte, C., Guivernau, M., Cassels, B.K. and Speisky, H. (1994) Anti-inflammatory and antipyretic effects of boldine. *Agents Actions*, **42**, 114–117.

Bannach, R., Valenzuela, A., Cassels, B.K., Núñez-Vergara, L.J. and Speisky, H. (1996) Cytoprotective and antioxidant effects of boldine on *tert*-butyl-hydroperoxide-induced damage to isolated hepatocytes. *Cell Biol. Toxicol.*, **12**, 89–100.

Cassels, B.K., Asencio, M., Conget, P., Speisky, H., Videla, L.A. and Lissi, E.A. (1995) Structure-antioxidative activity relationships in benzylisoquinoline alkaloids. *Pharmacol. Res.*, **31**, 103–107.

Cederbaum, A.I., Kukielka, E. and Speisky, H. (1992) Inhibition of rat liver microsomal lipid peroxidation by boldine. *Biochem. Pharmacol.*, **44**, 1765–1772.

De Las Heras, B., Ríos, J.L., Martínez-Mir, I. and Rubio, E. (1990) A study of the adrenergic activity of the alkaloids boldine and glaucine. *Pharmazie*, **45**, 443–444.

Gotteland, M., Espinoza, J., Cassels, B.K. and Speisky, H. (1995) Efecto de un extracto seco de boldo sobre el tránsito intestinal oro-cecal en voluntarios sanos. *Rev. Méd. Chile*, **123**, 955–960.

Gotteland, M., Jiménez, I., Brunser, O., Guzmán, L., Romero, S., Cassels, B.K. and Speisky, H. (1997) Protective effect of boldine in experimental colitis. *Planta Medica*, **63**, 311–315.

Hu, J., Speisky, H. and Cotgreave, I.A. (1995) The inhibitory effects of boldine, glaucine, and probucol on TPA-induced down regulation of gap junction function. *Biochem. Pharmacol.*, **50**, 1635–1643.

Ivorra, M.D., Chuliá, S., Lugnier, C. and D'Ocón, M.P. (1993) Selective action shown by two aporphines at α_1-adrenergic receptors and in the potential-operated calcium channel. *Eur. J. Pharmacol.*, **231**, 165–174.

Jiménez, I., Bannach, R., Gotteland, M., Valenzuela, A., Cassels, B.K. and Speisky, H. (1996) Cytoprotective effects of boldine on free radical-induced erythrocyte lysis. *Toxicology*, submitted.

Lanhers, M.C., Joyeux, M., Soulimani, R., Fleurentin, J., Sayag, M., Mortier, F., Younos, C. and Pelt, J.M. (1991) Hepatoprotective and anti-inflammatory effects of a traditional medicinal plant of Chile. *Peumus boldus*, *Planta Medica*, **57**, 110–115.

Madrero, Y., Elorriaga, M., Martínez, S., Noguera, M.A., Cassels, B.K., D'Ocon, P. and Ivorra, M.D. (1996) A possible structural determinant of selectivity of boldine and derivatives for the α_{1A}-adrenoceptor subtype. *Brit. J. Pharmacol.*, **119**, 1563–1568.

Madrero, Y., Elorriaga, M., Noguera, M.A., Cassels, B.K., D'Ocon, P. and Ivorra, M.D. (1997) Halogenated derivatives of boldine with high selectivity for α_{1A}-adrenoceptors in rat cerebral cortex. *Brit. J. Pharmacol.*, submitted.

Martínez, L.A., Ríos, J.L., Payá, M. and Alcaraz, M.J. (1992) Inhibition of non-enzymic lipid peroxidation by benzylisoquinoline alkaloids. *Free Radical Biol. Med.*, **12**, 287–292.

Morello, A., Lipchenca, I., Cassels, B.K., Speisky, H., Aldunate, J. and Repetto, Y. (1994) Trypanocidal effect of boldine and related alkaloids upon several strains of *Trypanosoma cruzi*. *Comp. Biochem. Physiol.*, **107C**, 367–371.

Speisky, H., Cassels, B.K., Lissi, E.A. and Videla, L.A. (1991a) Antioxidant properties of the alkaloid boldine in systems undergoing lipid peroxidation and enzyme inactivation. *Biochem. Pharmacol.*, **41**, 1575–1582.

Speisky, H., Squella, J.A. and Núñez-Vergara, L.J. (1991b) Activity of boldine on rat ileum. *Planta Medica*, **57**, 519–22.

Speisky, H., Cassels, B.K., Sanhueza, V., Nieto, S., Valenzuela, A. and Núñez-Vergara, L.J. (1993) Determination of boldine in plasma by HPLC. *J. Chromatogr.-Biomed. Appl.*, **612** , 315–319.

Speisky, H. and Cassels, B.K. (1994) Boldo and boldine: an emerging case of natural drug development. *Pharmacol. Res.*, **29**, 1–12.

Valenzuela, A., Nieto, S., Cassels, B.K. and Speisky, H. (1991) Inhibitory effect of boldine on fish oil oxidation. *J. Am. Oil Chem. Soc.*, **68**, 935–937.

Zetler, G. (1988) Neuroleptic-like, anticonvulsant and antinociceptive effects of aporphine alkaloids: bulbocapnine, corytuberine, boldine and glaucine. *Arch. Int. Pharmacodyn. Thér.*, **296**, 255–281.

12. METABOLIC ENGINEERING: A STRATEGY TO IMPROVE PLANT SECONDARY METABOLITE PRODUCTION

R. VERPOORTE[1], R. VAN DER HEIJDEN[1], J. MEMELINK[2] and H.J.G. TEN HOOPEN[3]

[1]*Division of Pharmacognosy, Leiden/Amsterdam Center for Drug Research, Leiden University, PO Box 9502, 2300 RA Leiden, The Netherlands*
[2]*Institute for Plant Molecular Sciences, Leiden University, The Netherlands*
[3]*Department of Biochemical Engineering, Delft University of Technology, Delft, The Netherlands*

INTRODUCTION

Plants produce a wide variety of secondary metabolites. These compounds, which vary for each plant species, play a role in the interactions of the plant with its environment, e.g. as attractants for pollinators, as antifeedants against predators or as antimicrobial active compounds (phytoalexin) against plant pathogens. Secondary metabolites play a major role in the quality of food plants and ornamentals, e.g. smell, taste, flavour and colour. Plant secondary metabolism is thus an important trait in connection with plant breeding.

Many plant secondary metabolites are used as fine chemicals, e.g. as drugs, dyes, insecticides, fragrances and flavours. A large number of these products are derived from rather rare plants, which is among others a reason to search for biotechnological production methods for such specialty chemicals. Taxol is an example. Plant cell cultures provide this sort of alternative approach to the production of plant secondary metabolites. In the past years all economically important plants have been brought into cell culture. However, in most cases the productivities found were too low to allow an economically feasible process. Research is therefore focusing on improving the production of secondary metabolites in plant cells.

Here we will discuss some aspects of plant cell biotechnology:

- the technological and economical feasibility;
- approaches used to improve production;
- prospects of metabolic engineering for improving yields;
- the role and the regulation of secondary metabolism.

This will be illustrated with examples from our research on *Catharanthus roseus*, *Tabernaemontana* species (both Apocynaceae) and *Cinchona* species (Rubiaceae). *C. roseus* is the source of ajmalicine, used as a drug which improves cerebral circulation, and the dimeric alkaloids vinblastine and vincristine, two important antitumor drugs. *Tabernaemontana* species produce alkaloids similar to those found in *C. roseus*, including

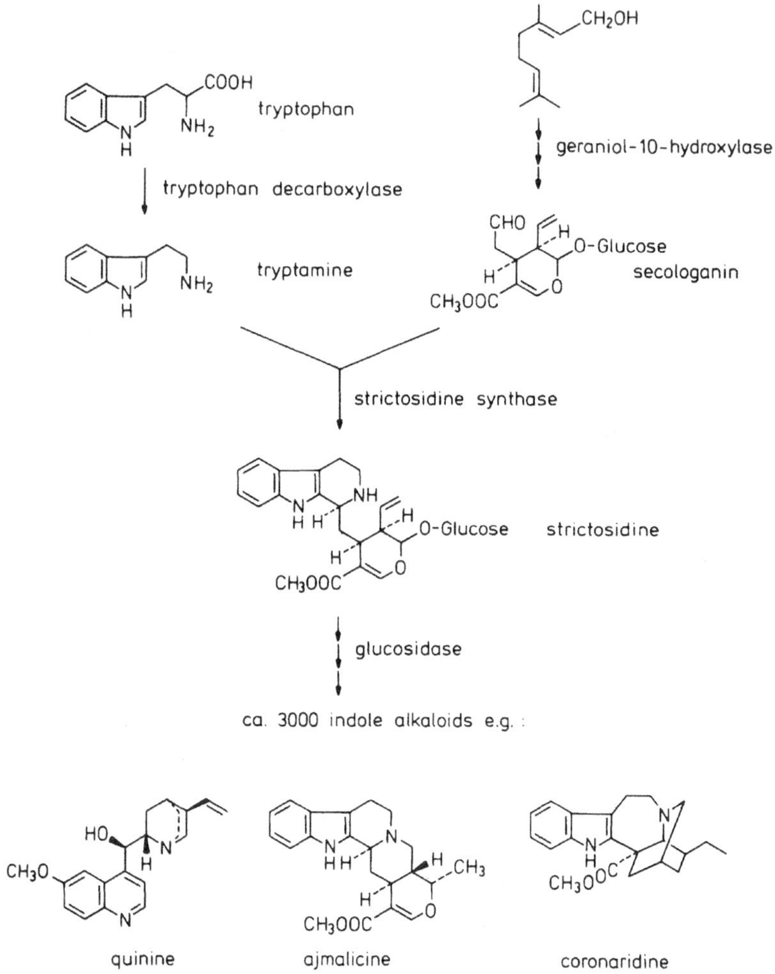

Figure 1 Early steps of terpenoid indole alkaloid biosynthesis

various dimeric alkaloids. From the bark of *Cinchona* species the quinoline alkaloids quinine, used as an antimalaria drug and as a bitter agent in beverages, and quinidine, an antiarhythmic drug, are isolated. Both are derived from the same biosynthetic pathway as the previously mentioned terpenoid indole alkaloids, in which strictosidine is an important intermediate (Figure 1).

LARGE SCALE CULTURE

Microorganisms are usually connected with biotechnological production. Of course this raises the question as to whether it is feasible to produce plant secondary metabolites in transgenic microorganims. In theory the answer is yes; however, in practice this is not realistic. It is feasible to express plant genes in microorganisms, but in case of the complex biosynthetic pathways of plant secondary metabolites, one needs to express a large number of genes in the microorganism – genes which in most cases are not yet known. Moreover, in plants the secondary pathways are regulated among others through compartmentation, i.e. first steps in plastids, next steps in the cytosol and final steps in the vacuole. In some cases, even different cells are involved. All these are situations which can not be realized in microorganisms. Only in cases of bioconversions involving one or two steps might transgenic microorganisms be an alternative. Therefore most efforts are focused on plant cell cultures as a possible production system.

The first question to be answered for a biotechnological production process is whether it is technically and economically feasible to produce secondary metabolites by means of plant cell cultures on a large scale.

Plant cells have been reported to be extremely shear sensitive, requiring special dedicated bioreactors. However, for an industrial application this is a major constraint, as the fermentation industry mostly runs stirred-tank type of bioreactors. The first thing we studied thus was the shear sensitivity of plant cells. Surprisingly it was found that, of four cell lines tested, two were clearly shear resistant, whereas the two others only showed reduced growth at very high stirrer speeds (Meijer *et al.* 1986; Meijer 1989). Similar findings were reported by Scragg and co-workers (Scragg *et al.* 1986; Leckie *et al.* 1990). The number of large scale production systems with plant cells reported in the literature confirm these results. Examples are the commercial production of shikonin in a 750 l stirred-tank bioreactor (Fuyita and Tabata 1987) and the growth of various plant cell cultures in 60 m^3 working volume stirred tanks by a German company (Westphal 1990).

Technologically the large scale growth of plant cells is thus feasible, but how about the economics? Several authors have presented cost price calculations (Drapeau *et al.* 1987; Fowler and Stepan-Sarkissian 1983; Goldstein *et al.* 1980; van Gulik *et al.* 1988; Verpoorte *et al.* 1991). In our calculations (van Gulik *et al.* 1988; Verpoorte *et al.* 1991) the *C. roseus*-ajmalicine production data were used to calculate the price for the production of this alkaloid (see Tables 1 and 2). Interestingly the lowest price is for a "classical" process in which a large amount of biomass is grown, production induced by a production medium and subsequently harvested for the product extraction. A system in which the product is released into the medium is more expensive, as the size of the required bioreactors is much larger. In the biomass production process the amount of extracellular medium is less than 10% – too small for a system where product released from the cells is to be recovered from the extracellular medium. Consequently in a production system based on product release, the biomass density has to be reduced. In this process the ratio biomass : medium is less favorable and the productivity per unit of bioreactor volume is thus less.

Obviously for compounds with a higher level of production the price will be lower. Using the same method of calculation for a compound like berberine the results are shown in Table 3.

Although these prices are quite high, they are in the same order of magnitude as the market prices of a number of plant-derived specialty chemicals. The improvement of

Table 1 Parameters used in cost price calculation for ajmalicine production by cell cultures of *Catharanthus roseus*

Design basis	
Yearly production	3000 kg
Product loss during isolation	20%
Effective production days/year	300
Growth parameters	
Specific growth rate	$0.029\ h^{-1}$
Initial biomass (D.W.)	$2.5\ kg/m^3$
Inoculation ratio	1 : 7
Dry weight yield from glucose	0.61 kg/kg
Maintenance energy requirement	$0.0066\ kg/kg \cdot h$
Maximum oxygen uptake rate	$0.0154\ kmol/m^3 \cdot h$
Single use of biomass	
Final biomass concentration (D.W.)	$40\ kg/m^3$
Final biomass concentration (F.W.)	$320\ kg/m^3$
Final ajmalicine content after 21 days production period	0.009 kg/kg
Repeated use of biomass	
Specific productivity, spontaneous product release	$2.36 \times 10^{-5}\ kg/kg \cdot h$
Ajmalicine concentration in biomass in case of forced release	0.009 kg/kg

Table 2 Breakdown of costs of different processes for large-scale production of ajmalicine by cell cultures of *Catharanthus roseus*, comparing a cell culture in which the alkaloid is to be extracted from the biomass with the same process but with ten times more productive cells, and the medium after spontaneous or forced release

	Product extraction from cells	*10 × improved*	*Product release*	*Permeabilization (DMSO)*
Production reactor volume (m^3)	6 × 145	6 × 15	6 × 250	6 × 335
Biomass reactor (m^3)	37	3.8	82	10
Storage tanks (m^3)	3 × 10	3 × 1.2	7 × 150	7 × 150
Depreciation costs in $/year	3×10^6	1.1×10^6	5.4×10^6	5.4×10^6
Medium costs in $/year	0.75×10^6	0.1×10^6	3.3×10^6	5.1×10^6
Energy costs in $/year	0.75×10^6	0.08×10^6	1.5×10^6	1.5×10^6
Total costs in $/year	4.5×10^6	1.3×10^6	10.2×10^6	12×10^6
Costs $/kg	1500	430	3400	4000

Table 3 Costprices calculated for the production of 3000 kg/year of berberine by means of plant cell cultures

Single use of biomass (*Coptis japonica*, 3.5 g/l)	360 $/kg
Spontaneous release of product, discontinuous process (*Thalictrum minus*, 1.4 g/l)	760 $/kg
Spontaneous release of product, continuous process (*Thalictrum minus*, 1.4 g/l)	850 $/kg
Immobilized cells (*Thalictrum minus*)	600 $/kg

the productivity of the cell cultures should thus be the main goal for research, in order to open the way to commercialization.

IMPROVING PRODUCTION

To improve production different approaches can be followed:

– screen and select for high producing cell lines
– improve growth and production media.

In the past these approaches were extensively used; in general one can say that an about 10–20 fold increase of production can be achieved by these means. However, in many cases success was not achieved – 20 times zero remains zero, e.g. compounds such as morphine, vinblastine and vincristine are not produced in a cell suspension culture. For those compounds which already have a high production these approaches might be useful, but in the cell suspension cultures of *Cinchona* species, for example, the alkaloid production remains very low, despite all efforts to optimize the media (Harkes *et al.* 1985; Hoekstra 1993; Hoekstra *et al.* 1990; Verpoorte *et al.* 1988, 1991; Wijnsma *et al.* 1986a). Furthermore, a major problem is the stability of high producing cell lines. In some cases, high producing cell lines obtained through an extensive screening procedure are not stable, though, for some plants, stable cell high producing cell lines have been obtained (Fuyita and Tabata 1987 see also for a review Ohta and Verpoorte 1992).

In recent years three other promising approaches have evolved:

– culturing of differentiated tissue cells
– elicitation of plant cells
– metabolic engineering.

Growth of Differentiated Tissue Cultures

Cell suspension cultures of *Cinchona* species hardly produce any alkaloids, but by changing the growth hormones in the media differentiation into shoot cultures was observed and alkaloids were produced. Certain combinations of growth hormones in the medium of the *Cinchona* cell cultures resulted in large cell aggregates, which were called compact globular structures (CGS). These CGS did produce considerable amounts of alkaloids. Microscopical analysis of the CGS showed that they have characteristics (tracheary elements) of the stem of the tree. The alkaloids were in fact also only found in the outer layers of the CGS, just like in the stem where the alkaloids are stored in the outer parts (the bark). Apparently the alkaloid biosynthesis is closely linked with morphological differentiation (Hoekstra 1993; Hoekstra *et al.* 1990). Similarly in *Tabernaemonatana pandacaqui* cultures it was noted that upon increasing the degree of differentiation, going from cell suspension to shoot culture, the amount of alkaloids increased; moreover the iboga-type of alkaloids could only be detected in more differentiated cultures (Sierra *et al.* 1991).

A specific form of organized cultures are the so-called "hairy root" cultures, obtained by transformation of the plant with *Agrobacterium rhizogenes.* In case of *Cinchona* species hairy roots have been obtained which do produce quinoline alkaloids (Hallard *et al.* 1996). Also for *C. roseus* such cultures could be obtained, and as expected they produced a similar alkaloid spectrum to the normal roots, with ajmalicine as major alkaloid (Hallard *et al.* 1996; Hoekstra 1993). As *A. rhizogenes* can also be used to introduce new genes into the plant, hairy roots are an interesting model system for studying metabolic engineering of secondary metabolism (see below). However, because of the difficulties involved with the large scale culture of such hairy roots, and the fact that the levels of alkaloids will be similar to those found in the plant, at the present we do not see this as the first choice for industrial production. In fact this holds true for all the differentiated cultures, their main importance being in studies of the regulation of secondary metabolism.

Elicitation of Secondary Metabolism

Addition of so called biotic elicitors, such as cell wall constituents of microorganisms or enzymes like cellulase or pectinase, to plant cell cultures induces the defense response against infections. For secondary metabolism this translates to the production of phytoalexins, low molecular weight compounds which have antimicrobial activity. For several systems the elicitation has shown to be an effective way to improve the production of certain secondary metabolites, e.g. sanguinarine in *Papaver somniferum* cell cultures (Eilert *et al.* 1985). In our model systems, quite different reactions were noted upon elicitation. In the case of *Tabernaemontana divaricata*, the alkaloid production is blocked, whereas considerable amounts of ursolic acid-type of triterpenes are formed, due to an effective channelling of the C_5-units into the triterpene pathway and inhibition of the sterol biosynthesis (van der Heijden *et al.* 1988, 1989). In *Cinchona* cell cultures a completely different pathway, leading to anthraquinones, is induced. These compounds need a C_5-unit and chorismic acid as precursors; both building blocks are also needed for the terpenoid indole alkaloid pathway. Alkaloid production was not induced by elicitation, the C_5-units being channelled away from the terpenoid/sterol biosynthesis into the anthraquinone biosynthesis (Ramos-Valdivia *et al.* unpublished results).

Earlier studies showed that these anthraquinones most likely act as phytoalexins in *Cinchona* species (Wijnsma *et al.* 1985, 1986b). Also *C. roseus* showed a different response – an increase of the tryptamine content and a slight increase of ajmalicine level was observed after elicitation with a *Pythium* elicitor preparation. The increase is due to a direct induction of the genes coding for tryptophan decarboxylase (TDC) and strictosidine synthase (STR), followed by an increase in the activity of the enzymes (Goddijn *et al.* 1992; Pasquali *et al.* 1992). Also the enzyme anthranilate synthase, the first enzyme in the pathway leading from chorismate to tryptophan, is induced after elicitation. Chorismate mutase, another important chorismate utilizing enzyme leading to phenylalanine and tyrosine and from there to a broad variety of secondary metabolites, is not affected (Poulsen and Verpoorte 1992). Surprisingly the enzyme isochorismate synthase is strongly induced with a concomitant formation of 2,3-dihydroxybenzoic acid (Moreno *et al.* 1994a,b). This compound proved to have antimicrobial activity, particularly in combination with UV-light. That combination also induced the activity of the enzymes TDC and STR.

Comparing the effects of elicitation for those three plants, which all share the trait of terpenoid-indole alkaloid biosynthesis, it is clear that despite the common features of this pathway, the regulation is apparently quite different.

Although these results offer very interesting possibilities for further studies of the regulation of the secondary metabolism in these systems, elicitation is not useful for the large scale production of the indole alkaloids. An increase of production can also be obtained by medium optimization or the use of fed-batch culture systems.

Metabolic Engineering

The genetic engineering of plants and plant cells is feasible nowadays; methods such as the *Agrobacterium* transformation system and the particle gun are quite successful methods for the introduction of new genes into plants. It should thus be possible to modify also secondary metabolism. Several possibilities can be envisaged:

- increasing the flux through a pathway, by increasing the activity of enzymes catalyzing limiting steps. This could be achieved by using the gene coding for that particular enzyme from the same plant, or a gene from another plant or organism which codes for an enzyme with a similar function. Even the construction of new, more efficient enzymes (protein engineering) can be envisaged, e.g. enzymes which are not sensitive anymore for feedback inhibition.
- constitutive expression of genes regulating (part of) a secondary metabolite pathway.
- blocking competitive pathways, e.g. competing for the same pool of C_5-building blocks (isoprenoids) in case of terpenoid indole alkaloids and anthraquinones in *Cinchona.*
- blocking catabolism of the desired product.

The first two approaches require sense genes, which express an active protein, allowing the biosynthetic machinery to work in a similar fashion as in the non-transformed plant or plant cells, but with a greater flux of carbon towards the desired product(s). The two latter approaches, require antisense genes which block certain steps in a pathway. Both the sense- and antisense approach have been shown to be effective in changing flower colours.

A major constraint for the metabolic engineering of plant secondary metabolite production is our lack of knowledge on secondary metabolism and how this is regulated.

Most secondary metabolite pathways are little studied, mostly only putative pathways have been proposed, based on the incorporation of various labeled intermediates. The knowledge on the level of enzymes is in most cases lacking, and consequently genes encoding steps in secondary metabolism, or genes regulating (parts of) the pathways are not known. Therefore to be able to apply metabolic engineering, first the biosynthetic pathways have to be studied in detail on the level of the products and the enzymes involved.

For the model systems in our studies, the first parts of the pathway are the same (see Figure 1). They all have strictosidine as the intermediate from which a broad variety of ca. 3000 indole and related alkaloids is derived. As targets for our studies we have chosen the two enzymes on the edge of primary and secondary metabolism, i.e. geraniol-10-hydroxylase (G10H) and tryptophan decarboxylase (TDC), and the enzyme linking the two pathways, i.e. strictosidine synthase (STR). Assays were developed for all enzymes; for TDC and STR, HPLC assays were developed, allowing the quantification of both substrate and products (Pennings *et al.* 1987, 1988, 1989a; Meijer *et al.* 1990, 1993a). With these assays the best source for the isolation of the enzymes was found to be *C. roseus*. The enzymes TDC and STR were purified (Pennings *et al.* 1989b,c, 1990), partly sequenced and antibodies were raised. STR was found to be present in 7 isoforms, 6 of which were purified and characterized (de Waal *et al.* 1995). Based on these results the encoding cDNAs could be cloned (Goddijn *et al.* 1992, Pasquali *et al.* 1992). In both cases only a single copy of the gene was present. The cDNAs were sequenced and with the aid of probes the regulation of the genes is now being further studied. The sequence found for the *tdc*-gene was identical to that reported by De Luca *et al.* (1989). The *str*-gene showed extensive homology with the one cloned from *Rauwolfia serpentina* (Kutchan *et al.* 1988). It was found that auxins repress the genes, whereas inoculation of the cells on an auxin-free medium resulted in a rapid induction of transcription. Similarly elicitors induced these genes in *C. roseus*. Constitutive expression of the *tdc*-gene in tobacco resulted in healthy plants which produced about 1% of tryptamine. It was shown that in these plants the activity of the enzyme anthranilate synthase was not increased, i.e. apparently in its normal tryptophan metabolism, tobacco can produce about 1% of tryptamine (Poulsen *et al.* 1993). It is interesting to note that in indole alkaloid-producing plants, levels of about 1% of total indole alkaloid are also found, if calculated on the total biomass. Tobacco cells in which both the *str*- and *tdc*-gene are introduced are able to produce strictosidine after feeding of secologanin (Hallard *et al.* 1997b).

Introduction of the *tdc*-gene into *C. roseus* cells resulted in an increased TDC activity and higher levels of tryptamine, but not in an increase of alkaloid levels. This was in agreement with other observations pointing to the terpenoid pathway as an important limiting factor in the alkaloid biosynthesis. This shows that it is feasible to overcome a limiting step, but that it directly results in meeting the next limiting step. Therefore further studies on the regulation of the expression of the genes already cloned are of interest, these may lead to the identification of genes regulating the complete pathway, or at least part thereof.

The first enzyme in the secologanin pathway is G10H, a membrane-bound cytochrome P-450 dependent enzyme which consists of two parts, the cytochrome P-450 containing protein, responsible for the selectivity of the enzyme, and a NADPH: cytochrome P-450 reductase, which serves all P-450 enzymes in the plant. Both enzymes could be purified. For the reductase part the gene was cloned and could be expressed in *E. coli* resulting in a functional enzyme (Meijer *et al.* 1993b). The cloning of the P-450 gene is hampered by the fact that a large number of closely related P-450 genes are present in the plant (Meijer *et al.* 1993c); this work is still in progess.

From these results it is clear that it is feasible to alter plant secondary metabolism by means of genetic engineering. But it must be realized that it requires extensive studies about the biosynthetic pathways, the enzymes involved and their regulation. Moreover, regulation does not only occur on the level of the genes, but also on the level of the enzymes and the products. Compartmentation and transport are further important factors involved in the biosynthetic pathways. Finally, catabolism is a factor which should not be overlooked as possibility to increase secondary metabolite production. In both *C. roseus* and *T. divaricata* we found that extensive breakdown of the alkaloids occurs, in fact equaling the de-novo biosynthesis (Dagnino *et al.* 1992, 1994; dos Santos *et al.* 1994; Schripsema *et al.* 1994).

Role of Secondary Metabolism for the Plant

The role of secondary metabolites is directly connected with their regulation. As discussed above, secondary metabolites are important in the interaction of the plant with its environment. Phytoalexins, such as the triterpenes and anthraquinones described above, are a clear example of such a role. In this case regulation consists of a signal-transduction chain that switches on the biosynthesis after infection of the plant with a microorganism. Antifeedant activity is probably connected with differentiation of tissues, resulting in the presence of protecting agents in vulnerable tissues, e.g. seedlings and young leaves.

How about the terpenoid indole alkaloids? We have studied the alkaloid production in *Cinchona* seedlings and plants in more detail (Aerts *et al.* 1990, 1991a,b,c, 1992). In seeds neither alkaloids nor the enzymes TDC and STR could be detected. Upon germination, however, these enzymes are rapidly induced and first tryptamine and then alkaloids are formed from pools of precursors apparently already present in the seeds. About five days after the radicle comes through the seedskin, the maximum alkaloid level is reached in the seedling. The metabolic effort is rather small, being less than 0.5% of the total metabolism during the development of the seedling. It was found that the alkaloids have a strong antifeedant activity against slugs. The level reached after five days gives complete feeding deterrence. The quinoline alkaloids are responsible for this effect. In the leaves of the plant the major alkaloids belong to the cinchophylline-type, i.e. semidimeric indole alkaloids. These alkaloids were found to have strong antifeedant activity against larvae of *Spodoptera exigua*, whereas the quinoline alkaloids did not have such an activity. Again the levels of the cinchophyllines in the young leaves were such that a complete deterrence was achieved (Aerts *et al.* 1992). Although the quinoline alkaloid exhibited allelopathic effects, the levels found around the roots of the plants are too low to cause such an effect in the field (Aerts *et al.* 1991c).

Thus in *Cinchona* species the alkaloids do play a major role in the defense against predators. From a common precursor strictosidine different products are made in the various plant parts, products which have a specific role in that part of the plant as antifeedant. In the lower parts and in the seedlings the quinoline alkaloids, active against ground-dwelling slugs, are produced; in the higher parts indole alkaloids, active against insects, are formed.

In *C. roseus* the glycoalkaloid strictosidine is present in rather high levels in young leaves. This led us to the idea that strictosidine might be more than just an intermediate. It is stored in vacuoles and the enzyme strictosidine glucosidase is found to be present in large excess connected with the tonoplast (Stevens *et al.* 1993a; Luijendijk 1995). Based on this we now think that strictosidine plays a role in plant defense. Upon wounding and cell rupture the glucosidase will rapidly convert strictosidine into a highly reactive dialdehyde, which might have strong biological activities. Indeed we were able to show that strictosidine in combination with glucosidase has antimicrobial activity against some plant pathogenic fungi, whereas strictosidine itself did not show such activity (Luijendijk *et al.* 1996). Strictosidine is thus thought to be a phytoanticipin, i.e. a compound already present in the plant, which is activated the very moment the plant is attacked by e.g. an insect.

CONCLUSIONS

As metabolic engineering is not only of interest for the production of specialty chemicals, but also in connection with improving the characteristics of crop plants, food and ornamental plants, studies on the regulation of secondary metabolism and its role for the plant will become increasingly important in the years to come. Phytochemistry, including biochemistry, will be the basis of such studies, which should open the way for metabolic engineering. The close collaboration between various disciplines, such as ecology, plant pathology, phytochemistry, biochemistry and molecular biology is an absolute necessity to make progress in this area.

Finally it must be noted that biodiversity is not only a rich source of interesting secondary metabolites, but also of enzymes for industrial use and of genes which, for example, might be useful to make other plants more resistant against pests and diseases.

REFERENCES

Aerts, R.J., van der Leer, T., van der Heijden, R. and Verpoorte, R. (1990) Developmental regulation of alkaloid production in *Cinchona* seedlings. *J. Plant Physiol.*, **136**, 86–91.

Aerts, R.J., Snoeijer, W., Aerts-Teerlink, O., van der Meijden, E., Verpoorte and R. (1991a) Control and biological implications of alkaloid synthesis in *Cinchona* seedlings. *Phytochemistry*, **30**, 3571–3577.

Aerts, R.J., Snoeijer, W., van der Meijden and E., Verpoorte, R. (1991b) Allelopathic inhibition of seed germination by *Cinchona* alkaloids? *Phytochemistry*, **30**, 2947–2951.

Aerts, R.J., de Waal, A., Pennings, E.J.M. and Verpoorte, R. (1991c) The distribution of strictosidine synthase activity and alkaloids in *Cinchona* plants. *Planta*, **183**, 536–541.

Aerts, R.J., Stoker, A., Beishuizen, M., Jaarsma, I., van de Heuvel, M., van der Meijden, E. and Verpoorte, R. (1992) Detrimental effects of *Cinchona* leaf alkaloids on larvae of the polyphagous insect *Spodoptera exigua*. *J. Chem. Ecol.*, **18**, 1955–1964.

Dagnino, D., Schripsema, J. and Verpoorte, R. (1992) Alkaloid metabolism in *Tabernaemontana divaricata* cell suspension cultures. *Phytochemistry*, **32**, 325–329.

Dagnino, D., Schripsema, J. and Verpoorte, R. (1994) Terpenoid indole alkaloid biotransformation of suspension cultures of *Tabernaemontana divaricata*. *Phytochemistry*, **35**, 671–676.

De Luca, V., Marineau, C. and Brisson, N. (1989) Molecular cloning and analysis of cDNA encoding a plant tryptophan decarboxylase: comparison with animal dopa decarboxylase. *Proc. Natl. Acad. Sci. USA*, **86**, 2582–2586.

De Waal, A., Meijer, A.H. and Verpoorte, R. (1995) Strictosidine synthase from *Catharanthus roseus*: purification and characterization of multiple forms. *Biochem. J.*, **306**, 571–580.

Dos Santos, R., Schripsema, J. and Verpoorte, R. (1994) Ajmalicine metabolism in *Catharanthus roseus* cell cultures. *Phytochemistry*, **35**, 677–681.

Drapeau, D., Blanch, H.W. and Wilke, C.R. (1987) Economic assessment of plant cell culture for the production of ajmalicine. *Biotechnol. Bioeng.*, **30**, 946–953.

Eilert, U., Kurz, W.G.W. and Constabel, F. (1985) Stimulation of sanguinarine accumulation in *Papaver somniferum* cell cultures by fungal elicitors. *J. Plant Physiol.*, **119**, 65–76.

Fowler, M.W. and Stepan-Sarkissian, G. (1983) Chemicals from plant cell fermentation. In A. Mizrahi and A.L. van Wezel (eds.), *Advances in Biotechnological Processes*, Vol. 2, Alan R. Liss, New York, pp. 135–158.

Fuyita, Y. and Tabata, M. (1987) Secondary metabolites from plant cells – Pharmaceutical applications and progress in commercial production. In C.E. Green *et al.* (eds.), *Plant Tissue and Cell Culture*, Alan R. Liss, New York, pp. 169–185.

Goddijn, O.J.M., de Kam, R.J., Zanetti, A., Schilperoort, R.A. and Hoge, J.H.C. (1992) Auxin rapidly down–regulates transcription of the tryptophan decarboxylase gene from *Catharanthus roseus*. *Plant Mol. Biol.*, **18**, 1113–1120.

Goldstein, W.E., Lazure, L.L. and Ingle, M.B. (1980) Product cost analysis. In E.J. Staba (ed.), *Plant Tissue Culture as a Source of Biochemicals*, CRC Press, Boca Raton, FL, pp. 191–234.

Hallard, D., Geerlings, A., Lopez–Cardoso, I., Hoge, J.H.C., van der Heijden, R. and Verpoorte, R. (1997a) Metabolic engineering of terpenoid indole and quinoline alkaloid biosynthesis in hairy root cultures. In P.M. Doran, (ed.), *Hairy Roots*, Harwood Academic Publishers, Berkshire, in press.

Hallard, D., van der Heijden, R., Verpoorte, R., Lopez–Cardoso, I., Pasquali, G., Memelink, J. and Hoge, J.H.C. (1997b) Suspension cultured transgenic cells of *Nicotiana tabacum* expressing tryptophan decraboxylase and strictosidine synthase cDNAs from *Catharanthus roseus* produce strictosidine upon feeding of secologanin. *Plant Cell Rep.*, **17**, 50–54.

Harkes, P.A.A., Krijbolder, L., Libbenga, K.R., Wijnsma, R. and Verpoorte, R. (1985) Influence of various media constituents on the growth of *Cinchona ledgeriana* tissue cultures and the production of alkaloids and anthraquinones therein. *Plant Cell Tissue Organ Culture*, **4**, 199–214.

Hoekstra, S.S. (1993) Accumulation of indole alkaloids in plant–organ cultures, Ph.D. Thesis, Leiden.

Hoekstra, S.S., Harkes, P.A.A., Libbenga, K.R. and Verpoorte, R. (1990) Effect of auxin on cytodifferentiation and production in compact globular structures of *Cinchona ledgeriana*. *Plant Cell Rep.*, **8**, 571–574.

Kutchan, T.M., Hampp, N., Lottspeich, F., Beyreuther, K. and Zenk, M.H. (1988) The cDNA clone for strictosidine synthase from *Rauvolfia serpentina*; DNA sequence determination and expression in *Escherichia coli*. *FEBS Letters*, **237**, 40–44.

Leckie, F., Scragg, A.H. and Cliffe, K.C. (1990) The effect of continuous high shear stress on plant cell suspension cultures. In H.J.J. Nijkamp, L.H.W. van der Plas and J. van Aartrijk, (eds.), *Progress in Plant Cellular and Molecular Biology*, Kluwer Academic Publishers, Dordrecht, pp. 689–693.

Luijendijk, T.J.C. (1995) Strictosidine glucosidase in indole alkaloid biosynthesis, Ph.D. Thesis, Leiden University.

Luijendijk, T.J.C., van der Meijden, E. and Verpoorte, R. (1996) Involvement of strictosidine as a defensive chemical in *Catharanthus roseus*. *J. Chem. Ecol.*, **22**, 1355–1366.

Meijer, J.J. (1989) Effects of hydrodynamic and chemical/osmotic stress on plant cells in a stirred bioreactor, Ph.D. Thesis, Technical University Delft.

Meijer, J.J., Van Gulik, W.M., Ten Hoopen, H.J.G. and Luyben, K.Ch.A.M. (1987) The influence of shear stress on growth and morphology of *Catharanthus roseus* in continuous culture. In O.M. Neijssel, R.R. van der Meer and K.Ch.A.M. Luyben, (eds.), *Proceedings 4th Eur. Congr. Biotechnol., Vol 2.*, Elsevier, Amsterdam, p. 409.

Meijer, A.H., Pennings, E.J.M., de Waal, A. and Verpoorte, R. (1990) Purification of cytochrome P-450-dependent geraniol-10-hydroxylase from a cell suspension culture of *Catharanthus roseus*. In H.J.J. Nijkamp, L.H.W. van der Plas and J. van Aartrijk, (eds.), *Progress in plant cellular and molecular biology*, Kluwer Academic Publishers, Dordrecht, pp. 769–774.

Meijer, A.H., de Waal, A. and Verpoorte, R. (1993a) Purification of cytochrome P-450-dependent geraniol-10-hydroxylase from cell cultures of *Catharanthus roseus*. *J. Chromatogr.*, **635**, 237–249.

Meijer, A.H., Lopes Cardoso, M.I., Voskuilen, J.Th., de Waal, A., Verpoorte, R. and Hoge, J.H.C. (1993b) Isolation and characterisation of a cDNA clone from *Catharanthus roseus* encoding NADPH: cytochrome P-450 reductase, an enzyme essential for reactions catalysed by cytochrome P-450 monooxygenases in plants. *Plant J.*, **4**, 47–60.

Meijer, A.H., Souer, E., Verpoorte, R. and Hoge, J.H.C. (1993c) Isolation of cytochrome P-450 cDNA clones from the higher plant *Catharanthus roseus* using a PCR strategy. *Plant Mol. Biol.*, **22**, 379–383.

Moreno, P.R.H., van der Heijden, R. and Verpoorte, R. (1994a) Elicitor-mediated induction of isochorismate synthase and accumulation of 2,3-dihydroxybenzoic acid in *Catharanthus roseus* cell suspension and shoot cultures. *Plant Cell Rep.*, **14**, 188–191.

Moreno, P.R.H., van der Heijden, R. and Verpoorte, R. (1994b) Induction of the secondary metabolism in *Catharanthus roseus* cell suspension cultures in response to UV irradiation and the addition of a benzoic acid derivative. *Heterocycles*, **39**, 457–465.

Ohta, S. and Verpoorte, R. (1992) Some accounts of variation (heterogeneity and/or instability) in secondary metabolite production by plant cell cultures. *Ann. Rep. Nat. Sci. Home Econ.*, **32**, 9–23.

Pasquali, G., Goddijn, O.J.M., de Waal, A., Verpoorte, R., Schilperoort, R.A., Hoge, J.H.C. and Memelink, J. (1992) Coordinated regulation of two indole alkaloid biosynthetic genes from *Catharanthus roseus* by auxin and elicitors. *Plant Mol. Biol.*, **18**, 1121–1131.

Pennings, E.J.M., Hegger, I., van der Heijden, R., Duine, J.A. and Verpoorte, R. (1987) Assay of tryptophan decarboxylase from plant cell cultures. *Anal. Biochem.*, **165**, 133–136.

Pennings, E.J.M., van der Heijden, R., van den Bosch, R.A., van der Leer, T. and Verpoorte, R. (1988) Rapid assay of tryptophan decarboxylase from plant cell cultures. In R.J. Robins and M.J.C. Rhodes, (eds.), *Manipulating Secondary Metabolism in Culture*, Cambridge University Press, Cambridge, pp. 79–82.

Pennings, E.J.M, van den Bosch, R.A., van der Heijden, R., Stevens, L.H., Duine, J.A. and Verpoorte, R. (1989a) Assay of strictosidine synthase from plant cell cultures by HPLC. *Anal. Biochem.*, **176**, 412–415.

Pennings, E.J.M., Groen, B.W., Duine, J.A. and Verpoorte, R. (1989b) Tryptophan decarboxylase from *Catharanthus roseus* is a pyridoxoquinoprotein. *FEBS Letters*, **255**, 97–100.

Pennings, E.J.M., Verpoorte, R., Goddijn, O.J.M. and Hoge, J.H.C. (1989c) Purification of tryptophan decarboxylase from cell cultures of *Catharanthus roseus*. *J. Chromatogr.*, **483**, 311–318.

Pennings, E.J.M., Giroud, C., Stevens, L. and Verpoorte, R. (1990) Purification of strictosidine synthase from a *Cinchona robusta* cell suspension culture. *Planta Med.*, **56**, 599.

Poulsen, C. and Verpoorte, R. (1992) Activities of chorismate utilizing enzymes and of enzymes involved in indole alkaloid biosynthesis in cell suspension cultures. *Plant Physiol. Biochem.*, **30**, 105–113.

Poulsen, C., Goddijn, O.J.M., Hoge, J.H.C. and Verpoorte, R. (1993) Anthranilate synthase and chorismate mutase activities in transgenic tobacco plants overexpressing tryptophan decarboxylase from *Catharanthus roseus. Transgenic Res.*, **3**, 43–49.

Schripsema, J., Dagnino, D., dos Santos, R. and Verpoorte R. (1994) Breakdown of indole alkaloids in suspension cultures of *Tabernaemontana divaricata* and *Catharanthus roseus. Plant Cell Tiss. Org. Cult.*, **38**, 301–307.

Scragg, A.H., Allen, E.J., Bond, P.A. and Smart, N.J. (1986) Rheological properties of plant cell suspension cultures. In P. Morris *et al.*, (eds.), Secondary Metabolism in Plant Cell Cultures, Cambridge University Press, Cambridge, pp. 178–194.

Sierra, M.I., van der Heijden, R., Schripsema, J. and Verpoorte, R. (1991) Alkaloid production in relation to differentiation in plant cell and tissue culture of *Tabernaemontana pandacaqui. Planta Med.*, **57**, 543–547.

Stevens, L.H., Blom, T.J.M. and Verpoorte, R. (1993a) Subcellular localization of tryptophan decarboxylase, strictosidine synthase and strictosidine glucosidase in suspension cultured cells of *Catharanthus roseus* and *Tabernaemontana divaricata. Plant Cell Rep.*, **12**, 573–576.

Stevens, L.H., Giroud, C., Pennings, E.J.M. and Verpoorte, R. (1993b) Purification and characterization of strictosidine synthase from a cell suspension culture of *Cinchona robusta. Phytochemistry*, **33**, 99–106.

Van der Heijden, R., Verheij, E.R., Schripsema, J., Baerheim Svendsen, A., Verpoorte, R. and Harkes, P.A.A. (1988) Induction of triterpene biosynthesis by elicitors in suspension cultures of *Tabernaemontana* species. *Plant Cell Rep.*, **7**, 51–54.

Van der Heijden, R., Threlfall, D.R., Verpoorte, R. and Whitehead, I.M. (1989) Regulation and enzymology of pentacyclic triterpenoid phytoalexin biosynthesis in cell suspension cultures of *Tabernaemontana divaricata. Phytochemistry*, **28**, 2981–2988.

Van Gulik, W.M, Meijer, J.J., van der Heijden, R., Verpoorte, R. and ten Hoopen, H.J.G. (1988) Feasibility of raubasine production by cell cultures of *Catharanthus roseus. Report of Biotechnology, Delft Leiden.*

Verpoorte, R., Schripsema, J. and van der Leer, T. (1988) Cinchona alkaloids. In A. Brossi, (ed.), *The Alkaloids, Vol. 34*, Academic Press, San Diego, pp. 332–391.

Verpoorte, R., van der Heijden, R., van Gulik, W.M. and ten Hoopen, H.J.G. (1991) Plant biotechnology for the production of alkaloids: present status and prospects. In A. Brossi, (ed.), *The Alkaloids, Vol. 40*, Academic Press, San Diego, pp. 1–187.

Westphal, K. (1990) Large-scale production of new biologically active compounds in plant cell cultures. In H.J.J. Nijkamp, L.H.W. van der Plas and J. van Aartrijk, (eds.), *Progress in Plant Cellular and Molecular Biology*, Kluwer Academic Publishers, Dordrecht, pp. 601–608.

Wijnsma, R., Go, J.T.K.A., van Weerden, I.N., Harkes, P.A.A., Verpoorte, R. and Baerheim Svendsen, A. (1985) Anthraquinones as phytoalexines in cell and tissue cultures of *Cinchona* species. *Plant Cell Rep.*, **4**, 241–244.

Wijnsma, R., Verpoorte, R., Harkes, P.A.A., van Vliet, T.B., ten Hoopen, H.J.G. and Baerheim Svendsen, A. (1986a) The influence of initial sucrose and nitrate concentrations on the growth of *Cinchona ledgeriana* cell suspension cultures and the production of alkaloids and anthraquinones therein. *Plant Cell Tissue Org. Cult.*, **7**, 21–29.

Wijnsma, R., van Weerden, I.N., Verpoorte, R., Harkes, P.A.A., Lugt, Ch.B., Scheffer, J.J.C. and Baerheim Svendsen, A. (1986b) Anthraquinones in *Cinchona ledgeriana* bark infected with *Phytophthora cinnamomi. Planta Med.*, **52**, 211–212.

INDEX

For Product Safety Concerns and Information please contact our EU representative GPSR@taylorandfrancis.com Taylor & Francis Verlag GmbH, Kaufingerstraße 24, 80331 München, Germany

Batch number: 10732772

Printed by Printforce, the Netherlands